HEALTH, NUTRITION AND FOOD DEMAND

Contents

Contributors

Ana M. Angulo is Assistant Professor in the Faculty of Economics, University of Zaragoza, Gran Vía, 2, 50005-Zaragoza, Spain. E-mail: aangulo@posta.unizar.es

Monia Ben Kaabia is Assistant Professor in the Faculty of Economics, University of Zaragoza, Gran Vía, 2, 50005-Zaragoza, Spain. E-mail: monia@posta.unizar.es

Oral Capps, Jr is Professor in the Department of Agricultural Economics, Texas A&M University, College Station, TX 77843-2124, USA. E-mail: ocapps@tamu.edu

Wen S. Chern is Professor in the Department of Agricultural, Environmental and Development Economics, The Ohio State University, 2120 Fyffe Rd, Columbus, OH 43210, USA. E-mail: chern.1@osu.edu

Jose M. Gil is Senior Researcher at the Agricultural Research Service (DGA), Apdo. 727, 50080-Zaragoza, Spain. E-mail: jmgil@posta.unizar.es

Azucena Gracia is Senior Researcher at the Agricultural Research Service (DGA), Apdo. 727, 50080-Zaragoza, Spain. E-mail: agracia@aragob.es

Harry M. Kaiser is Professor in the Department of Applied Economics and Management, Cornell University, 350 Warren Hall, Ithaca, NY 14853-7801, USA. E-mail: hmk2@cornell.edu

Kamhon Kan is Associate Research Fellow in the Institute of Economics, Academia Sinica, Nankang, Taipei, Taiwan. E-mail: kan@gate.sinica.edu.tw

Sung-Yong Kim, formerly of Texas A&M University, is Research Associate in the Korea Rural Economic Institute, 4-102 Hoigi-Dong Dongdaemoon-Gu, Seoul 130-710, Korea. E-mail: sykim@krei.re.kr

Henry W. Kinnucan is Professor in the Department of Agricultural Economics and Rural Sociology, Auburn University, Auburn, AL 36849, USA. E-mail: hkinnuca@acesag.auburn.edu

Dadi Kristofersson is PhD student in the Department of Economics and Social Sciences, Agricultural University of Norway, 1432 Ås, Norway. E-mail: dadi.kristofersson@ios.nlh.no

Tim Lloyd is Senior Lecturer in the School of Economics, University of Nottingham, University Park, Nottingham NG7 2RD, UK. E-mail: tim.lloyd@nottingham.ac.uk

Steven McCorriston is Reader in the Department of Economics, School of Business and Economics, University of Exeter, Streatham Court, Exeter EX4 4PU, UK. E-mail: S.McCorriston@exeter.ac.uk

David D. Mainland is currently retired but worked as Senior Economist in the Department of Agriculture and Food Economics, Scottish Agricultural College, Auchincruive, Ayr KA6 5HW, UK. E-mail: d.mainland@au.sac.ac.uk

Wyn Morgan is Senior Lecturer in the School of Economics, University of Nottingham, University Park, Nottingham NG7 2RD, UK. E-mail: wyn.morgan@nottingham.ac.uk

Øystein Myrland is Associate Professor in the Department of Economics and Management, University of Tromsø, Breivika, N-9037, Tromsø, Norway. E-mail: oysteinm@nfh.uit.no

Rodolfo Nayga, Jr is Associate Professor in the Department of Agricultural Economics, Texas A&M University, College Station, TX 77843-2124, USA. E-mail: Rnayga@tamu.edu

Véronique Nichèle is Economist in the Consumption Research Laboratory, Institut National de la Recherche Agronomique (INRA), 65 Boulevard de Brandebourg, 94205 Ivry-sur-Seine, France. E-mail: nichele@ivry.inra.fr

Laxmi Paudel is Graduate Research Assistant in the Department of Agriculture and Applied Economics, University of Georgia, Athens, GA 30602, USA. E-mail: lpaudil@agecon.uga.edu

Kyrre Rickertsen is Professor in the Department of Economics and Social Sciences, Agricultural University of Norway, 1432 Ås, Norway, and Senior Research Economist in the Norwegian Agricultural Economics Research Institute, Oslo. E-mail: kyrre.rickertsen@ios.nlh.no

John M. Santarossa is Adjunct Lecturer at the University of Glasgow and Economist in the Department of Agriculture and Food Economics,

Scottish Agricultural College, Auchincruive, Ayr KA6 5HW, UK. E-mail: j.santarossa@au.sac.ac.uk

Todd M. Schmit is Research Support Specialist in the Department of Applied Economics and Management, Cornell University, 312 Warren Hall, Ithaca, NY 14853-7801, USA. E-mail: tms1@cornell.edu

Jayachandran N. Variyam is Agricultural Economist in the Economic Research Service, US Department of Agriculture, 1800 M Street NW, Washington, DC 20036-5831, USA. E-mail: jvariyam@ers.usda.gov

Stephan von Cramon-Taubadel is Professor in the Department of Agricultural Economics, University of Göttingen, Platz der Göttinger Sieben 5, D-37073 Göttingen, Germany. E-mail: scramon@gwdg.de

Susanne Wildner completed her PhD in Agricultural Economics at the University of Kiel, Germany and currently works for a food-processing firm in Germany. She can be contacted via Stephan von Cramon-Taubadel.

Steven T. Yen is Associate Professor in the Department of Agricultural Economics, University of Tennessee, Knoxville, TN 37996-4518, USA. E-mail: syen@utk.edu

Preface

Over the last two decades, increasing concerns about health risks related to diets have had significant impacts on food-consumption patterns in the USA and Europe. The desire to improve diets in general and nutrient intakes in particular and the accompanying difficulties have received much attention among economists, nutritionists, health professionals, food producers and policy makers. The purpose of this book is to discuss the likely impacts of health information on the demand for various foods in various countries and among various socio-economic groups. Furthermore, effects of other types of information, such as food scares related to mad cow disease and advertising, are discussed. We hope that this book will be a useful reference for researchers and analysts in the food industry and government and that it can be used as a supplementary textbook in courses in applied microeconomics, consumer demand, health economics and food consumption and nutrition.

The book is based on the papers presented in the Mini-Symposium on 'Effects of Health Information on the Demand for Food: EU and US Experiences' organized as part of the XXIV International Conference of Agricultural Economists sponsored by the International Association of Agricultural Economists (IAAE), held in Berlin, Germany, 13–18 August 2000. Most of the European chapters contain results from the EU-supported research project 'Nutrition, Health and the Demand for Food'. Researchers from France, Germany, Norway, Scotland and Spain participated and we hope that the analyses from these countries provide a representative European perspective. As an additional source of information, each of the European chapters includes a brief review of food-demand studies in the respective countries and a discussion of changes in food-consumption patterns during the last 20 years of the 20th century.

Acknowledgements

Two anonymous referees were asked to review each chapter in a single-blind review process. We would like to express our appreciation to the following reviewers: Frank Asche, Stavanger University College, Norway; Xudong Feng, The Ohio State University, USA; Jose M. Gil, Agricultural Research Service (DGA), Zaragoza, Spain; Azucena Gracia, Agricultural Research Service (DGA), Zaragoza, Spain; Diane Hite, Mississippi State University, USA; Chung L. Huang, University of Georgia, USA; Kuo Huang, Economic Research Service, USDA, USA; Helen Jensen, Iowa State University, USA; Kamhon Kan, Academia Sinica, Taiwan; Henry Kinnucan, Auburn University, USA; Ernest Liu, The Ohio State University, USA; Øystein Myrland, University of Tromsø, Norway; Rudy Nayga, Texas A&M University, USA; Véronique Nichèle, INRA, Ivry-sur-Seine, France; John M. Santarossa, Scottish Agricultural College, UK; Jay Variyam, Economic Research Service, USDA, USA; Stephan von Cramon-Taubadel, University of Göttingen, Germany; and Steve Yen, University of Tennessee, USA.

We would also like to thank Judy Luke of The Ohio State University for her able assistance in editing and reformatting many chapters.

Introduction

1

WEN S. CHERN[1] AND KYRRE RICKERTSEN[2]

[1]*The Ohio State University, Columbus, Ohio, USA;*
[2]*Agricultural University of Norway, Ås, Norway*

There is a widely held belief that concerns about diet-related consequences for health have altered the landscape of food-consumption patterns in many industrialized countries. Analysts have frequently attempted to validate this belief and to provide empirical evidence on the subject. However, the economics literature has shown that it is very difficult to quantify precisely the impacts. Therefore, each chapter in this book attempts to discover whether information on health and nutrition have had any significant impacts on food demand. The book also includes a wide range of topics dealing with the methodology and econometric modelling of the impacts of health and nutrition information on the demand for various foods in different countries.

For the agricultural industry, policy makers and consumer groups, it is important to understand and quantify the impacts of health concerns on food demand because of the critical role of consumer preference in determining the future course of agricultural production, marketing and trade. We believe that a comparison of research findings between the USA and Europe will provide useful insights into the measurement of health-risk concerns and health information and, furthermore, their impacts on food demand. It is our hope that analysts from other parts of the world can adopt the methodologies used in this book.

The lessons from the USA and Europe on the changing patterns of nutrition and food intakes can be used to aid the food and nutrition policies in other developed countries, such as Japan, and in developing countries, such as Brazil, China and India. This topic is of growing importance given the dietary changes already occurring in the newly industrialized countries of East Asia, such as Singapore, South Korea and Taiwan. By using appropriate tools, such as nutrition-education campaigns, these countries may be able to lessen the adverse health impacts of dietary changes that occur with rapid income growth. For the remainder of this chapter, we shall summarize the main contributions of each chapter.

American food choices have been affected by health and nutrition information related to diets as well as by the traditional demand variables, such as income and prices. In Chapter 2, Wen Chern first reviews the historical trends of food intakes and food expenditures by American households for the last 20 years of the 20th century. He points out that, over this period, American households have improved their diets by eating more cereals, poultry, fruits and vegetables, and drinking more low-fat milk. Furthermore, Americans have been eating less red meat, particularly beef, and drinking less whole milk. On the other hand, Americans love cheese, ice cream and fast food, so have had a hard time cutting down the consumption of sugar, fats and oils, and have not managed to add much more seafood to their dinner plates. In accord with human nature, Americans are often inconsistent with respect to healthy eating. Chern attempts to quantify the impacts of health information related to fat and cholesterol on food consumption in the USA. He reviews and suggests appropriate conceptual frameworks on how to incorporate health concerns in a neoclassical demand model. Chern discusses the methodology of constructing the monthly fat and cholesterol information indices based on the numbers of articles published in medical journals and in the popular press, such as the *Washington Post*. One very innovative feature of this methodology is the assumption that any published article has only a finite duration as a useful source of health information to consumers. This methodology is used in constructing the similar European indices used in several European studies included in this book. His econometric model and results clearly show that consumers' health-risk concerns, as measured by his fat- and cholesterol-information index, have affected American food choices in the direction of healthier diets. Medical research has certainly played a very important role in helping Americans to improve their diets.

The effects of health information on the demand for foods have rarely been studied using European data. In Chapter 3, Rickertsen and von Cramon-Taubadel summarize the effects of health information that were found in a European Union (EU) project including France, Germany, Norway, Scotland and Spain. These countries vary considerably as regards diet and mortality rates from diet-related illness, such as heart disease. For example, the consumption of vegetable products is higher in Mediterranean than in northern Europe, and the mortality rates from coronary heart disease are high in Scotland and Norway and low in Spain and France. A two-stage demand model is estimated using a separability structure, commodity aggregation and functional form as similar as possible across countries. However, there are important differences in the data and sample periods used. As in many previous studies, the health-information index is based on the Medline database and a new index, based on the number of articles dealing with fats, cholesterol and the diet in English-language medical journals, is developed. Price and expenditure elasticities are calculated. Rickertsen and von Cramon-Taubadel find that it is difficult to discover any European pattern. There is a wide spread in the values of the own-price elasticities for all goods except for the group 'other goods' and a surprisingly high number of expenditure-elastic food groups and types of meat in different countries. It is impossible to draw clear conclusions regarding the impact of health information on the demand for food in general and meat in particular in the EU. A consistent positive or negative effect of health information on the demand for any specific goods is not found. Several of the significant health-information effects are unexpected, indicating that the common model may be too restrictive. Most of the participating countries have been able to produce more plausible results, as discussed in later chapters, by deviating from the model presented in Chapter 3.

The next three chapters cover US experiences dealing with health, nutrition and food demand in a cross-sectional context. In Chapter 4, Jayachandran Variyam deals with the intakes of several key macro-nutrients by American adults. Specifically, he attempts to estimate the impacts of educational attainment on the intakes of total fat, saturated fat, cholesterol and fibre. His study deviates from the previous literature in the field by estimating the impacts of education and other social and demographic factors on different segments of the intake distribution rather than means. He does so by employing the quantile regression method. Quantile regression provides a useful method when we are interested in the impacts among demographic groups with varying intakes and thus different potential health risks. His database is obtained from the Continuing Survey of the Food Intakes by Individuals (CSFII) for 1994–1996 provided by the US Department of Agriculture (USDA). This is one of the most important databases in the USA for doing nutrient-intake analyses. Since the study's focus is on the impact of education, Variyam reports only the regression results related to the education variable. His findings suggest that education not only affects the intakes of the key macronutrients in the USA, but also has stronger impacts on the households in the upper tail of the intake distribution. That is, education appears to have a stronger impact on the intakes of fat, saturated fat and cholesterol in the 75th and 90th percentiles than in the 10th or 25th percentiles. This information is important because the health risks are higher for those people in the 75th or 90th percentiles of the intake distribution.

In Chapter 5, Sung-Yong Kim, Rodolfo Nayga and Oral Capps investigate the impacts of label use on diet quality in the USA. Health information is assumed to be provided by food labels, and thus using food labels is similar to getting health information on food. The study is based on the healthy eating index (HEI) developed by the USDA. The USDA has been attempting to monitor the American dietary trends and needed a measure of dietary quality. The HEI was developed to measure dietary quality for individuals or the nation as a whole. Kim *et al.* model the HEI as a function of label use by the types of food label. They also specify a probit model for estimating the probability of label use simultaneously with the dietary quality measured by the HEI. They use data from the 1994–1996 USDA CSFII and the Diet and Health Knowledge Survey (DHKS). These are important surveys conducted by the USDA for monitoring the nutrition status of people in the USA. Thus this study uses the same database as that used by Variyam in Chapter 4. The main findings include the significant impacts of label uses on dietary quality. They also reinforce the previous finding in Chapter 4 that education is important for achieving a more healthy diet. This chapter also provides a detailed investigation of the patterns of label uses and their impacts on dietary quality among different demographic groups. Despite the importance of demographic characteristics, the results related to demographic variables are not always transparent. For example, it is not clear why non-Hispanic people are less likely to use label information. Furthermore, why would male users of nutrient claims have higher HEIs than female label users, but male non-label users have lower HEIs than female non-users?

Kamhon Kan and Steven Yen, in Chapter 6, model health information, health knowledge and egg consumption in a recursive simultaneous-equations system. Basically, they assume that information affects health knowledge, which, in turn, affects egg consumption in both consumption participation and quantities consumed. Thus there are four equations in the system. As in Chapters 4 and 5, they use data from the 1994–1996 CSFII and data from the DHKS. However, Kan and Yen construct the information and knowledge variables differently from the previous authors. Health information is a binary variable based on whether or not the survey respondent has heard about the health effect of cholesterol. The knowledge variable, on the other hand, is measured by the relative importance of choosing a low-cholesterol diet simply because eggs contain more dietary cholesterol than most other foods. Many typical demographic variables, such as education, age, sex, race and ethnicity, are included. Since these survey data do not contain information on prices, no prices are included and this is, of course, a typical problem of using the USDA CSFII data. Among the main findings are: (i) health information as defined does affect health knowledge; (ii) health knowledge affects only the participation in egg consumption, not the level of consumption; and (iii) education affects information positively, but affects knowledge and participation negatively. Similarly to the attempt in Chapter 5, Kan and Yen treat health information and knowledge as endogenous in the food-intake decision. This methodology will be evaluated further in the last chapter of this book.

The next four chapters deal with the European experiences. Even though there are many studies of German food demand, none of them, to our knowledge, have been published in English. In Chapter 7, Susanne Wildner and Stephan von Cramon-Taubadel fill this gap. Food demand evolved in different environments in East Germany (EG) and West Germany (WG) prior to 1989. Nevertheless, the importance of meat in the food budgets of both EG and WG was considerable prior to reunification and remains so today. The share of meat and processed meat in food expenditure was 25% in EG and 21% in WG in 1998. A demand model is used to test the possible linkages between the dissemination of health information and the demand for different types of meat. Meat is expected to be especially responsive to health information and many health campaigns are geared towards reducing the intake of animal fats. The health-information index discussed in Chapter 3 is also used in Chapter 7. Furthermore, two additional indices are developed. The first index is based on publications in English by German scientists, while the second index also includes German-language publications. A linear approximate almost ideal demand system (AIDS) is estimated. Five groups of meat (beef and veal, pork, poultry, sausage and other processed meats, and fish and fish products) are studied. The analysis is based on monthly data covering the 1991 to 1998 period. The demand of a 'standard household' consisting of a married couple with two children and middle income is studied. They find that, with the exception of poultry in EG, the two German indices have no significant impact on meat demand in either EG or WG. The health-information index discussed in Chapter 3 is significant in three of the five equations in both EG and WG. Significant health-information effects appear to be stronger in WG than in EG. Fish and fish products appear to have benefited at the expense of pork from the dissemination of cholesterol- and fat-related information. These results are plausible. Fish is regarded as a healthy food while pork is considered less healthy due to its cholesterol content and the poorer composition of its fatty acids.

Chapter 8 is about France. Even though the French mortality rates from cardiovascular diseases are among the lowest in the world, this group of diseases is still the primary cause of mortality, causing 32 deaths out of 100. Véronique Nichèle estimates the impact of information about fat and cholesterol on food demand. The main differences from the French model described in Chapter 3 are related to the selected demand system and data set. The quadratic almost ideal demand system (QUAIDS) is estimated in Chapter 8. This system is a generalization of the AIDS, which is consistent with non-linear Engle curves. Furthermore, pooled microdata from the French National Food Survey are used. Household data enhance the empirical analysis in several ways. First, food products are aggregated into 15 food groups according to high and low fat content. Secondly, the inclusion of sociodemographic variables allows the investigation of the impact of household characteristics on the demand. Thirdly, the estimated price effects are more reasonable than the rather unexpected results for France found in Chapter 3. The quadratic expenditure terms are less important than the linear expenditure terms; however, they are significant for 13 of the 15 food groups, indicating that the QUAIDS specification is appropriate. Substantial health-information effects are found. Fat and cholesterol information has a negative impact on the demand for beef, other meats, eggs and butter and, to a lesser extent, on the demand for pork and vegetables. A positive impact is found on the demand for meat products, milk, yoghurt, oils and grain products. The demand for poultry, fish, cheese and fruits is relatively unaffected by health information. Most of the results are plausible. The negative effects for beef, pork, other meats, eggs and butter are as expected. However, the negative effect on vegetable demand is unexpected. The demand for oils benefits from increased health information at the expense of the demand for butter. The positive effect on milk demand is related to the change in the composition of milk demand. The consumption of whole milk has decreased while the consumption of low-fat and non-fat milk has increased.

A Spanish study is presented in Chapter 9. We note first that it is likely that consumers make their decisions not only in terms of final food products but also in terms of food nutrients. Ana Angulo, Jose Gil, Azucena Gracia and Monia Ben Kaabia develop and estimate a demand system with food quantities as dependent variables and income, prices, sociodemographic variables and nutrient content of different food groups as explanatory variables. Their approach is related but not identical to Lancaster's approach (see Chapter 2). A demand system based on the generalized addilog demand system (GADS) is developed. They use the Spanish Quarterly National Expenditure Survey for 1995 and study the demand for six broad food groups (cereals and potatoes; meats; fish; fruits and vegetables; dairy products and eggs; and oils, sugar and other food products), including six nutrients (carbohydrates, lipids, proteins, vitamins, minerals and fibre) in the model. A likelihood-ratio test suggests that nutrient composition of food is important and should be included. As expected, total expenditure elasticities are positive and own-price elasticities are negative. Furthermore, the inclusion of nutrients causes the expenditure elasticities for relatively healthy products to increase and, except for dairy products and eggs, the own-price elasticities to become more elastic. The own-nutrient elasticities suggest that the demand for cereals and potatoes is positively affected by the content of fibre and minerals, while the effect of carbohydrates is negative. The amount of proteins, minerals and vitamins in meat has a positive impact on demand, while the content of lipids has a negative effect. The demand for fish is positively influenced by the content of proteins and vitamins. The demand for fruits and vegetables is positively affected by the content of every included nutrient. Dairy products and eggs are positively affected by their content of proteins and minerals while lipids have a negative effect. Finally, the demand for oils, sugar and other products is negatively affected by their content of carbohydrates and lipids.

Recall that, in our summary of Chapter 3, no clear conclusions regarding the effects of health information on the demand for food in the EU emerged. The effects varied from one European country to another and were often insignificant and frequently unexpected. Autocorrelation, however, is a major problem in the estimation of all the country-specific time-series models and different correction methods were applied in Chapter 3. In Chapter 10, Kyrre Rickertsen and Dadi Kristofersson investigate the fragility in estimated health-information effects for choice of autocorrelation correction. The two-stage AIDS, the Norwegian data and the health-information index described in Chapter 3 are also used in Chapter 10. Six correction mechanisms are applied: the general Berndt–Savin correction, a restrictive version of this correction, the general partial adjustment model, a restricted partial adjustment model, the general Berndt–Savin correction used on the general partial adjustment model and the Bewley transformation. The general Berndt–Savin correction in combination with the general partial adjustment model is preferred. The estimated price, total expenditure and health-information effects are quite sensitive to choice of correction. Some health effects, however, are robust across different specifications. A positive effect on the demand for poultry is found. An unexpected negative effect on the demand for fish and an unexpected positive effect on the demand for fats and oils indicate that there are additional problems in the model.

The last four chapters deal with the relationships between consumer health information and producers, marketing and government responses. In Chapter 11, Henry Kinnucan, Øystein Myrland and Laxmi Paudel compare the impacts of health information related to cholesterol and advertising on the meat sector in the USA. Their study extends the scope of the analysis presented in the preceding chapters to deal with not only the demand side but also the supply side. They argue that consumers respond to negative information, such as the health impacts of cholesterol, much more strongly than the positive information about a food product, such as beef, generated from commercial advertising. They construct an industry-wide model for beef, pork and poultry in the USA. The model links the retail to the farm sectors through the simultaneous determination of the equilibrium quantities and prices at the retail and farm levels. While the supply and own-price demand elasticities are taken from previous studies, they estimate the elasticities associated with health information and advertising in the present study. For health information, they compare the index of cholesterol information developed by Brown and Schrader and the fat- and cholesterol-information index developed by Kim and Chern (similar to the one discussed in Chapter 2). They use the results obtained by using the Kim and Chern index for a simulation. From the elasticity analysis, they find that the impacts of health information are much larger than the impacts from advertising, thus supporting their conceptual hypothesis of a greater response of consumers to negative information. They specifically find that health information resulted in decreases in beef consumption and increases in poultry consumption during the sample period of 1976–1993. Furthermore, even though health information did not have a direct impact on pork consumption, the indirect effects, due to the induced price changes, resulted in decreases in pork consumption. Therefore, they argue that the analysis of the welfare impact of health information cannot be complete without analysing the total effect from the entire commodity sector, including the supply side. The most interesting finding is that advertising effects on meat markets are relatively small as compared with the effects of health information. The implication is that the meat industry cannot expect to be able to counter the negative impacts of health information about the high content of fat and cholesterol in red meats, particularly beef, by commercial advertising.

In Chapter 12, Todd Schmit and Harry Kaiser investigate the impacts of dietary cholesterol concerns on the demand for eggs in the USA. This is the second study in this book dealing with egg demand. Unlike the study described in Chapter 6, Schmit and Kaiser use aggregate disappearance data and model both the cholesterol information and generic advertising effects. One of the most interesting aspects of this chapter is the quantification of the opposite impacts of the changes in cholesterol-awareness information from negative to positive as a result of a landmark study published in 1997. The study changed the perception of eggs in the eyes of news media and thus the public as well. Schmit and Kaiser use a cholesterol-information index similar to that constructed by Brown and Schrader. The numbers of cholesterol-related articles were based on the *Reader's Guide to Periodical Literature*. Their results show that the cholesterol information decreased egg consumption during 1987–1996, but increased egg consumption during 1997–2000. Similarly to the approach used in Chapter 11, Schmit and Kaiser's model also includes farm supply, intermediate marketing split and the wholesale demand for eggs in the USA. Therefore, they are able to draw inferences from the impacts of health information not only on egg demand, but also on farm prices of eggs. During the 1997–2000 period, positive cholesterol information on eggs increased the average farm price of eggs by 16.5%. Interestingly, this study reinforces the findings obtained in Chapter 11 that health information has a much more dominant effect on consumer demand than commercial advertising. As a result, the authors find that recent increases in egg consumption in the USA can be attributed mostly to the changing perception of eggs in terms of the decreasing significance of the impact of dietary cholesterol on health. Advertising had only a small impact on egg-demand enhancement.

The situation appears to be different in Scotland. Advertising campaigns and health information have not been very successful in changing the Scottish diet. In Chapter 13, John Santarossa and David Mainland use the AIDS. They find that the health-information index discussed in Chapter 3 has no significant effects on Scottish food demand and propose a different approach, relying on taxation of unhealthy nutrients to achieve some desirable objectives. The authors propose to tax amounts that are in excess of the recommended levels given in official guidelines for intakes of various fat components. To calculate the optimal tax for various nutrients, they focus on the nutritional content of various foods. The suggested taxes imply small price changes for most food groups, but the prices of dairy products, eggs, and fats and oils would increase by 4, 11 and 24%, resulting in reductions in expenditure shares of 5, 13 and 11%, respectively. The authors find that the proposed taxes reduce the consumption of fat by 20%. Specifically, reductions of 21% for saturated and monounsaturated fats and 26% for polyunsaturated fats are observed. The energy intake from fats is also reduced from 38 to 30%, which is below current UK guidelines.

Chapter 14 deals with the market responses to food scares in the UK. Bovine spongiform encephalopathy (BSE) first came to the public's attention in the mid-1980s; however, at the time there was little concern that BSE represented a threat to human health. It was not until the mid-1990s that BSE was fully recognized as a crisis following the first human deaths from variant Creutzfeld–Jakob disease (CJD). Consumption of red meat fell immediately by 40% in the UK and bans on imports of beef from the UK were imposed. Since then, the link between BSE and variant CJD in humans has been confirmed and, to date, 90 deaths have been recorded in the UK. In Chapter 14, Tim Lloyd, Steven McCorriston and Wyn Morgan investigate the impact of BSE on price adjustment in the UK beef market by using a co-integration framework. The study provides an interesting case-study on how to incorporate the consumer health information into market equilibrium prices of the affected food product. Lloyd *et al.* consider price adjustment at the retail, wholesale and producer stages. Prices fell at all marketing stages in the 1990s and the spreads between prices have not remained constant. Retail prices have fallen by 18%, while wholesale and producer prices have fallen by around 40% each, suggesting that the costs of the BSE crisis has fallen primarily on farmers. The price series are supplemented by a food publicity index, counting the number of articles in broadsheet newspapers concerning the safety of meat. Articles relating to BSE dominate the index, although other topics are also covered. The authors conclude that the index has affected the evolution of UK beef prices. A one-standard-error shock in the log of the index resulted in long-run price reductions of 1.80, 2.80 and 3.11 pence kg^{-1} in the retail, wholesale and producer prices, respectively.

All the studies presented in this book employ demand or sector models for investigating the impacts of consumer health concerns and information on food demand and marketing strategies in the USA and Europe or for designing effective government responses to achieve the desirable outcomes of healthier diets. Some further evaluations of the methodology, models, databases and policy implications are provided in the last chapter of this book.

Health, Nutrition and Demand for Food: an American Perspective

2

WEN S. CHERN

The Ohio State University, Columbus, Ohio, USA

Introduction

On many counts, Americans have been eating healthier diets. The percentage of calorie intake from fat is slowly dropping and per capita consumption of fruits and vegetables is increasing (Frazao, 1999). Furthermore, consumer knowledge about nutrition and health is also slowly improving. However, statistics also show that more and more Americans are becoming overweight and the rate of obesity continues to increase. Dietitians noted that the desire for high-fat and high-sodium foods often outweighs nutrition and health concerns (Frazao, 1999). Quantifying consumer health information and knowledge and their impacts on food-consumption behaviour is not a simple task.

The objectives of this chapter are to discuss several methodological issues on how to incorporate the demand for health into food-demand analysis and to present a case-study of the impact of health-risk concerns on the demand for food in the USA. Understanding the linkage between the demand for health and the demand for food is important because of the increasing importance of developing a satisfactory model to investigate the impact of the consumer's health-risk concerns on food consumption behaviour. The subject is important to agricultural producers and marketers because the impact of health-risk concerns suggests a different operational strategy from that of demand affected simply by income and prices, as implied by the neoclassical demand theory.[1]

In recent years, major agricultural producers of beef, pork, poultry and dairy products appear to have accepted the perception that consumers have altered their consumption behaviour in response to the information concerning diet–health relationships. As a result, beef and dairy producers have spent millions of dollars in commercial advertising to counter any perception of negative health attributes in their products. Interestingly, the economics literature offers no conclusive empirical evidence of the impacts of health-risk concerns on consumption behaviour. Beef is perhaps one of the most investigated food items for analysing the structural change in demand caused by health-risk concerns. Earlier studies using time-series data have attributed structural shifts to increasing health-risk concerns about cholesterol (Braschler, 1983; Chavas, 1983; Moschini and Meilke, 1984). However, Chalfant and Alston (1988), using a non-parametric analysis, found no structural changes in beef demand in the USA and Australia due to concerns about fat and cholesterol. More recently, Robenstein and Thurman (1996) also showed no evidence of the relationship between health-risk and the demand for red meat using daily time-series data of futures price from 1983 to 1990. The earlier studies of beef demand could not identify the causes of structural changes in beef demand because the health-risk concerns were not quantified in these studies. In this study, we shall review some of the recent literature related to various methods of measuring health-risk concerns by information proxy variables used for econometric estimation.

Since the neoclassical demand theory assumes that the consumer maximizes a utility function subject to a budget constraint, the demand function derived from such a model is based on a stable utility function, i.e. given taste and preference. The health-risk concern differs from the demographic variables in a demand analysis because it affects the consumer's taste and preference. On the other hand, demographic variables are often used to account for household variation in consumption. More precisely, the consideration of these demographic variables is necessary either because we use household-level data or because we use time-series data in a period when household composition has changed. According to the neoclassical demand theory, it is awkward, if not outright incorrect, to incorporate the taste and preference into the demand function because the demand function itself is derived from assuming taste and preference as given. In this chapter, we shall review the approaches employed by researchers that extend the neoclassical model to incorporate health into a demand model and point out some of the unresolved issues.

Specifically, this chapter will address two major issues. One deals with the methodology for incorporating the consumer's health information into a food-demand model for analysing consumption behaviour. The other issue is related to the method of creating proxy variables to quantify both the amount and changing trends of consumer health information over time. We shall focus on the health-risk concerns about fat and cholesterol. Several alternative measures of consumer information on fat and cholesterol will be discussed. The empirical application involves estimation of a linear approximate version of the almost ideal demand system (LA/AIDS) for ten selected food items. The monthly time-series data created from the Consumer Expenditure Survey conducted by the Bureau of Labor Statistics (BLS) in the USA are used for the case-study. The empirical results provide important implications for assessing the patterns of the consumer's dietary changes and public policies for medical research and consumer dietary education in the USA.

Trends of Food Consumption and Expenditure in the USA

Table 2.1 shows the annual per capita food consumption in kilograms in the USA for selected years during 1980–1999. These data were extracted from aggregate 'disappearance' or 'availability' data. The data clearly show that Americans have been eating less beef, about the same amount of pork and more poultry during this period. The data for 1999 suggest a reversal of the declining consumption trends for beef and pork, reflecting

Table 2.1. Annual per capita consumption of major food items in the United States.[a] (From Putnam and Allshouse, 1997; Putnam *et al.*, 2000; and unpublished tables obtained from the Economic Research Service of the USDA (31 August 2001).)

Food item	1980 (kg)	1985 (kg)	1990 (kg)	1995 (kg)	1999 (kg)
Meats[b]	83.2	84.5	81.7	86.8	91.5
Beef	34.7	35.9	30.7	30.5	31.3
Pork	25.8	23.4	22.4	23.7	24.4
Poultry	21.5	23.8	27.5	31.7	35.1
Other meats[c]	1.3	1.5	1.1	1.0	0.9
Fish	5.6	6.8	6.8	6.8	6.9
Eggs	15.8	14.9	13.7	13.7	14.9
Cheese	7.9	10.2	11.2	12.4	12.5
Fluid milk (plain)	101.3	97.8	94.8	89.3	85.9
Whole	64.3	54.3	39.8	32.9	31.3
2% fat	24.8	31.1	35.6	31.9	29.3
Light to fat-free	12.2	12.4	19.4	24.4	25.3
Vegetables and fruits	169.6	282.0	293.6	311.1	326.2
Fresh vegetables	67.7	70.8	75.4	78.7	87.1
Fresh fruits	47.5	50.2	52.8	57.2	60.1
Processed vegetables	84.9	91.6	99.9	105.0	103.9
Processed fruits	69.4	69.4	65.4	70.2	75.0
Flour and cereals	65.6	70.9	82.6	87.3	91.6
Sugar	37.9	28.4	29.2	29.7	30.8
Fats and oils	27.4	30.6	29.6	30.3	30.1

[a]Original data are in lbs and they were converted using the conversion factor of 1 kg = 2.2046 lb.
[b]Meats are in 'retail cut equivalent', not 'boneless and trimmed equivalent'.
[c]Other meats include veal and lamb.

at least in part a result from the concerted efforts by the beef and pork industries in advertising and promotion. Of course, the recent 'mad cow' and foot-and-mouth disease epidemics in Europe will probably soften any upward surge in the consumption of beef and pork in the USA.

Despite the urge by nutritionists to eat more fish and seafood due to their rich content of beneficial omega-3 fatty acid, the per capita consumption of fish remained flat during 1980–1999. As for eggs, with their high dietary cholesterol content, the trend has been downwards until recent years. In 1999, the consumption level increased to the 1985 level of 14.9 kg per person per year. Notably, Americans have been drinking less whole milk and more low-fat milk, apparently due to their health concerns. But the consumption of cheese has been rapidly increasing. This is a strong case showing that sometimes factors other than health-risk concerns are more important for consumption decisions.

Another healthier eating trend observed is for fruits and vegetables. Americans have been eating more of these foods, which are highly recommended by nutritionists and health professionals. Specifically, there was a 92% increase of the annual per capita consumption of fruits and vegetables from 169.6 kg in 1980 to 326.2 kg in 1999. Furthermore, the consumption of flour and cereals increased over this period. On the other hand, the consumption of sugar, as well as fats and oils, has been very steady from 1985 to 1999, despite the urge to cut down their consumption as part of many campaigns for better diets.

Table 2.2 presents the comparative statistics on food expenditures, income and household size in the USA for 1980, 1992 and 1999. The first thing to note is that the food-at-home share of the total food budget has been declining from 69% in 1980 to 63% in 1999. Among the foods consumed at home, the expenditure shares of beef and pork have been decreasing, while that for poultry remained constant during 1992–1999. Therefore, the expenditure trends for beef and pork follow their consumption trends discussed earlier. However, the trend of increasing poultry consumption did not translate into a trend of increasing expenditure, apparently as a result of decreasing poultry prices. On the contrary, the constant consumption trend of fish and seafood yielded an increasing trend in expenditure on them, apparently due to the increasing prices of fish and seafood over these years. Another interesting observation is that American consumers have been spending less on dairy products. But they are willing to spend relatively more on fruits and vegetables. Finally, the most notable increase in expenditure share goes to miscellaneous foods, which consist of many prepared foods. This segment of the food market has been growing very rapidly, resulting from an increasing demand for convenience foods by American consumers.

Overall, these historical trends show that there are indicators of Americans' food-consumption habits moving towards healthier diets.

Table 2.2. Annual per-household food expenditures and shares in the USA. (From Paulin (1998), and unpublished summary tables of Dairy Survey results for 1999 provided by the Bureau of Labor Statistics (29 August 2001).)

Item	1980	1992	1999
Income before tax	$17,985	$33,407	$43,292
Household size	2.7	2.5	2.5
Food expenditure	$2,504	$3,902	$4,580
Food at home	$1,727	$2,600	$2,875
Cereal and bakery products	$222	$411	$448
Meat, poultry, fish and eggs	$594	$687	$749
Dairy products	$232	$302	$322
Fruits and vegetables	$256	$429	$500
Other food at home	$422	$772	$856
Food away from home	$777	$1,302	$1,705
Share of food at home (%)	100.0%	100.0%	100.0%
Cereal and bakery products	12.9%	15.8%	15.6%
Meat, poultry, fish and eggs	34.4%	26.4%	26.0%
Beef	13.2%	8.1%	7.6%
Pork	7.3%	6.0%	5.4%
Other meats	4.6%	3.6%	3.4%
Poultry	4.5%	4.7%	4.7%
Fish and seafood	2.8%	2.9%	3.7%
Eggs	1.9%	1.1%	1.1%
Dairy products	13.5%	11.6%	11.2%
Fresh milk and cream	7.1%	5.1%	4.2%
Other dairy products	6.4%	6.5%	7.0%
Fruits and vegetables	14.8%	16.5%	17.4%
Fresh vegetables	4.2%	4.9%	5.2%
Fresh fruits	4.3%	4.9%	5.3%
Processed vegetables	2.8%	2.9%	3.0%
Processed fruits	3.5%	3.9%	3.9%
Other food at home	24.4%	29.7%	29.8%
Sugar and sweets	3.6%	3.9%	3.9%
Fats and oils	2.9%	2.8%	2.9%
Miscellaneous	8.8%	14.8%	14.6%
Non-alcoholic beverages	9.2%	8.2%	8.4%

However, there are also other factors that continue to drive American consumers to consume foods with high fat and cholesterol contents.

Demand for Health and Demand for Food

The diet–health relationships have received increasing attention from consumers, health professionals, food marketers, and agricultural and health policy makers. Increasing demand for health may affect the food-demand structure. It is therefore important to incorporate health in a food-demand model.

Grossman (1972) is among the first to analyse the demand for health. In his seminal paper, he constructed a model of the demand for the commodity 'good health' based on the household production theory developed by Becker (1965). Specifically, Grossman specifies an intertemporal utility function as:

$$U = U(\phi_0 H_0, \ldots, \phi_n H_n, Z_0, \ldots, Z_n) \qquad (2.1)$$

where H_0 is the initial stock of health, H_i is the cumulated stock of health in the ith time period, ϕ_i is the service flow of health stock in the ith period and Z_i is the consumption of other commodities in the ith period. Grossman then assumes that the health stock depends on health investment according to the following scheme:

$$H_{i+1} - H_i = I_i - \delta_i H_i \qquad (2.2)$$

where I_i is gross investment and δ_i is the depreciation rate of the health stock during the ith period. Under this framework, both I_i and Z_i are household produced goods. The household productions are specified as:

$$I_i = I_i(M_i, TH_i; E_i) \qquad (2.3)$$

$$Z_i = Z_i(X_i, T_i; E_i) \qquad (2.4)$$

where M_i is medical care, X_i is the goods input, TH_i and T_i are time inputs and E_i is the stock of human capital. The stock of human capital can be affected by education and other knowledge accumulated from other sources such as information. Grossman's primary interest is the link between health stock and medical care. He argues that the consumer demands medical care not because medical care directly provides the utility to the consumer, but rather it provides useful input to enhance the stock of health, which directly contributes to the consumer's utility function. This is, of course, the essence of Becker's household production theory.

Grossman's model can be easily extended to include food as another explicit input to the production of health stock. Thus, Equation 2.3 can be rewritten as:

$$I_i = I_i(D_i, M_i, TH_i; E_i) \qquad (2.5)$$

where D_i is the diet chosen by the consumer. For simplicity, let us disregard the intertemporal relationship. Under Becker's household production theory, the consumer maximizes his/her utility function in Equation 2.1 subject to the full income constraint of budget and time resources. The equilibrium conditions would yield the demand functions for 'good health' and other household-produced commodities.

The demand for health would be a function of the price of 'good health', prices of other commodities, income and the value of time. The prices of 'good health' and commodities are, of course, not observable. There exist 'shadow prices' for these commodities in the model. As demonstrated by Grossman, the model has been very useful for deriving many insightful comparative-statistics results for understanding the complex relationships of the demand for health. However, very few empirical studies can implement the Becker-type model exactly, due to the unavailability of data related to the time input.

Theoretically, we can also derive the demand functions for medical care and diet as a function of all the exogenous variables in the model. Such a model can be generally expressed as:

$$D_i = D_i\left(PD_i, PM_i, PX_i, Y_i, W_i; E_i\right) \tag{2.6}$$

where PD_i is the price of diet (or the prices of foods), PM_i is the price of medical care, PX_i is the price of other market goods, Y_i is income, W_i is the value of time (which may be approximated by wage rate) and E_i is the stock of human capital. We can, of course, make a further assumption that the stock of human capital is a function of education, health knowledge or health information, so that Equation 2.6 can be estimated empirically. It is noted, however, that the derivation of Equation 2.6 is not a straightforward matter in Becker's household production theory.

Another approach is to adopt the characteristics model developed by Lancaster (1971). Lancaster argues that the consumer draws utility from the characteristics embodied in the goods he/she purchases. Therefore, utility should be a function of these characteristics or attributes of the goods, not the goods themselves. Using this concept, we can argue that a diet has positive characteristics such as good taste and sources of energy and other nutrients, but it also has negative characteristics such as the health-risk associated with the consumption of fat and cholesterol. Thus, in general, we can define the utility function as:

$$U = U(NC, PC, C) \tag{2.7}$$

where NC is the negative attribute, such as health-risk, PC is the positive attribute and C represents all other attributes. Since all these attributes come from the market goods, they can be obtained directly from them:

$$NC = NC(X; E) \tag{2.8}$$

$$PC = PC(X; E) \tag{2.9}$$

$$C = C(X; E) \tag{2.10}$$

where X is market goods, including food, and E is the stock of human capital. In Lancaster's framework, the consumer maximizes the utility function expressed in Equation 2.7 subject to the budget constraint and the input–output relationships in Equations 2.8–2.10. These input–output relationships between attributes and market goods are similar to Becker's household production functions. Note that for some attributes, such as calorie content, the input–output coefficients are essentially fixed, while for other attributes, and especially perceived attributes such as the linkage to blood cholesterol level, the input–output relationships may be highly dependent on the stock of human capital, E. The resulting equilibrium conditions would yield the demand functions for the characteristics of attributes, NC, PC and C. The demand determinants typically include the prices of all attributes and income. It is interesting to note that Lancaster's model has many similarities to Becker's model, but is much simpler as it does not address time allocation as part of the consumer's choice problem. Similarly to the prices of Becker's household-produced commodities, the prices of good characteristics are not observable. However, these prices can be estimated directly from the prices of market goods through the well-known hedonic price model.

Theoretically, we can also derive the demand functions for market goods such as food in Lancaster's framework. Such demand functions may be expressed as:

$$X = X(PX; Y; E) \tag{2.11}$$

where PX represents the prices of market goods and Y is income. Note that Equation 2.11 is much simpler than Equation 2.6, and it is also much easier to estimate.

Chern *et al.* (1995) adopted this general framework to develop a more elaborate model based on the following expected utility function:

$$\mu = \mu(R, PF, Z) \tag{2.12}$$

where R is the perceived health-risk index related to fat consumption, PF is the positive aspects of fat consumption and Z is the expenditure for all other goods. They assume that the health-risk index depends on fat consumption, given the consumer's characteristics and belief in health-risk:

$$R = R(F; E, \phi) \tag{2.13}$$

where F is the total fat consumption, E is the stock of human capital (measured by demographic variables) and ϕ is the belief in health-risk associated with fat consumption.

Furthermore, the total fat consumption, F, and the positive aspects of fat consumption, PF, are directly related to the type of food consumed:

$$\begin{aligned} F &= F(D) \\ PF &= PF(D) \end{aligned} \tag{2.14}$$

where D is the diet consumed.

In Chern *et al.*'s (1995) model, ϕ is assumed to be a random variable with a certain probability distribution characterized by the mean μ_ϕ and variance σ_ϕ^2. Maximization of the expected utility function in Equation 2.12, subject to budget constraint and the additional constraints of Equations 2.13 and 2.14, yields the following demand function for diet or food:

$$D = D(PD, Y, E, \mu_\phi, \sigma_\phi^2) \tag{2.15}$$

where PD is the price of food. Chern *et al.* (1995) estimated this demand function for fats and oils in the USA. In their model, the mean and variance of the health-belief random variable capture the impact of health-risk concerns related to fat and cholesterol. While their approach is noble and innovative, the empirical implementation using a Bayesian procedure to quantify the probability distribution of the health-risk belief is very cumbersome.

Thus, most empirical studies of the impacts of the demand for health on the demand for food adopted either Equation 2.6, following Becker's approach, or Equation 2.11, following Lancaster's framework. These specifications may be viewed as reduced-form demand functions because the original model does not yield these functions directly. Furthermore, the demand for health is often assumed to be directly related to either the consumer's health knowledge (internal) or the health-risk information provided for the consumer (external).

Recently, there have been attempts to treat the health knowledge in Equation 2.6 or the health information in Equation 2.11 as endogenous by arguing that both health input (food demand) and health knowledge are affected by the same set of demographic variables (Kenkel, 1990; Nayga, 2000; Kim *et al.*, 2001). However, the theories of Becker and Lancaster do not provide an adequate theoretical framework for deriving such joint demand functions for goods and consumer health knowledge/information. This modelling issue requires further investigation.

Measurement of Health-risk Concern

The next methodological issue is how to quantify the consumer's health-risk concern or beliefs. For empirical studies, one often makes an assumption that health concern is affected by either knowledge or information. Surveys have been conducted to quantify the knowledge of diet–health relationships. For example, the Diet and Health Knowledge Survey (DHKS) of the US Department of Agriculture collected data on how well the surveyed households answered certain nutrition- and health-related questions in the questionnaire. These data have been used to analyse the impacts of health knowledge on food-purchasing behaviour (Variyam *et al.*, 1996; Nayga, 1997). The Food and Drug

Administration (FDA) has regularly conducted a health-knowledge survey to monitor changes in consumers' health knowledge and beliefs. One of the problems in using these survey data is that the ability to answer certain selected diet–health questions tends to be highly correlated with education. It is therefore very difficult to sort out the causal relationships between the knowledge and health information generated by government programmes in health and medical research, and consumer education and outreach. Another problem is that these surveys tend to cover very generic diet–health questions, which do not provide a specific measure of knowledge in a particular health-risk concern such as that relating to fat and cholesterol.

It is expected that consumer responses to any health concern about diet–health-risk will take place very slowly. People do not adjust and change their dietary patterns abruptly and suddenly unless they face a life-threatening situation. Time-series data would be the best to capture such changes. One major problem in using time-series data, however, is how to construct an appropriate time series for health knowledge or information. Previous researchers have used a time-trend variable, but that has later been shown to produce misleading results (Chern *et al.*, 1995). The time-trend variable is almost always highly correlated with other explanatory variables in a time-series model, making it very difficult to attribute the impacts of time entirely to either health knowledge or information.

Brown and Schrader (1990) made a seminal contribution in constructing a cholesterol-information index based on the numbers of published articles in medical journals. Their index has been used in several other studies (Capps and Schmitz, 1991; Yen and Chern, 1992). By assuming that an article will have a constant effect as a source of information after it is published, the index, in essence, simply accumulates the numbers of articles over time. Therefore, it often performs as a time trend in a demand model.

Recently, attempts have been made to create alternative indices, following Brown and Schrader's approximation based on the numbers of articles published in medical journals or popular newspapers (Chern and Zuo, 1995; Kim and Chern, 1997, 1999). These researchers assume that, after an article is published, its effectiveness as a source of consumer information diminishes over time. One such index is defined as:

$$I_i = \sum_{i=0}^{n} s_i N_{t-i} \tag{2.16}$$

where I is the health-information index, s is the weight of time periods, N is the number of articles and n is the number of the lagged periods for which an article will remain as a useful source of information. The time $t = 0$ is used to denote the initial period when an article is published. Chern and Zuo (1995) use the following third-degree polynomial weight function to create the weights:

$$s_i = \gamma_0 + \gamma_1 i + \gamma_2 i^2 + \gamma_3 i^3 \tag{2.17}$$

where i is the number of the ith lagged period. Assume further that the maximum weight would occur in the period m. Chern and Zuo (1995) show that, with certain regularity conditions, the weights can be derived analytically as a function of the two key parameters, n and m. Chern and Zuo (1997) found that their fat- and cholesterol-information indices, as constructed within a reasonable range of n and m, performed fairly well in their empirical demand model for fresh milk in the USA.

For our empirical study, we updated the indices developed by Chern and Zuo (1997) to 1997. Specifically, we used the Medline database in the Lexis/Nexis data system. The set of key words used in the search included: 'fat and cholesterol and (heart disease or arteriosclerosis)'. Thus, we added the word 'fat' to the list previously used by Brown and Schrader (1990). The search was done month by month for the period 1965–1997. Adjustments were subsequently made so that some undated articles could be assigned with an appropriate month of publication within the year. We also conducted a similar search using the database for the *Washington Post*. This database has only been available since September 1978.

Figures 2.1 and 2.2 show the results of the searches in terms of the numbers of articles related to fat and cholesterol published in all English-language medical journals in the world contained in Medline and articles in the *Washington Post*, respectively. The numbers of articles in Medline (Fig. 2.1) show an upward trend, but the monthly fluctuations are very substantial. The data from the *Washington Post* (Fig. 2.2) clearly show that the numbers of articles peaked during 1989–1990. They also exhibit a highly fluctuating monthly trend. Using the cubic weighting scheme described previously with the specific assumption of $n = 24$ and $m = 1$, the fat- and cholesterol-information indices were computed using the data series from Medline and the *Washington Post*. These indices are shown in Figs 2.3 and 2.4. These indices clearly exhibit a trend distinctively different from a simple time trend. The index from the *Post* shows that consumer fat and cholesterol information peaked in 1989–1990, sharply declined until late 1996 and then rapidly increased again. These two indices appear to reflect a similar general trend since they perform almost the same in the demand system estimated in this study, as discussed later.

These fat- and cholesterol-information indices have been used by many researchers in the USA, Japan, Korea and Europe, especially Norway. However, the methodology has met with many criticisms. Many

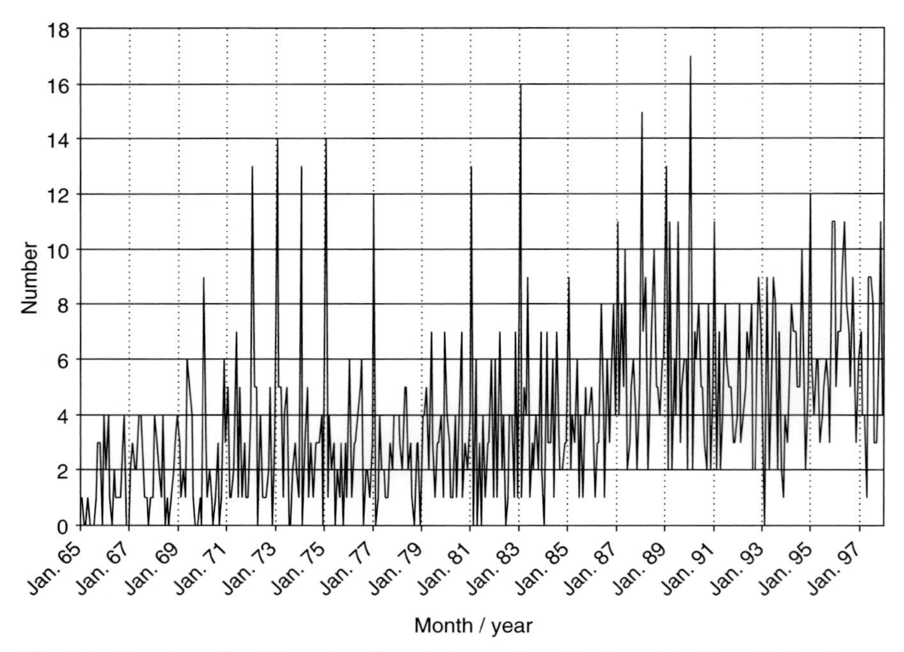

Fig. 2.1. Number of articles related to cholesterol and fat from Medline, 1965–1997.

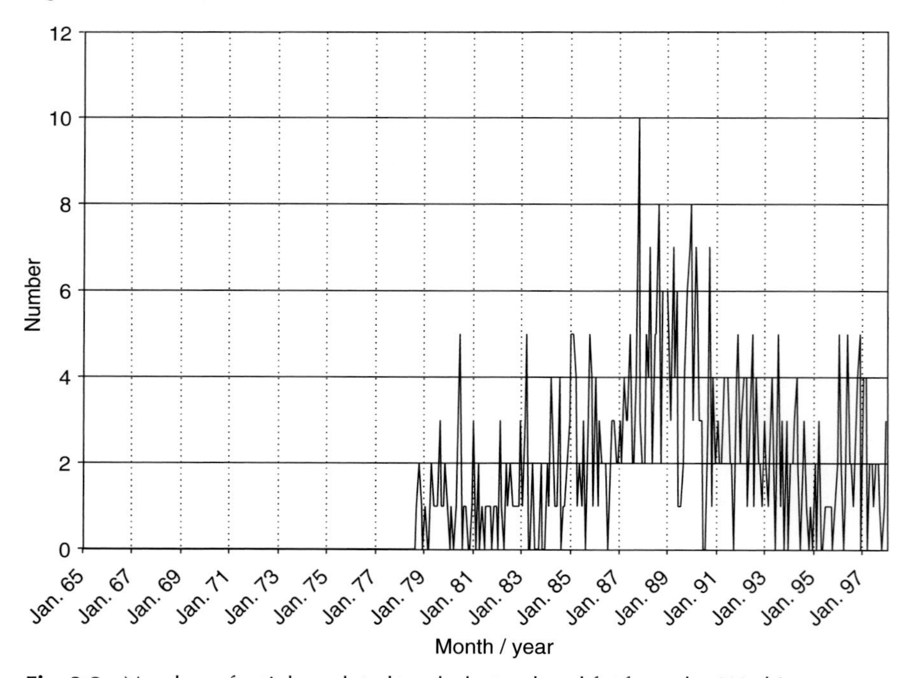

Fig. 2.2. Number of articles related to cholesterol and fat from the *Washington Post,* 1978–1997. No data available for 1965–1977.

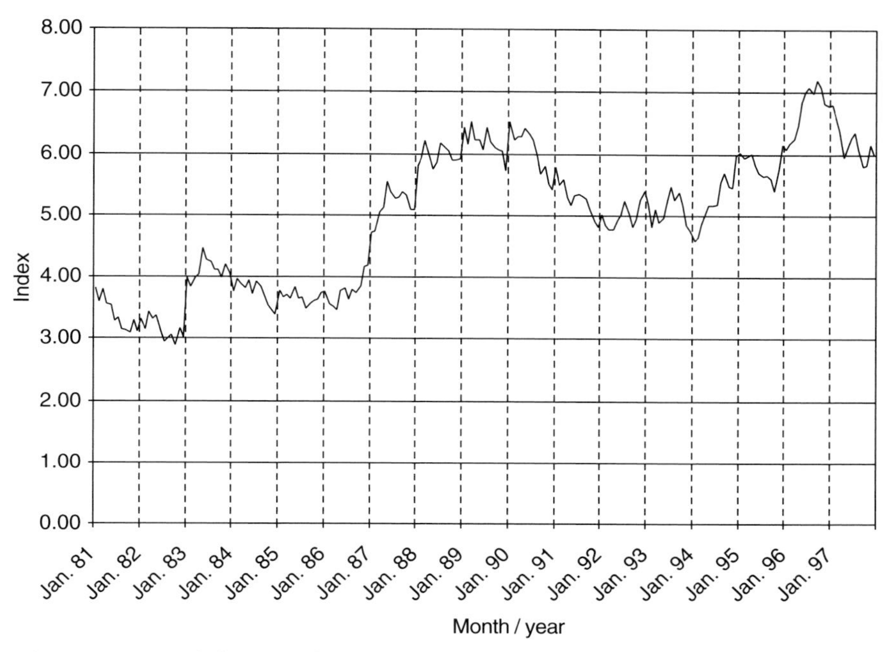

Fig. 2.3. Fat- and cholesterol-information index from Medline, 1981–1997.

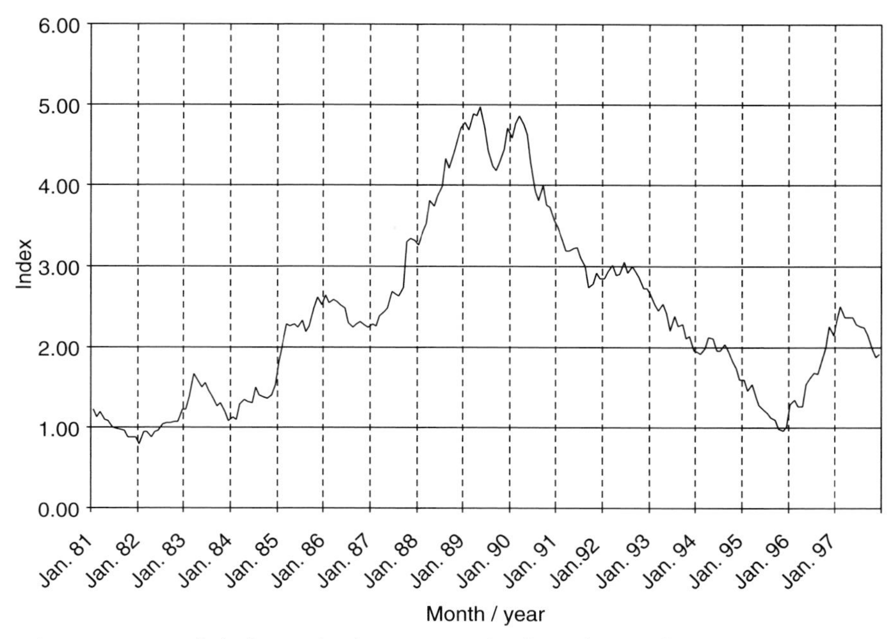

Fig. 2.4. Fat- and cholesterol-information index from the *Washington Post*, 1981–1997.

have questioned the justification for using the numbers of the published journal articles as a measure of consumer health information. It is true that consumers do not read these journals directly. But these articles are the basic sources of information used by health professionals, newspapers and magazines. Similarly, one may ask, 'How can we use the articles published in the *Washington Post* as a proxy for all news and information accessible to the US consumer?' Of course, only a small fraction of the population reads the *Washington Post*. However, we argue that its coverage may reflect the national trend during the sample period. With respect to the weighting scheme, it would be most desirable if the weighting parameters, n and m, could be estimated directly in the demand model. It is, in fact, possible that, given the duration of the lag, i.e. n, the parameters γ_i in Equation 2.17 can be estimated directly and, therefore, the parameter m will no longer be needed. These estimation procedures definitely require further investigation.

An Empirical Application

As a case-study, a food-demand model incorporating the demand for health is estimated. Specifically, we adopt the demand function expressed in Equation 2.11 based on Lancaster's characteristics-modelling framework. Equation 2.11 is used as the basic model for investigating the impacts of health-risk concerns about fat and cholesterol on the demand for ten food items in the USA. Thus, the demand for a food is specified as a function of the prices of foods, income (or total expenditure on these ten foods) and the stock of household human capital. The stock of human capital, in turn, is assumed to be dependent upon health-risk concerns, measured by the fat and cholesterol index described earlier, and a set of demographic variables.

The model

The almost ideal demand system (AIDS) developed by Deaton and Muellbauer (1980) is used for the case-study. Following Deaton and Muellbauer, the Marshallian demand functions in the budget form for the ten foods under consideration can be expressed as:

$$
\begin{aligned}
w_i &= \alpha_i + \sum_{j=1}^{10} \gamma_{ij} \log p_j + \beta_i \log(Y / P) \\
\log P &= \alpha_0 + \sum_{k=1}^{10} \alpha_k \log p_k + 1/2 \sum_{j=1}^{10} \sum_{k=1}^{10} \gamma_{kj} \log p_k \log p_j
\end{aligned}
\tag{2.18}
$$

where p_i terms are prices, w_i terms are expenditure shares and Y is the total expenditure on the ten foods under consideration. The ten food items are: (1) beef, (2) pork, (3) other meats, (4) poultry, (5) seafood, (6) eggs, (7) dairy and related products, (8) fresh vegetables, (9) fresh fruits and (10) fats and oils. These ten food groups are selected because they are mostly likely to be affected by the health-risk concerns about fat and cholesterol. The ten foods accounted for more than 50% of Americans' expenditure on foods at home (Table 2.2).

The fat- and cholesterol-information index is included in the demand system through the demographic translation, as are several other demographic variables. Specifically:

$$\alpha_i = \alpha_i^* + \sum_{j=1}^{11} d_{ij} D_j + d_{i12} HS + d_{i13} AGE + d_{i14} EAR \\ + d_{i15} FS + d_{i16} I \qquad (2.19)$$

where D_j terms are monthly dummy variables, HS is household size, AGE is the age of the household head, EAR is the number of wage earners, FS is the food stamps received by the average household, and I is the fat- and cholesterol-information index. Note that the demographic variables are the averages over the households in the samples within the month. Furthermore, we tried two alternative information indices – one based on Medline and the other from the *Washington Post*. The two variables performed equally well and produced similar elasticities. This may be due to the significance of the general trend, rather than specific peaks and fluctuations of the health-information indices in the econometric model. Both indices seem to capture a very similar general trend of fat and cholesterol information during this period. Only the results using the Medline index will be reported in this study.

The AIDS model is very popular, in part because its linear approximate version is very easy to estimate. Deaton and Muellbauer (1980) suggested replacing the general price in Equation 2.18 with the Stone index. However, Moschini (1995) points out that the Stone index is not invariant to the changes in unit of measurement, and thus it may affect the approximation properties of the model. In this study we compare the following two indices suggested by Moschini – the Tornqvist index:

$$\log P_t = 1/2 \sum_i (w_{it} + w^0{}_i) \log (p_{it} / p_i^0) \qquad (2.20)$$

and the simplified Laspeyres index:

$$\log P_t = \sum_i w^0{}_i \log p_{it} \qquad (2.21)$$

where w_i^0 is the expenditure share of good i in the base period, p_{it} is the price of good i in period t and p_i^0 is the price of good i in the base period. As it turned out, both price indices performed equally well and produced very similar elasticity estimates. The Tornqvist index involves using the current budget shares, which are also the dependent variables in the demand system, raising the issue of endogeneity. Since the Laspeyres index avoids the problem of endogeneity in the model, only the results based on the Laspeyres index are presented and discussed here.

Data

Monthly time-series data are used for the case-study. These data were obtained from the Consumer Expenditure Survey conducted by the BLS. The expenditure data are weekly household expenditure on the ten food items generated by averaging the expenditures from all the households in the sample. There are, on average, 800 households in each month. The sample period covers 1981–1995, yielding 180 monthly observations. All the demographic variables were generated in the same fashion, and thus they are all monthly averages from the sample. The price data are the consumer price index (CPI) by food group. Since the BLS collected all the expenditure and CPI data, they match exactly.

Regression results

The LA/AIDS model using the Laspeyres price index was estimated using the iterative non-linear seemingly unrelated regression (SUR) estimator (ITSUR). The equation of fats and oils was dropped in the estimation, due to the use of budget shares as dependent variables, causing a singularity problem in the variance and covariance matrix. The theoretical restrictions of homogeneity and symmetry conditions were imposed in the estimation to ensure that the estimated model was consistent with utility-maximizing consumer behaviour.

The regression results are not reported here, but are available from the author. Among the nine equations estimated in the model, eggs (i = 6) and fresh vegetables (i = 8) have the best fit, with their R^2 values of 0.88 and 0.84, respectively. The other meats (i = 3) and seafood (i = 5) have the poorest fit with an R^2 of only 0.31 and 0.35, respectively. All the other foods have a reasonably high R^2. The estimated coefficients of price and expenditure all appear to be reasonable. We shall learn more about the estimated price and income effects when their elasticities are presented later. The seasonal dummies have many significant coefficients, indicating strong seasonality in the consumption of these food items. The demographic variables are also very significant in many cases. Since the focus of this study is on the impacts of health-risk information on the demand for food, we pay particular attention to the estimated coefficients for the fat- and cholesterol-information variables. It is noted that six out of the nine coefficients associated with this information variable are statistically significant at the 5% level. These results confirm that the information index does indeed work very well in the model.

Estimated elasticities

Table 2.3 presents the expenditure and the uncompensated Marshallian price elasticity matrix computed at the sample means. Their standard errors are also computed. The results show that beef has the highest expenditure elasticity, and dairy and related products have the lowest. The results also show that these foods all have relatively high price elasticities. Furthermore, all expenditure and own-price elasticities are statistically significant.

Table 2.4 presents the estimated elasticities and their estimated standard errors for the fat- and cholesterol-information index and other demographic variables. With respect to the health-risk concern about fat and cholesterol, the results show that the information index has a negative impact on the demand for beef, pork, poultry, eggs, and fats and oils, and a positive impact on other meats, seafood, dairy and related products, fresh vegetables and fresh fruits. Among these, the estimated impacts for beef, other meats and seafood are not statistically significant. These results appear to be very plausible for the most part. It is somewhat surprising to obtain the negative impact on poultry. It is often expected that the health concern about fat and cholesterol should have induced the substitution of red meats (e.g. beef and pork), by poultry. As noted earlier, the increasing consumption trend of poultry was matched by its constant expenditure share, resulting from declining relative prices. Thus, for poultry, the price effect may have outweighed any impact related to health concerns. In the meat group, the results show that the health concern had the greatest impact on pork, then poultry and less on beef. (The impact on other meats is positive but insignificant.) Perhaps, the beef commercial campaigns have been working, as we have observed a reversal of the beef-consumption trend in recent years.

The result for dairy and related products may not be too surprising. The positive impact of the health concern may reflect the increasing consumption of low-fat milk, cheese, yoghurt and ice cream. For the

Table 2.3. Estimated expenditure and uncompensated price-elasticity matrix.[a]

		Prices									
Food	Expenditure	P1	P2	P3	P4	P5	P6	P7	P8	P9	P10
Beef	1.537***	−0.398*	−0.070	0.096	0.086	−0.269**	−0.004	−0.538***	−0.180**	−0.108	−0.151***
	(0.116)	(0.182)	(0.087)	(0.112)	(0.081)	(0.099)	(0.029)	(0.153)	(0.059)	(0.057)	(0.043)
Pork	1.229***	−0.076	−0.880***	−0.023	−0.099	0.016	−0.045	−0.053	0.062	0.116	−0.246***
	(0.127)	(0.137)	(0.125)	(0.106)	(0.085)	(0.106)	(0.030)	(0.142)	(0.063)	(0.062)	(0.063)
Other meats	0.876***	0.358	0.006	−1.146*	−0.120	−0.171	0.042	−0.406	0.267*	−0.156	0.45***
	(0.124)	(0.266)	(0.156)	(0.533)	(0.154)	(0.198)	(0.068)	(0.313)	(0.105)	(0.086)	(0.118)
Poultry	1.054***	0.264	−0.109	−0.121	−0.915***	0.017	−0.066	−0.32	0.395***	0.105	−0.304***
	(0.137)	(0.168)	(0.110)	(0.137)	(0.155)	(0.141)	(0.043)	(0.193)	(0.079)	(0.075)	(0.085)
Seafood	1.301***	−0.761**	0.024	−0.247	0.003	−1.034**	0.055	0.377	0.158	0.080	0.044
	(0.230)	(0.287)	(0.193)	(0.246)	(0.197)	(0.321)	(0.073)	(0.325)	(0.140)	(0.127)	(0.139)
Eggs	0.726***	0.158	−0.124	0.128	−0.181	0.167	−0.252***	−0.616**	−0.275**	−0.033	0.301**
	(0.127)	(0.181)	(0.119)	(0.191)	(0.136)	(0.164)	(0.071)	(0.221)	(0.088)	(0.080)	(0.110)
Dairy and related products	0.580***	−0.224	0.043	−0.107	−0.076	0.142	−0.063*	−0.335*	−0.159***	−0.074	0.273***
	(0.075)	(0.115)	(0.068)	(0.100)	(0.071)	(0.085)	(0.026)	(0.167)	(0.048)	(0.044)	(0.049)
Fresh vegetables	0.829***	−0.237*	0.103	0.217**	0.362***	0.121	−0.074**	−0.404***	−0.903***	−0.052	0.039
	(0.117)	(0.104)	(0.070)	(0.081)	(0.068)	(0.087)	(0.025)	(0.114)	(0.070)	(0.049)	(0.062)
Fresh fruits	1.057***	−0.14	0.151*	−0.139*	0.094	0.064	−0.018	−0.281*	−0.071	−0.744***	0.024
	(0.105)	(0.107)	(0.072)	(0.070)	(0.068)	(0.083)	(0.023)	(0.113)	(0.051)	(0.066)	(0.044)
Fats and oils	0.916***	−0.085	−0.111	0.163	−0.112	0.022	0.039	0.294	0.018	0.018	−1.162***
	(0.130)	(0.185)	(0.114)	(0.180)	(0.124)	(0.148)	(0.048)	(0.229)	(0.093)	(0.077)	(0.125)

[a]Figures in parentheses are estimated standard errors; *** indicates significance at the 1% level, ** at the 5% level and * at the 10% level.

Table 2.4. Estimated elasticities for information and demographic variables.[a]

Food	Information index	Household size	Age	Number of earners	Food stamps income
Beef	−0.061	−0.220	−1.008**	−0.158	−0.016
	(0.043)	(0.258)	(0.383)	(0.163)	(0.018)
Pork	−0.243***	0.365	−0.741	−0.413*	0.051*
	(0.047)	(0.284)	(0.421)	(0.179)	(0.020)
Other meats	0.036	0.246	0.809	−0.112	−0.021
	(0.055)	(0.281)	(0.415)	(0.189)	(0.020)
Poultry	−0.144**	0.287	−0.054	−0.419*	0.022
	(0.052)	(0.307)	(0.458)	(0.196)	(0.022)
Seafood	0.059	−1.117*	1.736*	−0.152	0.042
	(0.095)	(0.513)	(0.764)	(0.327)	(0.036)
Eggs	−0.131*	1.161***	−1.080*	−0.538**	−0.06**
	(0.052)	(0.283)	(0.425)	(0.182)	(0.020)
Dairy and related products	0.118***	−0.004	0.341	0.397***	−0.019
	(0.031)	(0.166)	(0.249)	(0.106)	(0.012)
Fresh vegetables	0.106*	−0.282	0.305	0.342*	−0.001
	(0.042)	(0.263)	(0.398)	(0.165)	(0.019)
Fresh fruits	0.197***	0.001	0.102	0.020	0.015
	(0.039)	(0.235)	(0.352)	(0.149)	(0.017)
Fats and oils	−0.044	0.100	0.045	0.061	−0.005
	(0.052)	(0.292)	(0.441)	(0.186)	(0.021)

[a]Figures in parentheses are estimated standard errors; *** indicates significance at the 1% level, ** at the 5% level and * at the 10% level.

dairy sector, further disaggregation of dairy products would be necessary in order to estimate more clearly the impacts of health concerns on individual products, such as whole milk.

The estimated impacts of health concern on fresh fruits and vegetables are of course very plausible. As noted in Table 2.2, both the quantities and expenditure shares of fruits and vegetables were increasing during the sample period. Americans have been consuming and paying more for fruits and vegetables in response to their health concerns about fat and cholesterol.

The results for fats and oils suggest that the concerns regarding fat and cholesterol have a negative impact on the demand for those foods with high fat content. However, the impact is not statistically significant. Chern *et al.* (1995) have found that health-risk concerns induced substitutions between animal fats and vegetable oils. However, they also found that the overall impact resulted in a net increase in the total consumption of fats and oils.

Conclusions

This chapter reviews the conceptual models used to incorporate the demand for health in food-demand analysis. Becker's household production theory offers a more rigorous conceptualization of the demand for health than the characteristics model developed by Lancaster. But the model of the Lancaster type is easier to implement in empirical studies. It links health-risk information directly to the characteristics of food. There are also other important methodological issues that remain unresolved in the treatment of the demand for health and the demand for food in empirical applications. These subjects require further investigation.

Significant progress has been made on the measurement of health-risk information as a proxy for health-risk concern/knowledge. Various information indices related to fat and cholesterol have been developed, using the numbers of articles published in medical journals over time (Medline) or in popular newspapers such as the *Washington Post*. Employing specific lag-distribution schemes, we are able to show that consumer information on fat and cholesterol may not exhibit an ever-increasing trend. This methodology of measuring consumer health information is yet to be accepted by the economics and health professions.

The case-study involves the estimation of the LA/AIDS for ten foods with either high or low fat content in the USA. The focus of the study is to evaluate the performance of the fat- and cholesterol-information indices in the demand model. The results show that health-risk concerns have had a significant impact on food-consumption behaviour. Specifically, health-risk concern is shown to increase the consumption of fresh fruits and vegetables and dairy products, but to reduce the consumption of meats, eggs, and fats and oils. These changing patterns of food consumption are certainly desirable from the health point of view. This study is among the first to show such strong evidence of the impacts of health concern on so many food products in a single model. The study also validates the appropriateness of using the numbers of articles published in medical journals as a measure of consumer health information. The findings also imply that medical research around the world has had a significant impact on the consumer's dietary pattern in the USA. The other chapters by European authors in this book will reveal other econometric results of the impact of health-risk concerns on food consumption in Europe.

Notes

[1] If the demand were affected by the relative prices, the appropriate operational strategy would be to reduce the costs of production for the firms to be competitive in the market-place.

References

Becker, G.S. (1965) A theory of the allocation of time. *Economic Journal* 75, 493–517.

Braschler, C. (1983) The changing demand structure for pork and beef in the 1970s: implications for the 1980s. *Southern Journal of Agricultural Economics* 15, 105–110.

Brown, D.J. and Schrader, L.F. (1990) Information on cholesterol and shell egg consumption. *American Journal of Agricultural Economics* 73, 548–555.

Capps, O. Jr and Schmitz, J.D. (1991) A recognition of health and nutrition factors in food demand analysis. *Western Journal of Agricultural Economics* 16, 21–35.

Chalfant, J.A. and Alston, J.H. (1988) Accounting for changes in tastes. *Journal of Political Economy* 96, 391–410.

Chavas, J.P. (1983) Structural change in the demand for meat. *American Journal of Agricultural Economics* 65, 148–153.

Chern, W.S. and Zuo, J. (1995) Alternative measures of changing consumer information on fat and cholesterol. Paper presented in the *Annual Meeting of the American Agricultural Economics Association*, Indianapolis, Indiana, 6–9 August.

Chern, W.S. and Zuo, J. (1997) *Impacts of Changing Health Information of Fat and Cholesterol on Consumer Demand: Application of New Indices*. Discussion Paper No. 443, Institute of Social and Economic Research, Osaka University, Osaka, Japan, 35 pp.

Chern, W.S., Loehman, E.T. and Yen, S.T. (1995) Information, health-risk beliefs, and the demand for fats and oils. *Review of Economics and Statistics* 76, 555–564.

Deaton, A. and Muellbauer, J. (1980) An almost ideal demand system. *American Economic Review* 70, 312–326.

Frazao, E. (1999) *America's Eating Habits*. Agriculture Information Bulletin Number 750, US Department of Agriculture, Washington, DC, 473 pp.

Grossman, M. (1972) On the concept of health capital and the demand for health. *Journal of Political Economy* 80, 223–255.

Kenkel, D. (1990) Consumer health information and the demand for medical care. *Review of Economics and Statistics* 72, 587–595.

Kim, S.R. and Chern, W.S. (1997) Indicative measures of health-risk information on fat and cholesterol for US and Japanese consumers. *Consumer Interest Annual* 43, 84–89.

Kim, S.R. and Chern, W.S. (1999) Alternative measures of health information and demand for fats and oils in Japan. *Journal of Consumer Affairs* 33, 92–109.

Kim, S.R., Nayga, M. Jr and Capps, O. Jr (2001) Health knowledge and consumer use of nutritional labels: the issue revisited. *Agricultural and Resource Economics Review* 30, 10–19.

Lancaster, K. (1971) *Consumer Demand: a New Approach*. Columbia University Press, New York, 177 pp.

Moschini, G. (1995) Units of measurement and the Stone index in demand system estimation. *American Journal of Agricultural Economics* 77, 63–67.

Moschini, G. and Meilke, K.D. (1984) Parameter stability and the US demand for beef. *Western Journal of Agricultural Economics* 9, 271–282.

Nayga, R.M. (1997) Impact of sociodemographic factors on perceived importance of nutrition in food shopping. *Journal of Consumer Affairs* 31, 1–9.

Nayga, R.M. Jr (2000) Nutrition knowledge, gender, and food label use. *Journal of Consumer Affairs* 34, 97–112.

Paulin, G.D. (1998) *The Changing Food-at-home Budget: 1980 and 1992 Compared.* Monthly Labor Review. Bureau of Labor Statistics, Washington, DC.

Putnam, J.J. and Allshouse, J.E. (1997) *Food Consumption, Prices, and Expenditures, 1970–95.* Statistical Bulletin No. 939, Economic Research Service, US Department of Agriculture, Washington, DC, 146 pp.

Putnam, J., Kantor, L.S. and Allshouse, J. (2000) Per capita food supply trends: progress toward dietary guidelines. *Food Review* 23, 2–14.

Robenstein, R.G. and Thurman, W.N. (1996) Health-risk and the demand for red meat: evidence from futures markets. *Review of Agricultural Economics* 18, 629–641.

Variyam, J.N., Blaylock, J. and Smallwood, D. (1996) A probit latent variable model of nutrition information and dietary fibre intake. *American Journal of Agricultural Economics* 78, 628–639.

Yen, S.T. and Chern, W.S. (1992) Flexible demand systems with serially correlated errors: fat and oil consumption in the United States. *American Journal of Agricultural Economics* 74, 689–697.

Health, Nutrition and Demand for Food: a European Perspective

3

KYRRE RICKERTSEN[1] AND
STEPHAN VON CRAMON-TAUBADEL[2]

[1]*Agricultural University of Norway, Ås, Norway;*
[2]*University of Göttingen, Göttingen, Germany*

Introduction

Since at least 1963, the link between cholesterol in the diet and heart disease has been researched and become increasingly well documented (Brown and Schrader, 1990). As medical research has stressed the relationship between diet and health, government agencies have been trying to educate the public about the advantages of healthful diets. Therefore, it is of interest to evaluate the impact of health research and information on behaviour using actual market data. The purpose of this study is to evaluate, using a well-established methodology, the impact of health information on the demand for food in Europe, with a special emphasis on meats and fish.

Previous studies have focused on effects on the demand for eggs (e.g. Brown and Schrader, 1990); beef, pork, poultry and fish (e.g. Capps and Schmitz, 1991); edible fats (e.g. Yen and Chern, 1992; Gray *et al.*, 1998); and fresh milk (Chern and Zuo, 1995), using data from the USA and Canada. Kaabia *et al.* (2001) and Rickertsen *et al.* (2003) studied the effects of health information on the demand for meat in Spain and the Nordic countries, respectively.

France, Germany, Norway, Scotland and Spain, which represent a wide range of geographical and cultural differences, are included in the study. These countries vary considerably as regards both diet and mortality rates from diet-related illness, such as heart disease. For example, the per capita consumption of vegetable products is much higher in Mediterranean than in northern Europe, while in a ranking of 31 countries presented by the Scottish Home Office and Health Department (1993), Scotland was ranked first, Norway 11th, Germany

21st, Spain 28th and France 30th in mortality rates from coronary heart disease for individuals between 40 and 60 years of age.

Most information concerning the impact of diet on health is quite international in nature, at least in the industrialized countries, and is available to consumers through medical advice published in mass media, by governmental campaigns or directly by the medical profession. Given the differences and similarities between the countries included in this study, it is interesting to investigate whether health information has significant effects on the consumption of food in these countries and how these effects, if any, differ from country to country.

Trends in Consumption

A variety of authors have concluded that dietary patterns are becoming more homogeneous across European countries – and industrialized countries in general – over time (Blandford, 1984; Gil *et al.*, 1995; Herrmann and Röder, 1995). Nevertheless, considerable differences in dietary patterns and in consumer responses to changes in the main determinants of demand remain between countries in Europe (Angulo *et al.*, 2000). It is difficult to compare levels of consumption in the European countries we have studied because of differences in official data sources. Our consumption data from France, Germany and Norway are constructed from disappearance data, while the Scottish and Spanish data are based on household budget surveys. For most food items, per capita consumption derived from household surveys is lower than consumption derived from disappearance data. These differences can be substantial, as shown for Norway (see Rickertsen and Kristofersson, Chapter 10, this volume). Loss and waste in processing and distribution cause many of the observed differences. The Food and Agriculture Organization (FAO) publishes disappearance data for all the countries; however, the quality of these data is questionable and we shall not present them.

Trends in consumption may be more comparable regardless of whether we use household survey or disappearance data. Annual percentage changes in per capita consumption for the 1980–1998 period are presented in Table 3.1. The decline in fluid milk consumption is general, but the picture is more mixed for other products. Beef consumption has been going down, except in Norway, and the consumption of poultry and fish has increased, except in Spain. Pork consumption has been going up, except in Germany. With the exception of France, the consumption of eggs has declined, while cheese consumption has increased, except in Scotland. Fruits and vegetables are divided into fresh and processed in the French and Spanish data. There are no clear trends in the consumption of fruits, vegetables, cereals or fats across Europe. Sugar consumption is falling in all countries, except in Norway.

Table 3.1. Annual percentage changes in per capita consumption of food products in selected European countries for the period 1980–1998.

Food item	France[a]	Germany[b]	Scotland[c]	Spain	Norway
Meats	0.11	–0.44	–0.53	–0.19	0.81
Beef	–1.34	–2.77		–1.29	0.25
Pork	0.48	–0.21		2.27	0.52
Lamb					0.08
Poultry	1.24	2.31		–1.76	5.68
Other meats	0.04	–0.68		0.32	0.22
Fish	3.14[d]	1.24	0.57	–1.11	0.76
Eggs	0.26	–1.31	–4.64	–2.16	–0.25[f]
Cheese	1.54	2.24	–0.26	0.25	1.03
Fluid milk	–0.53	–0.48	–1.51	–0.54	–1.40
Whole		–1.05			–8.20
Low-fat		1.20			7.49[e]
Non-fat		–4.63			0.25
Vegetables and fruits	–0.21	0.45	–1.07	–0.99	0.51[f]
Vegetables	0.05	0.93	–2.38	–0.74	1.61[f]
Fruits	–0.21	0.03	2.64	–1.79	0.15[f]
Processed vegetables and fruits	2.24			3.57	
Potatoes	–1.47				0.09[f]
Grains and cereals	–0.44	0.53	–0.93	–0.92	0.81
Sugar	–3.59	–0.65	–5.13	–4.48	0.47
Fats and oils	0.16	0.71	–2.50	–1.66	–0.88

[a]French data only available until 1997.
[b]West German data for 1980 and German data for 1998 are used.
[c]Scottish data cover the period 1979–1998.
[d]Data only available until 1990.
[e]Data from 1985 to 1998.
[f]Data from 1979 to 1998.

Health Information

In studying the impact of health information on observed consumption, a major challenge is to construct a variable that measures how informed consumers are about the links between diet and health. As in previous studies, the hypothesis underlying this study is that results from basic medical research are disseminated through scientific medical journals. Medical personnel, consumer advocates, journalists and representatives of the food industry read these journals and can transfer the resulting knowledge to consumers. Consumers' behaviour regarding cholesterol will therefore change as scientific information accumulates and is disseminated.

Several indices have been constructed to approximate the effects of health information on consumers' consumption decisions. The indices used in published studies are generated by counting the number of articles in the Medline database. Medline contains abstracts of articles published in roughly 4000 major medical journals. The records in Medline are indexed in any one of 67 languages. In the database, 47% of the records stem from European journals and 44% from North American journals (Pratt, 1994). Hence, Medline is a very rich database and contains much of the world's knowledge on the links between diet, cholesterol and heart disease.

However, there are some problems associated with using a count of articles as a proxy for the practically unobservable level of health information. First, the number of articles that is identified by a search in Medline – and hence the nature of the resulting health-information index – has been found to change substantially depending on what keywords are chosen to define the search. Secondly, simple counts ignore the fact that some articles, published in leading journals, may have a much stronger and longer-lasting impact on health knowledge and consumer behaviour than others that are published in less well known or prestigious journals. Furthermore, some articles/journals may be more influential in some countries than in others. Thirdly, while the assumption that medical research publications influence consumer behaviour is reasonable, the actual process of dissemination of the information in these publications is likely to be highly complex. Little or no attempt has been made in the pertinent literature to distinguish between the effects of personal and non-personal information, neutral and market-oriented information, or information that corroborates and information that contradicts the link between cholesterol and heart disease. The latter problem is further complicated by the fact that the discussion of possible linkages is changing over time. The focus has gradually changed from dietary fat and cholesterol to 'good' and 'bad' cholesterol and, further, to the omega-3 fatty acids found in fish-oils and the *trans*-fats found in certain vegetable oils.

Even though none of these problems is easily solved, health information is clearly important for consumption decisions and should be accounted for. Moreover, since different indices show quite similar patterns over time, we believe that they may be used as reasonable proxies for the level of health information possessed by consumers.

Brown and Schrader (1990) (B&S) assume that health information accumulates and thus they created a health-information index based on the cumulative number of published articles. This index has been widely used in North American studies. B&S initially used quarterly data for the 1966–1987 period and later extended the resulting index. They use the keywords 'cholesterol and (heart disease or arteriosclerosis)' and discard as irrelevant articles that focus on smoking, alcohol abuse, etc.

Each article that provides evidence of a link adds one unit to the index, and each article that provides evidence in the opposite direction subtracts one unit. Furthermore, B&S assume a 6-month lag between publication and the date an article first affects consumption.

The B&S index may suffer from several weaknesses. First, only articles published in US (and later Canadian) journals are included. The exclusion of European and other non-American journals is an obvious weakness for European applications. Secondly, the cumulative nature of the index may be a problem. It seems reasonable to assume that an article's influence will decline over time and may even vanish if the message is not repeated. The assumption that information has an infinite lifetime is especially questionable when the focus of this information is changing substantially over time. Finally, the B&S index closely resembles a time trend, which may lead to spurious results.

Chern and Zuo (1995) (C&Z) have constructed a new monthly index that covers the period 1980–1994 and which has been extended by Kim and Chern (1999) to cover 1965–1994. The C&Z index differs from the B&S index in several respects. First, the C&Z index includes 'fat(s)' as an additional keyword. Secondly, C&Z include articles from all English-language journals in their search. Thirdly, C&Z assume that all articles have a carry-over and a decay effect. They experiment with quadratic and cubic distributed lag functions and various lifespans for articles (12 and 24 months) to model these effects. Fourthly, they do not distinguish between articles that support and articles that question the link to health because the number of questioning articles is small (1% of all articles in the period 1966–1969, 6% in 1970–1979 and 4% in 1980–1989). Furthermore, consumers may be especially responsive to 'negative' information, such as that linking saturated-fat and cholesterol intake to heart diseases, and a simple subtraction rule may over-emphasize 'positive' information. Finally, C&Z do not manually discard 'irrelevant' articles in constructing their index.

Rickertsen and Slåen (Slåen, 1999) constructed the adjusted global index (GIA), which is used in the present study. The GIA is a modified C&Z index that covers all English-language journals. First, the keyword 'diet' is included to make an explicit link between diet and heart disease. The decision to include this keyword was based on experiments that consisted of counting the number of hits elicited by various strings of keywords for the month of January 1994. The keywords 'fat(s) and cholesterol and (heart disease or arteriosclerosis)' resulted in 83 hits, while the apparently minor change to '(fat(s) or cholesterol) and (heart disease or arteriosclerosis)' increased the number of hits to 516. Our selected keywords '(fat(s) or cholesterol) and (heart disease or arteriosclerosis) and diet' resulted in 155 hits and found most of the relevant articles without including an excessive number of irrelevant articles. Secondly, a total of 4708 abstracts were read to remove 528 articles that either were duplicates or judged to be 'irrelevant' because they focused on veterinary research, experimental methods, smoking and cholesterol, cholesterol and eye, joint, skin or gall-bladder disease, and snoring and cholesterol.

Tokoyama and Chern have simultaneously developed an (unpublished) index (FATMED2) based on the above keywords, using a polynomial distributed lag, but not discarding 'irrelevant' articles. The global index before adjustment (GI) and after adjustment (GIA) and the FATMED2 index are presented in Fig. 3.1. The number of 'irrelevant' articles has increased since 1980. The GIA and FATMED2 follow a similar pattern, although the development in FATMED2 is somewhat smoother.

The Almost Ideal Demand System Incorporating Health Information

To investigate the effects of health information, we assume weak separability and estimate a two-stage model for each country. The first stage distinguishes between five aggregates of food (meats and fish; dairy and eggs; oils and fats; cereals, fruits, vegetables and potatoes; and other foods and beverages) and one non-food group. The second stage breaks down meats and fish into beef and veal, pork, lamb, poultry, sausage and other processed meat, and fish and fish products.

Deaton and Muellbauer's (1980) almost ideal demand (AID) model is frequently used in applied food-demand analysis. The linear approximate version of the model is used here. The ith good's expenditure share, w_i, is given by:

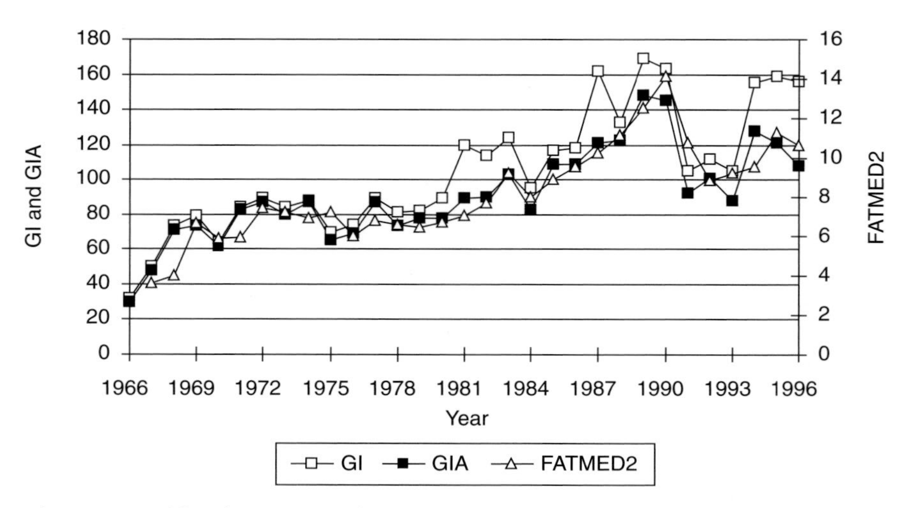

Fig. 3.1. Health-information indices.

$$w_i = \alpha_i + \sum_j \gamma_{ij} \ln p_j + \beta_i [\ln x - \sum_j w_{j(1995)} \ln p_j] \quad (3.1)$$

where p_i is the per unit price of good i, and x is per capita total expenditure. The 1995 expenditure shares replace actual expenditure shares in Stone's price index, so the resulting index satisfies the properties of a price index (Moschini, 1995).

The effects of health information are introduced through modifications of the constant term, α_i, in Equation 3.1. In the models for France, Norway and the first stage for Spain, annual data are used. In the second stage for Spain quarterly data are used. The German and Scottish models are estimated using monthly data. Because of these different data frequencies, it is not possible to incorporate the effects of health information and seasonality in the same way in each of the country models. Based on experiments with data and lag structures, the following specifications were selected.

France: $$\alpha_i = \alpha_{i0} + \alpha_{i1} t + \lambda_i \log GIA_t \quad (3.2)$$

Germany: $$\alpha_i = \alpha_{i0} + \sum_{d=1}^{11} \delta_{id} D_d + \sum_{h=0}^{6} \omega_{ih} GIA_{t-h} \quad (3.3)$$

Norway: $$\alpha_i = \alpha_{i0} + \lambda_i \sum_{h=0}^{2} \omega_h GIA_{t-h} \quad (3.4)$$

Scotland: $$\alpha_i = \alpha_{i0} + \sum_{h=0}^{10} \omega_{ih} GIA_{t-h} \quad (3.5)$$

Spain: $$\alpha_i = \alpha_{i0} + \lambda_i GIA_t \text{ (stage 1)}$$
$$\alpha_i = \alpha_{i0} + \sum_{d=1}^{3} \delta_{id} D_d + \sum_{h=0}^{4} \omega_{ih} GIA_{t-h} \text{ (stage 2)} \quad (3.6)$$

Here t is a trend variable and D_d terms are seasonal dummy variables. Unrestricted distributed lags are applied in the models for Germany, Scotland and Spain, while the lag weights are restricted to sum to 1 ($\Sigma_h \omega_h = 1$ where h denotes the number of lags) in the Norwegian model. Adding up implies the following restrictions:

$$\sum_i \alpha_{i0} = 1, \sum_i \alpha_{i1} = \sum_i \beta_i = \sum_i \lambda_i = 0, \sum_i \delta_{id} = 0 \ \forall d,$$
$$\sum_i \gamma_{ij} = 0 \ \forall j, \sum_i \omega_{ih} = 0 \ \forall h \quad (3.7)$$

Homogeneity and symmetry are imposed on the system through the restrictions:

$$\sum_j \gamma_{ij} = 0 \quad \forall i \quad \text{and} \quad \gamma_{ij} = \gamma_{ji} \tag{3.8}$$

Marshallian own- (e_{ii}) and cross-price (e_{ij}) elasticities are calculated according to Asche and Wessells (1997), such that:

$$e_{ii} = -1 + \left(\frac{\gamma_{ii}}{w_i}\right) - \beta_i \quad \text{and} \quad e_{ij} = \left(\frac{\gamma_{ij}}{w_i}\right) - \beta_i \left(\frac{w_j}{w_i}\right), \quad i \neq j \tag{3.9}$$

The corresponding Hicksian elasticities (e_{ij}*) are calculated by using the general relationship $e_{ij}^* = e_{ij} + w_j E_i$, with E_i, the expenditure elasticities, given by:

$$E_i = 1 + \frac{\beta_i}{w_i} \tag{3.10}$$

Health information elasticities, η_i, are calculated as:

$$\text{Norway:} \qquad \eta_i = \lambda_i \frac{GIA}{w_i} \tag{3.11}$$

$$\text{France:} \qquad \eta_i = \frac{\lambda_i}{w_i} \tag{3.12}$$

$$\text{Germany, Scotland and Spain:} \qquad \eta_i = \sum_h \omega_{ih} \frac{GIA}{w_i} \tag{3.13}$$

All elasticities are calculated at the 1995 values of expenditure shares and health-information indices.

Data and Empirical Implementation

The data employed are, for the most part, collected by national statistical agencies in the participating countries. In the following, the few exceptions are noted. Annual national accounts data for 1959–1997 are used to estimate the French model. Monthly household data for the period 1976–1997 (stage one) and 1966–1997 (stage two) are used for the German model. Results for a typical household consisting of a married couple with two children and a middle income are presented. Since lamb is a minor product in Germany, it is excluded from the study. Annual national accounts data for 1966–1995 are used to estimate stage one of the Norwegian model. Stage two of the Norwegian model is estimated using disappearance data collected by the Norwegian Agricultural Economics Research Institute and consumer price index data. No Norwegian price data exist for the group 'sausage and other processed meat' and the group is excluded from the study. The Scottish model is estimated using monthly household expenditure survey data collected by the National Food Survey for 1977–1996. Annual national accounts data for the 1966–1995 period are used to estimate stage one of the Spanish model. The second stage of this model is based on quarterly household survey data on expenditures for 1985:3 to 1995:4, while prices are extracted from the consumer price index.

Autocorrelation is frequently a major problem in time-series estimations. First-order autocorrelation was tested for equation by equation, using the Breusch–Godfrey test. This test is only strictly valid in a single-equation framework; hence, the *P* values presented in Table 3.2 must be interpreted with some caution. The first column for each country gives the *P* values prior to any correction for autocorrelation. The *P* values show the lowest significance level at which the null hypothesis of no autocorrelation can be rejected. It is rejected at the 5% level if the corresponding *P* value is below 0.05. First-order autocorrelation is apparently a problem in the French, German and Norwegian models. Results (not shown in the table) of a multivariate

Table 3.2. Results of equation-by-equation tests for first-order autocorrelation before/after correction, *P* values.[a]

	France	Germany	Norway	Scotland
Stage one				
Meats and fish	**0.00**/0.99	**0.00**/0.09	**0.01**/0.82	0.12/0.18
Dairy products and eggs	0.14/0.09	0.07/0.65	**0.01**/0.85	0.10/0.17
Carbohydrates	**0.00/0.00**	**0.00/0.04**	**0.00**/0.57	0.08/0.13
Fats and oils	**0.00**/0.40	**0.00**/0.54	**0.02**/0.96	0.10/0.17
Other foods	0.15/0.63	**0.00**/0.99	**0.00**/0.82	0.10/0.34
Other goods	**0.00/0.00**	[b]	**0.00**/0.57	0.10/0.22
Stage two				
Beef and veal	0.19/0.58	**0.00**/0.92	0.25/0.38	0.08/0.19
Pork	0.44/0.51	**0.00**/0.44	0.40/0.26	0.10/0.19
Lamb	**0.00**/0.42	[c]	0.26/0.53	**0.04**/0.19
Poultry	0.07/0.57	**0.02/0.02**	**0.00**/0.06	0.08/0.18
Processed meat	0.44/0.57	[b]	[c]	0.09/0.18
Fish	0.71/0.17	**0.00**/0.47	0.22/0.21	0.07/0.19

[a] Significant values at the 5% level are printed in bold type.
[b] The equation was deleted from the estimation and the hypothesis was not tested.
[c] 'Lamb' in Germany and 'Processed meat' in Norway were not included in the analysis.

Breusch–Godfrey test (Godfrey, 1988) indicate that autocorrelation is also a problem in the Spanish models. The food group cereals, fruits, vegetables and potatoes is referred to as 'carbohydrates' in the tables.

All of the models, except for the German model, are estimated using maximum-likelihood methods. As discussed below, three-stage least squares is used in the German case. Due to the differences in the data frequencies employed, different corrections for autocorrelation are adopted in the different countries. The French models are estimated in first-difference form and the total expenditure at stage one is instrumented by income to control for endogeneity. The Scottish models are estimated using a correction similar to Berndt and Savin's (1975) general correction for autocorrelation.

The German team applied the Bewley (1979) transformation. For Norway and Spain a partial adjustment model was selected. These models can be specified in terms of modifications of the constant terms, α_{i0}, in Equations 3.3, 3.4 and 3.6 such that:

$$\text{Germany:} \quad \alpha_{i0} = \alpha_{i00} + \sum_{j=1}^{n-1} \theta_{ij} \Delta w_{j,t-1} + \sum_{j=1}^{n-1} \phi_{ij} \Delta w_{j,t-12} \tag{3.14}$$

$$\text{Norway and Spain:} \quad \alpha_{i0} = \alpha_{i00} + \sum_{j=1}^{n-1} \theta_{ij} w_{j,t-1} \tag{3.15}$$

The Berndt and Savin (1975) correction is used when the Norwegian models are estimated to remove system autocorrelation. In the German case, the models are estimated using $w_{j,t-1}$ and $w_{j,t-12}$ as instruments for $\Delta w_{j,t-1}$ and $\Delta w_{j,t-12}$.

The second column for each country in Table 3.2 gives the P values after these corrections. The autocorrelation is apparently reduced. Results (not shown in the table) of the system test indicate that the autocorrelation is removed from the Spanish model as well.

Price and Total Expenditure Effects

Complete tables of uncompensated price and expenditure elasticities are presented in the Appendix. Here we focus on the own-price elasticities presented in Table 3.3 and the expenditure elasticities presented in Table 3.4 to see if there are evident patterns across countries. It is quite difficult to see any European pattern. There is a wide spread in the values of the own-price elasticities for all goods except for the group 'other goods'. However, there are no significantly positive own-price elasticities and all the French and Scottish own-price elasticities are significantly negative. The demands for various meats are

Table 3.3. Uncompensated own-price elasticities at 1995 values.[a]

	France	Germany	Norway	Scotland	Spain
Stage one					
Meats and fish	**–1.40**	**–0.65**	**–0.97**	**–0.95**	–0.06
Dairy products and eggs	**–1.20**	–0.29	0.06	**–1.09**	–0.23
Carbohydrates	**–1.25**	**–0.95**	**–1.37**	**–0.79**	0.11
Fats and oils	**–1.19**	–0.58	**–1.29**	**–0.31**	1.63
Other foods	**–1.46**	0.02	**–0.58**	**–0.88**	**–0.23**
Other goods	**–1.11**	**–1.01**	**–0.94**	**–1.11**	**–0.88**
Stage two					
Beef and veal	**–1.72**	**–3.09**	**–0.70**	**–0.60**	**–0.79**
Pork	**–1.25**	**–1.48**	**–0.57**	**–1.03**	**–0.59**
Lamb	**–1.89**	[b]	**–0.56**	**–0.83**	–0.67
Poultry	**–1.13**	**–1.10**	**–0.45**	**–1.36**	**–0.71**
Processed meat	**–1.10**	**–0.81**	[b]	**–1.31**	**–0.47**
Fish	**–1.44**	**–1.10**	**–0.91**	**–0.59**	**–0.92**

[a]Significant values at the 5% level are printed in bold type.
[b]'Lamb' in Germany and 'Processed meat' in Norway were not included in the analysis.

Table 3.4. Total expenditure elasticities at 1995 values.[a]

	France	Germany	Norway	Scotland	Spain
Stage one					
Meats and fish	**0.89**	–0.06	**0.94**	**0.90**	**0.90**
Dairy products and eggs	**0.73**	**0.18**	0.71	**1.72**	**0.60**
Carbohydrates	**0.86**	**0.51**	**1.48**	**0.93**	0.21
Fats and oils	**1.01**	–0.47	0.72	**0.47**	0.23
Other foods	**1.00**	–0.05	**1.36**	**1.05**	**0.41**
Other goods	**1.03**	**1.28**	**0.95**	**0.69**	**1.14**
Stage two					
Beef and veal	**0.82**	0.31	**0.98**	**1.40**	**1.31**
Pork	**1.17**	**1.67**	**1.00**	**0.67**	0.63
Lamb	**1.25**	[b]	0.40	**1.25**	**2.87**
Poultry	**0.93**	**0.75**	**1.61**	**1.96**	**0.56**
Processed meat	**1.04**	**0.94**	[b]	**0.53**	**0.42**
Fish	**1.15**	**0.72**	**1.16**	**0.63**	**1.28**

[a]Significant values at the 5% level are printed in bold type.
[b]'Lamb' in Germany and 'Processed meat' in Norway were not included in the analysis.

surprisingly price elastic in France and also partly in Germany and Scotland. The demands for all types of meat are price inelastic in Norway and Spain.

None of the total expenditure elasticities is significantly negative. The French and Scottish expenditure elasticities are again all significant. The expenditure elasticities for poultry, fish, other meats and other goods are significant in all countries. We note a surprisingly high number of expenditure-elastic food groups and types of meat in the various countries. However, we have to remember that the presented elasticities are conditional on the separability structure and different from the corresponding unconditional elasticities.

The Effects of Health Information

The health-information elasticities for 1995 are presented in Table 3.5. It is impossible to detect any clear pattern. Health information does not have a consistent positive or negative effect on the demand for any specific good across all countries. Focusing on statistically significant elasticities, the most surprising elasticity at stage one is the numerically large and positive effect on the demand for fats and oils in Norway. One possible, but not entirely satisfying, explanation for this finding is that this group consists of 'healthy' fats (e.g. olive oil) as well as 'unhealthy' fats (e.g. lard). At stage two, most of the significant effects are surprising.

Table 3.5. Health-information elasticities.[a]

	France	Germany	Norway	Scotland	Spain
Stage one					
Meats and fish	–0.01	–0.06	0.03	**–0.03**	**–0.05**
Dairy products and eggs	0.00	0.03	–0.05	**–0.06**	**–0.10**
Carbohydrates	–0.01	0.01	–0.02	0.04	–0.04
Fats and oils	–0.02	–0.08	**0.67**	–0.01	**–0.15**
Other foods	0.00	–0.04	**0.10**	**0.07**	–0.02
Other goods	0.00	0.01	–0.01	**0.04**	0.02
Stage two					
Beef and veal	0.01	**–0.23**	–0.02	–0.02	**0.09**
Pork	0.00	**0.09**	0.04	–0.00	–0.09
Lamb	–0.01	[b]	–0.09	**0.08**	–0.06
Poultry	0.00	0.00	**0.51**	–0.13	–0.00
Processed meat	–0.00	0.01	[b]	–0.03	0.02
Fish	–0.01	–0.01	**–0.08**	0.00	–0.03

[a]Significant values at the 5% level are printed in bold type.
[b]'Lamb' in Germany and 'Processed meat' in Norway were not included in the analysis.

Positive effects on beef demand in Spain and lamb demand in Scotland and a negative effect on fish demand in Norway are not as might be expected a priori. No significant effects are found in France.

Conclusions

The impact of health information on food demand varies considerably from country to country and is often insignificant and occasionally unexpected. Based on these results, it is not possible to draw clear conclusions regarding the impact of health information on the demand for food in general and meat in particular in the EU. Several problems with the data, as well as more fundamental problems related to the measurement of health information, may at least partly explain our unexpected results.

First, the sample period varies from 1959–1997 (France) to 1985:3–1995:4 (the Spanish meat model). There is reason to believe that the relatively volatile level of health information in the late 1980s and the 1990s had different effects on demand than the more trended development in previous years. Furthermore, household survey data are used in some countries while aggregate market data are used in others. These sample differences will affect the empirical results but could not be avoided, given data availability in the participating countries.

Secondly, the results may not be robust with respect to choices regarding specification and estimation. In this regard, it is interesting to note that most of the participating country specialists have been able to produce more plausible results, as discussed in later chapters, by deviating somewhat from the 'common' model that has been presented here. An important question is whether these deviations and the 'improved' results that they generate represent an inevitable outcome of data mining, and should thus be discounted, or whether they are indications that it is possible to 'home in' on the truth by using theory, common sense and exploratory techniques. Furthermore, the numerical values of the estimated elasticities are quite sensitive to the method used to correct for autocorrelation. This sensitivity is illustrated using Norwegian data (see Rickertsen and Kristofersson, Chapter 10, this volume), where it is demonstrated that the health-information elasticities are sensitive to choice of correlation mechanism for autocorrelation. A related problem is that the health-information index employed is seasonal (see Wildner and von Cramon-Taubadel, Chapter 7, this volume), which may lead to spurious results in the models that use monthly (Germany, Scotland) or quarterly (Spain) data.

Thirdly, commodity aggregation may cause some problems, especially at the first stage, because perceived healthy products (e.g.

chicken) are aggregated with less healthy products (e.g. beef). The oils and fats group is also problematic in this regard, as mentioned above. For this reason, it is perhaps unreasonable to expect consistent results from the first stage of the demand model, as the estimated coefficients and elasticities represent 'hybrids' that combine effects of different sign and magnitude. In general, it is apparent that national statistics are not collected with the needs of demand-system estimation in mind, and even less so with a view to providing a basis for comparative inter-EU demand analysis.

Finally, as discussed above, counting articles in medical journals is likely to produce only an approximate measure of the level of health information. Some articles may be very influential while others have little or no influence on consumer behaviour. No explicit attempt is made to model how information moves from medical journals to consumers, and how consumers then translate it into demand behaviour. It is reasonable to expect that these processes will have changed over time as the relative importance of different media (i.e. print vs. television) and the degree of consumer sensitivity to topics such as diet and health (i.e. as a function of the demographic structure of the population) have changed.

Acknowledgements

Financial support for this research has been received from the EU, contract no. FAIR5–CT97–3373, and is greatly appreciated. The following have contributed to the EU research project entitled 'Nutrition, Health and the Demand for Food', from which the results are presented in this chapter: David D. Mainland (coordinator) and John Santarossa (Scotland); Lars Erik Aarstad, Dadi Kristofersson, Kyrre Rickertsen and Bjørn Slåen (Norway); Ana M. Angulo, Jose M. Gil, Azucena Gracia and Monia Ben Kaabia (Spain); Stephan von Cramon-Taubadel and Susanne Wildner (Germany); Jérôme Adda, Pierre Combris and Véronique Nichèle (France). We would like to thank Wen S. Chern for help with the construction of the health-information indices and two anonymous reviewers for helpful comments.

References

Angulo, A.M., Gil, J.M. and Gracia, A. (2000) *Calorie Intake and Income Elasticities in EU Countries: a Convergence Analysis using Cointegration.* University of Zaragoza, Spain.

Asche, F. and Wessells, C.R. (1997) On price indices in the almost ideal demand system. *American Journal of Agricultural Economics* 79, 1182–1185.

Berndt, E.R. and Savin, N.E. (1975) Evaluation and hypothesis testing in singular equation systems with autoregressive disturbances. *Econometrica* 43, 937–957.

Bewley, R.A. (1979) The direct estimation of the equilibrium response in a linear dynamic model. *Economic Letters* 61, 357–361.

Blandford, D. (1984) Changes in food consumption patterns in the OECD area. *European Review of Agricultural Economics* 11, 43–65.

Brown, D. and Schrader, L.F. (1990) Cholesterol information and shell egg consumption. *American Journal of Agricultural Economics* 72, 548–555.

Capps, O. Jr and Schmitz, J. (1991) A recognition of health and nutrition factors in food demand analysis. *Western Journal of Agricultural Economics* 16, 21–35.

Chern, W.S. and Zuo, J. (1995) Alternative measures of changing consumer information on fat and cholesterol. Paper presented at the American Agricultural Economics Association's Annual Meeting, Indianapolis.

Deaton, A. and Muellbauer, J. (1980) An almost ideal demand system. *American Economic Review* 70, 312–326.

Gil, J.M., Gracia, A. and Pérez y Pérez, L. (1995) Food consumption and economic development in the European Union. *European Review of Agricultural Economics* 22, 385–399.

Godfrey, L.G. (1988) *Misspecification Tests in Econometrics*. Cambridge University Press, Cambridge, 252 pp.

Gray, R., Malla, S. and Stephen, A. (1998) Canadian dietary fat substitutions, 1955–93, and coronary heart disease costs. *Canadian Journal of Agricultural Economics* 46, 233–246.

Herrmann, R. and Röder, C. (1995) Does food consumption converge internationally? Measurement, empirical tests and the influence of policy. *European Review of Agricultural Economics* 22, 400–414.

Kaabia, M.B., Angulo, A.M. and Gil, J.M. (2001) Health information and the demand for meat in Spain. *European Review of Agricultural Economics* 28, 499–517.

Kim, S. and Chern, W.S. (1999) Alternative measures of health information and demand for fats and oils in Japan. *Journal of Consumer Affairs* 33, 92–109.

Moschini, G. (1995) Units of measurement and the Stone index in demand system estimation. *American Journal of Agricultural Economics* 77, 63–68.

Pratt, G.F. (1994) A brief hitchhiker's guide to MEDLINE. *Database* 17, 41–49.

Rickertsen, K., Kristofersson, D. and Lothe, S. (2003) Effects of health information on Nordic meat and fish demand. *Empirical Economics* (in press).

Scottish Home Office and Health Department (1993) *The Scottish Diet*. Her Majesty's Stationery Office, Edinburgh, Scotland.

Slåen, B. (1999) Effekten av helseinformasjon på etterspørselen etter mat. MSc thesis (in Norwegian), Agricultural University of Norway, Ås, Norway.

Yen, S.T. and Chern, W.S. (1992) Flexible demand systems with serially correlated errors: fat and oil consumption in the United States. *American Journal of Agricultural Economics* 74, 689–697.

Appendix: Price and Expenditure Elasticities by Country

All elasticities are calculated at the 1995 values of expenditure shares and health-information indices. Bold type indicates significance at the 5% level.

Table A3.1. Expenditure and Marshallian price elasticities in France, stage 1.

Elasticity with respect to	Meats and fish	Dairy products and eggs	Fats and oils	Carbo-hydrates	Other foods	Non-durables
Total expenditure	**0.89**	**0.73**	**1.01**	**0.86**	**1.00**	**1.03**
Price: meats and fish	**–1.40**	0.03	0.01	0.01	0.08	**0.03**
Price: dairy products and eggs	0.01	**–1.20**	0.04	**–0.08**	0.02	**0.01**
Price: fats and oils	0.00	0.02	**–1.19**	–0.02	–0.04	0.01
Price: carbohydrates	0.01	**–0.14**	–0.12	**–1.25**	0.06	0.02
Price: other foods	0.06	0.06	–0.22	0.07	**–1.46**	**0.02**
Price: non-durables	**0.43**	**0.51**	0.48	**0.41**	**0.34**	–1.11

Table A3.2. Expenditure and Marshallian price elasticities in France, stage 2.

Elasticity with respect to	Beef and veal	Pork	Poultry	Fish	Processed meat	Lamb
Expenditure	**0.82**	**1.17**	**0.93**	**1.15**	**1.04**	**1.25**
Price: beef and veal	**–1.72**	**0.31**	0.05	**0.44**	**0.22**	**0.58**
Price: pork	**0.13**	**–1.25**	0.06	–0.00	**–0.10**	0.07
Price: poultry	0.05	0.09	**–1.13**	–0.09	0.00	0.22
Price: fish and fish products	**0.27**	–0.01	–0.04	**–1.44**	–0.02	0.04
Price: sausage and processed meat	**0.30**	**–0.36**	0.03	–0.08	**–1.10**	–0.26
Price: lamb	**0.16**	0.06	0.10	0.02	–0.04	**–1.89**

Table A3.3. Expenditure and Marshallian price elasticities in Germany, stage 1.

Elasticity with respect to	Meats and fish	Dairy products and eggs	Fats and oils	Carbo-hydrates	Other foods	Non-durables
Expenditure	–0.07	**0.18**	–0.48	**0.50**	–0.05	**1.28**
Price: meats and fish	**–0.67**	**0.16**	0.00	0.10	0.07	0.41
Price: dairy products and eggs	**0.29**	–0.30	**0.16**	**0.28**	–0.08	**–0.52**
Price: fats and oils	0.08	**0.85**	–0.58	**0.58**	0.46	–0.91
Price: carbohydrates	0.08	**0.16**	**0.06**	**–0.96**	–0.05	0.20
Price: other foods	0.15	–0.09	0.10	–0.06	0.01	–0.06
Price: non-durables	**–0.22**	**–0.24**	**–0.20**	**–0.19**	**–0.22**	**–1.02**

Table A3.4. Expenditure and Marshallian price elasticities in Germany, stage 2.

Elasticity with respect to	Beef and veal	Pork	Poultry	Fish	Processed meat
Expenditure	0.33	**1.67**	**0.74**	**0.72**	**0.94**
Price: beef and veal	**–3.12**	**2.08**	0.16	0.09	0.45
Price: pork	**0.85**	**–1.46**	–0.05	–0.02	**–0.98**
Price: poultry	0.12	0.06	**–1.10**	**–0.06**	**0.24**
Price: fish and fish products	0.22	0.04	**–0.17**	**–1.10**	0.30
Price: sausage and processed meat	0.05	**–0.14**	**0.06**	**0.05**	**–0.81**

Table A3.5. Expenditure and Marshallian price elasticities in Norway, stage 1.

Elasticity with respect to	Meats and fish	Dairy products and eggs	Fats and oils	Carbo-hydrates	Other foods	Non-durables
Expenditure	**0.94**	0.71	0.72	**1.48**	**1.36**	**0.95**
Price: meats and fish	**–0.97**	**–0.16**	**–0.11**	**0.10**	**0.17**	0.03
Price: dairy products and eggs	**–0.26**	0.06	0.02	0.13	**–0.44**	–0.22
Price: fats and oils	**–1.50**	0.13	**–1.29**	**–1.14**	1.10	1.97
Price: carbohydrates	0.09	0.07	**–0.11**	**–1.37**	**–0.26**	0.10
Price: other foods	**0.08**	**–0.17**	0.05	**–0.12**	**–0.58**	**–0.62**
Price: non-durables	0.00	–0.02	**0.01**	**0.03**	–0.03	**–0.94**

Table A3.6. Expenditure and Marshallian price elasticities in Norway, stage 2.

Elasticity with respect to	Beef and veal	Pork	Poultry	Fish	Lamb
Expenditure	**0.98**	**1.00**	**1.61**	**1.16**	0.40
Price: beef and veal	**–0.70**	**–0.34**	0.03	0.10	–0.07
Price: pork	**–0.33**	**–0.57**	–0.02	**–0.13**	0.05
Price: poultry	0.06	**–0.43**	**–0.45**	–0.21	**–0.59**
Price: fish and fish products	0.09	**–0.27**	–0.02	**–0.91**	–0.06
Price: lamb	–0.08	**0.41**	**–0.19**	0.02	**–0.56**

Table A3.7. Expenditure and Marshallian price elasticities in Scotland, stage 1.

Elasticity with respect to	Meats and fish	Dairy products and eggs	Fats and oils	Carbo-hydrates	Other foods	Non-durables
Expenditure	**0.90**	**1.72**	**0.47**	**0.93**	**1.05**	**0.69**
Price: meats and fish	**–0.95**	**0.04**	**–0.08**	–0.02	–0.03	**0.13**
Price: dairy products and eggs	–0.06	**–1.09**	**–0.05**	**–0.05**	**–0.07**	**–0.25**
Price: fats and oils	**–0.60**	–0.01	**–0.31**	–0.04	0.04	**0.45**
Price: carbohydrates	–0.14	–0.00	**–0.04**	**–0.79**	–0.01	0.06
Price: other foods	–0.13	0.03	–0.01	–0.01	**–0.88**	–0.05
Price: non-durables	**0.41**	**–0.07**	**0.10**	0.04	0.01	**–1.11**

Table A3.8. Expenditure and Marshallian price elasticities in Scotland, stage 2.

Elasticity with respect to	Beef and veal	Pork	Poultry	Fish	Processed meat	Lamb
Expenditure	**1.40**	**0.67**	**1.96**	**0.63**	**0.53**	**1.25**
Price: beef and veal	**–0.60**	**–0.23**	–0.02	**–0.18**	–0.08	**–0.30**
Price: pork	–0.16	**–1.03**	–0.00	–0.02	**0.32**	**0.20**
Price: poultry	–0.23	–0.25	**–1.36**	**–0.60**	0.35	0.13
Price: fish and fish products	–0.06	–0.02	–0.10	**–0.59**	0.08	0.05
Price: sausage and processed meat	0.10	**0.37**	0.15	0.11	**–1.31**	–0.03
Price: lamb	**–0.37**	0.09	0.07	–0.06	–0.15	**–0.83**

Table A3.9. Expenditure and Marshallian price elasticities in Spain, stage 1.

Elasticity with respect to	Meats and fish	Dairy products and eggs	Fats and oils	Carbo-hydrates	Other foods	Non-durables
Expenditure	**0.90**	**0.60**	0.23	0.21	**0.41**	**1.14**
Price: meats and fish	–0.06	0.04	**–0.16**	–0.03	0.07	**–0.76**
Price: dairy products and eggs	**0.13**	–0.23	0.06	–0.05	–0.03	–0.48
Price: fats and oils	**–1.41**	**0.21**	1.63	**–0.72**	0.04	0.01
Price: carbohydrates	**0.02**	**–0.01**	**–0.10**	0.11	**–0.15**	–0.08
Price: other foods	**0.15**	**–0.01**	**0.01**	**–0.19**	**–0.23**	–0.13
Price: non-durables	**–0.10**	**–0.03**	**–0.01**	**–0.07**	**–0.05**	**–0.88**

Table A3.10. Expenditure and Marshallian price elasticities in Spain, stage 2.

Elasticity with respect to	Beef and veal	Pork	Poultry	Fish	Processed meat	Lamb
Expenditure	**1.31**	0.63	**0.56**	**1.28**	**0.42**	**2.87**
Price: beef and veal	**−0.79**	0.07	−0.07	−0.07	−0.30	−0.16
Price: pork	**0.23**	**−0.59**	−0.16	**−0.59**	0.37	0.10
Price: poultry	0.00	**−0.13**	**−0.71**	0.06	0.09	**0.14**
Price: fish and fish products	−0.03	**−0.23**	**−0.05**	**−0.92**	−0.18	**0.12**
Price: sausage and processed meat	−0.02	0.13	0.04	0.08	**−0.47**	**−0.18**
Price: lamb	**−0.65**	−0.05	0.00	0.12	−1.63	−0.67

4 Double Impact: Educational Attainment and the Macronutrient Intake of US Adults

JAYACHANDRAN N. VARIYAM
Economic Research Service, USDA, Washington, DC, USA

Introduction

The benefits of education, from its influence on earnings to its impact on a variety of 'non-monetary' outcomes, such as health, crime and the environment, have been the subject of extensive research in economics and related social sciences (Mincer, 1974; Juster, 1975; Ashenfelter and Krueger, 1994; Behrman and Stacey, 1997). Precise estimates of the educational effects on these outcomes are crucial for guiding policy debates on investment in education by individuals and government. The role of educational attainment in people's health and health-related behaviours is gaining attention due to persistent evidence of schooling as 'the most important correlate of good health', according to Fuchs (1979).

Interest in the economics of dietary behaviour in general and the role of education and health information in particular has increased due to two reasons. First, the economic stakes of dietary behaviour are high: $71 billion per year (1995 dollars) in medical costs, lost productivity and the value of premature deaths due to diet-related illnesses, such as heart disease and cancer, which are the leading causes of death in the USA (Frazao, 1999). Secondly, the amount of scientific and public information linking diet and health has increased sharply over recent years. The government's role in providing information and nutrition education for the public has seen a parallel increase through nutrition-labelling regulations and the popularization of the Dietary Guidelines for Americans and the Food Guide Pyramid. The increased governmental role is motivated by the recognition that, among the numerous factors influencing dietary choices, nutrition information is the most amenable to modification (Thomas, 1991).

While nutrition information is a key 'input' into the production of a healthy diet, the acquisition and use of information are driven by consumer's human capital, often proxied by educational attainment.[1] Under household production theory, which provides the framework for many of the empirical analyses of health behaviour, information is the key pathway through which education influences health behaviour and health outcomes (Strauss and Thomas, 1996). The strong positive relationship between nutrition information and educational level is a well-established empirical fact (Levy *et al.*, 1993; Gould and Lin, 1994; Variyam *et al.*, 1998). The strength of this relationship, coupled with the theoretical underpinning of the household production theory, has led many previous studies on dietary behaviour to use educational attainment to capture the extent of nutrition information available to households and individuals (Ippolito and Mathios, 1990; Bhattacharya and Currie, 2001).

In this chapter, we estimate the effect of educational attainment on the dietary intakes of US adults. Many previous studies have estimated the effect of education on nutrient intakes and diet quality. However, to our knowledge, all these studies have restricted themselves to estimating the effect of education at the conditional mean of the intake distributions (Nayga, 1994; Adelaja *et al.*, 1997). By focusing only on the marginal effect at the conditional mean, these estimates presuppose that the influence of education is identical along the entire range of intake distribution. The contribution of this chapter is to estimate the influence of education at different points along the entire conditional distribution of intakes and to test whether the marginal effects at these points are identical.

Estimating the marginal effect of education at different points along the conditional distribution of intakes is important because our interest in dietary intakes stems from the need to better understand those at risk of excessive or inadequate intakes. But the risk of inadequacy or excess is greater at the tails of the intake distributions rather than the mean. Tables 4.1 and 4.2 illustrate this point. Table 4.1 reports the mean and selected percentiles of the daily intake of four major macronutrients among men and women 18 years or older, obtained from the US Department of Agriculture's 1994–1996 Continuing Survey of the Food Intakes by Individuals (CSFII). These univariate statistics were estimated using combined 3-year sampling weights. Table 4.2 reports the recommended daily intakes for these same macronutrients. Comparing these tables, it is clear that in most cases both the mean and median intakes are within the daily intake levels recommended by health authorities. For example, the recommended total fat intake for men between 19 and 50 years of age is less than or equal to 96.7 g fat day^{-1} based on a 2900-calorie diet. The mean and median intakes of total fat for men during 1994–1996 are about 92 and 84 g, both well within this recommended level. However, intake at the 90th percentile is

Table 4.1. Sample statistics for the distribution of macronutrient intake among US adults, 1994–1996.[a]

Nutrient	Units	Percentile					Mean
		10	25	50	75	90	
Men							
Total fat	g	43.2	60.2	83.8	114.5	146.0	91.8
Saturated fat	g	13.3	19.4	27.9	38.8	50.9	31.0
Cholesterol	mg	121.4	181.6	283.2	423.2	594.8	329.7
Fibre	g	8.0	11.6	16.2	22.9	29.8	18.1
Women							
Total fat	g	27.2	39.7	55.7	73.9	94.6	59.2
Saturated fat	g	8.4	12.3	18.1	25.0	32.6	19.7
Cholesterol	mg	71.6	113.2	176.5	272.9	389.3	209.6
Fibre	g	5.9	8.6	12.4	16.9	22.3	13.5

[a]Weighted estimates. Men $N = 4882$ and women $N = 4714$.

Table 4.2. Recommended daily intakes of selected macronutrients. (From Lin *et al.*, 1999: 6, Table 4.)

Sex and age	Nutrient				
	Energy (cal)	Total fat[a] (g)	Saturated fat[b] (g)	Cholesterol (mg)	Fibre (g)
Men					
19–20	2900	≤ 96.7	<32.2	≤ 300	Age + 5
21–24	2900	≤ 96.7	<32.2	≤ 300	33.4
25–50	2900	≤ 96.7	<32.2	≤ 300	33.4
51+	2300	≤ 76.7	<25.6	≤ 300	26.5
Women					
19–20	2200	≤ 73.3	<24.4	≤ 300	Age + 5
21–24	2200	≤ 73.3	<24.4	≤ 300	25.3
25–50	2200	≤ 73.3	<24.4	≤ 300	25.3
51+	1900	≤ 63.3	<21.1	≤ 300	21.9

[a]Based on the recommendation that no more than 30% of total calories come from fat.
[b]Based on the recommendation that less than 10% of total calories come from saturated fat.

146 g, considerably above the healthful level. Similarly, for women, the cholesterol intake even at the 75th percentile, 273 milligrams (mg), is below the recommended daily intake of 300 mg, whereas at the 90th percentile the intake is 389 mg, well above the healthful level.

Since the risk of excess or inadequacy of nutrient intake is located at the tails rather than at the mean, the marginal influence of a characteristic such as education at the conditional mean may provide an incomplete picture of its effect on risky dietary behaviour. Therefore, it is important to examine whether the marginal influence is different depending on the location along the intake distribution. If we were studying a nutrient whose excessive consumption is a health issue, policy makers would be more concerned if the educational impact were predominantly at the upper end of the distribution as opposed to the lower end of the distribution or the conditional mean. If the effects are larger at parts of the distribution with greater risk of inadequacy or excess, the 'average' estimates (marginal effect at the conditional mean) may undervalue the benefits of education.

In this chapter, we employ the quantile regression method to better characterize the impact of educational attainment on parts of the conditional distribution of dietary intakes other than the conditional mean. The quantile regression approach relaxes the assumption that the effects of explanatory variables are constant along the whole distribution of the dependent variable and allows such effects to vary over the entire range of dietary intakes. Using national food-consumption survey data, we estimate the effects of educational attainment on the intakes of total fat, saturated fat, cholesterol and fibre by US adults at five separate quantiles. We focused on these four macronutrients because of their links to cardiovascular disease, obesity and certain types of cancer (National Research Council, 1989).

Empirical Framework

Our study, similarly to most previous empirical work on the effect of education on health and health-related behaviours, is framed within the context of Grossman's (1972) health-demand model. According to this model, education affects health outcomes by raising technical and allocative efficiencies of health-input use (Grossman and Kaestner, 1997). Technical efficiency causes the more educated to produce a larger health output from a given level of health inputs. Allocative efficiency causes the more educated to acquire and use more information about the true effects of inputs on health. To estimate the marginal effect of education on intakes, we employed regression functions of the form:

$$Y_{ij} = \beta_0 + \beta_1 S_i + \beta_2 S_i^2 + \beta_3 X_{i3} + \ldots + \beta_K X_{iK} + \varepsilon_{ij}, \quad i = 1, \ldots, N, \quad j = 1, \ldots, M \tag{4.1}$$

where Y_{ij} is the intake of the jth nutrient of the ith individual, S_i is the educational attainment of the ith individual measured by the number of years of schooling completed, $X_{i3}, \ldots, X_{iK}$ are other sociodemographic and anthropometric characteristics associated with the ith individual, and ε_{ij} is the error term. These empirical regression functions approximate the reduced-form health-input demand functions generated from the household production model (Grossman, 1972; Strauss and Thomas, 1996). In a fully specified structural model of health-input demand, educational attainment is properly treated as an endogenous variable, but this introduces estimation issues, such as the availability of relevant instruments, that are outside the scope of this chapter. Nayga (2000a,b) discusses the role of education as a causal variable in health production and approaches to modelling the endogeneity of education.

Besides education, other sociodemographic factors included in our estimated regressions include factors that influence an individual's health-production efficiency and investment decisions. For example, if the demand for health is inelastic and if health stock depreciates at an increasing rate with age, then health investment will increase with age (Grossman, 1972). In terms of nutrient intake, this implies that older individuals will be more likely to have better diets than younger individuals. Income appears in the reduced-form demand function through the budget constraint and it may also influence information-processing efficiency (Ippolito and Mathios, 1990). Based on previous empirical studies and the need to isolate the informational effect of education, other variables to control for include household size, region, degree of urbanization, and race and ethnicity. Given that dietary intakes are measured over short periods and do not represent long-run intakes, it is also important to include anthropometric measures, such as height and weight.[2]

Data

The nutrient-intake data were obtained from the US Department of Agriculture's 1994–1996 CSFII (Tippett and Cypel, 1997). Each year of the 3-year CSFII comprised a nationally representative sample of non-institutionalized persons residing in the USA. Dietary data for selected sample persons from a screened sample of 9664 households were collected on 2 non-consecutive days through in-person interviews using 24-h recalls. For both days, 15,303 sample persons provided information on food intakes, giving a 2-day response rate of 76.1%. From the sample persons providing complete 2-day dietary intake records, we selected men and women 24 or above in age. We restricted the sample to those at or above age 24 so that an individual could have attained the maximum possible level of recorded educational attainment (17 years) at that age. By combining the food records with a nutrient database, CSFII provides information on the intakes of a variety of macronutrients, vitamins and minerals. We used the mean daily intakes of total fat, saturated fat, cholesterol and fibre as our dependent variables.

Table 4.3 lists the explanatory variables used in the regressions. These fall into three groups: household characteristics, personal characteristics and survey-related variables. Educational attainment, household income, household size, age, height and weight are continuous

Table 4.3. Explanatory variables, means and sample size.[a]

Variable	Men	Women
Household characteristics		
Gross annual income ($10,000)	3.99	3.64
	(2.69)[b]	(2.65)
Household size	2.83	2.73
	(1.47)	(1.50)
Region (northeast omitted)		
Midwest	0.24	0.25
South	0.36	0.36
West	0.21	0.20
Urbanization (city omitted)		
Suburb	0.46	0.44
Non-metropolitan	0.27	0.25
Personal characteristics		
Level of education (years)	12.71	12.53
	(3.27)	(3.06)
Age (years)	51.53	51.73
	(16.55)	(16.38)
Height (in)	69.72	63.97
	(3.02)	(2.84)
Weight (lb)	184.36	152.92
	(34.30)	(35.69)
Race–ethnicity (Non-Hispanic white omitted)		
Non-Hispanic black	0.09	0.13
Non-Hispanic other	0.03	0.03
Hispanic	0.08	0.08
Sample size	4346	4003

[a]Squares of income, education, age, height and weight were included in all regressions. Additionally, 14 dummy variables were also used in all regressions: two for the survey years (1995, 1996; base = 1994), six for the survey seasons (spring, summer and autumn for each of the 2 days of intake; base = winter), two indicating whether the intake was recorded on a weekend (one dummy variable for each day of the 2-day intake) and four indicating respondent's opinion whether each day of the recorded intake was less than or more than his or her usual intake (base = intake was usual).
[b]Standard deviations are reported in parentheses.

variables. The remaining variables are dummy indicator variables. As noted earlier, we used the years of schooling completed and its square in all regressions. To account for other possible non-linear effects, the quadratic form was also used for income, age, height and weight.[3] Height and weight were included to control for the influence of body mass on the amount of food intake. While previous studies have often used the body mass index (BMI) for this purpose, estimating a single coefficient for BMI implies a restriction on the coefficients for the BMI components, height and weight. Therefore, we left height and weight in the unrestricted form.

Besides the variables listed in Table 4.3, 14 additional dummy variables were included as explanatory variables in all regressions. These were two for the survey years (1995, 1996; base=1994), six for the survey seasons (spring, summer and autumn for each of the 2 days of intake; base = winter), two indicating whether the intake was recorded on a weekend (one dummy variable for each day of the 2-day intake) and four indicating the respondent's opinion whether each day of the recorded intake was less than or more than his or her usual intake (base = intake was usual). Since the intake is the mean of a 2-day dietary recall, these variables are used to account for some of the random day-to-day variation in intakes.

Quantile Regression

Koenker and Bassett (1978) introduced quantile regression as a generalization of the sample quantiles to conditional quantiles expressed as a linear function of explanatory variables. This is analogous to the ordinary least squares (OLS) regression model, which expresses the conditional mean in a linear form. However, by permitting *any* conditional quantile to be expressed in linear form, quantile regression enables one to describe the entire conditional distribution of the dependent variable, given a set of regressors. The familiar least absolute deviation (LAD) estimator is a special case of quantile regression that expresses the conditional median as a linear function of covariates.

Quantile regression's ability to characterize the whole conditional distribution is most potent when there is heteroscedasticity in the data (Deaton, 1997). When the data are homoscedastic, the set of slope parameters of conditional quantile functions at each point of the dependent variable's distribution will be identical with each other and with the slope parameters of the conditional mean function. In this case, the quantile regression at any point along the distribution of the dependent variable reproduces the OLS slope coefficients, and only the intercepts will differ. However, when the data are heteroscedastic, the set of slope parameters of the conditional quantile functions will differ from each other as well as from the OLS slope parameters. Therefore, estimating conditional quantiles at various points of the distribution of the dependent variable will allow us to trace out different marginal responses of the dependent variable to changes in the explanatory variables at these points.

Two additional features of the quantile regression model are relevant to our application (Buchinsky, 1998). First, the classical properties of efficiency and minimum variance of the least-squares estimator are obtained under the restrictive assumption of independently, identically and normally distributed (i.i.d.) errors. When the distribution of errors is non-normal, the quantile regression estimator may be more efficient than the least-squares estimator. Secondly, the quantile regression estimator is 'robust' when the dependent variable has outliers or the error distribution is 'long-tailed'. Since the objective function from which the quantile regression estimator is derived is a weighted sum of absolute deviations, the parameter estimates are less sensitive to a few large or small observations at the tails of the distribution. The distributional statistics reported in Table 4.1 show that the mean intakes of macronutrients are consistently higher than the median (50th percentile) intakes. This suggests that the intakes are asymmetrically distributed, with some influential observations at the upper tails. Quantile regression ensures that the parameter estimates are less sensitive to such observations, compared with OLS.

Estimation

The quantile regression model for Equation 4.1 can be written as:

$$y_i = \boldsymbol{x}_i'\beta_\theta + \varepsilon_{\theta i}, \qquad Q_\theta\left(y_i \middle| \boldsymbol{x}_i\right) = \boldsymbol{x}_i'\beta_\theta \tag{4.2}$$

where $\boldsymbol{x}_i$ denotes a $(K + 1) \times 1$ vector of the explanatory variables, β_θ is the corresponding vector of coefficients and $Q_\theta(y_i|\boldsymbol{x}_i)$ denotes the θth conditional quantile of y_i $(0 < \theta < 1)$. From Equation 4.2, the quantile regression estimator of β_θ is obtained by solving:

$$\min_{\beta_\theta} \frac{1}{N}\left\{ \sum_{i:y_i \geq \boldsymbol{x}_i'\beta_\theta} \theta \left| y_i - \boldsymbol{x}_i'\beta_\theta \right| + \sum_{i:y_i < \boldsymbol{x}_i'\beta_\theta} (1-\theta) \left| y_i - \boldsymbol{x}_i'\beta_\theta \right| \right\} \tag{4.3}$$

When $K = 0$ and $\boldsymbol{x}_i$ is a (1×1) vector that includes only the intercept for all i, this minimization problem reduces to an estimator of the sample θth quantile. The minimization problem in Equation 4.3 has a linear programming representation, which is guaranteed to have a solution in a finite number of simplex iterations (Buchinsky, 1998). Several estimators for the asymptotic covariance matrix for $\hat{\beta}_\theta$ obtained from the above minimization are available, but, for obvious reasons, those that rely on the assumption of i.i.d. errors are of limited value (Deaton, 1997). Buchinsky (1995) has shown that the design-matrix bootstrap estimator provides a consistent estimator for the covariance matrix under very general conditions. In the design-matrix bootstrap, quantile regression is re-estimated on a sample of N observations $(y_i^*, \boldsymbol{x}_i^*)$, $i = 1, \ldots, N$, drawn randomly from the original sample. The process is repeated B times to obtain bootstrap estimates $\hat{\beta}^*_{\theta b}$, $b = 1, \ldots, B$. *The* covariance matrix of $\hat{\beta}_\theta$ is then obtained as the covariance of $\hat{\beta}^*_\theta$ computed from the B bootstrap estimates with $\hat{\beta}_\theta$ as the pivotal value. Additional details regarding the estimation of the quantile regression model and the estimation of the asymptotic covariance matrix of the parameters are discussed in Buchinsky's (1998) methodological survey.

Results

Conditional quantile functions for the intake of the four macronutrients were estimated at five representative quantiles, separately for men and women.[4] Tables 4.4–4.7 report the slope-coefficient estimates for education, with their absolute t values given in the parentheses. Since education was entered in the quadratic form, the slope estimates were calculated as $\hat{\beta}_1 + 2\hat{\beta}_2 S$, where $\hat{\beta}_1$ and $\hat{\beta}_2$ are the coefficients of level of education and its square and S is the level of education reported in the first column of each table. For comparison with the quantile estimates, the second column in each table presents the OLS slope estimates. The standard errors for the quantile regression estimates were obtained using the design-matrix bootstrap with 500 replications. The standard errors for the OLS estimates were computed using White's method. All regressions included an intercept and 31 additional explanatory variables. These included all the variables listed in Table 4.3 and the survey-related variables discussed above, as well as the squares of income, age, height and weight. The F statistics reported in Tables 4.4–4.7 are for a test of the equality of slope estimates across the five quantiles, with their P values given in parentheses. The R^2 values for the quantile regression are squared correlation between observed and predicted intakes. The R^2 values are generally low, indicating the high level of unexplained variation in the intakes computed from 2 days of recall data, but are in line with those of previous studies (e.g. Adelaja *et al.*, 1997).

In general, the results are in the expected direction with education exerting a negative effect on fat, saturated fat and cholesterol intakes, and a positive effect on fibre intake. These findings imply that diet quality improves with education and are consistent with previous

Table 4.4. Estimated effect of education on total fat intake.[a]

	Men							Women						
		Quantile							Quantile					
Years of education	OLS	0.10	0.25	0.50	0.75	0.90	F test[b]	OLS	0.10	0.25	0.50	0.75	0.90	F test[b]
10	−0.19	0.32	0.25	−0.13	−0.44	−0.72	1.51	0.41**	0.35	0.41*	0.26	0.52*	0.30	0.54
	(0.85)	(1.50)	(1.00)	(0.47)	(1.40)	(1.40)	(0.20)	(2.60)	(1.80)	(2.10)	(1.50)	(2.10)	(0.77)	(0.71)
12	−0.63**	0.27	−0.05	−0.74**	−1.19**	−1.43**	5.21	0.10	0.07	0.18	0.05	0.11	−0.20	0.39
	(2.90)	(1.10)	(0.24)	(2.80)	(3.80)	(2.90)	(0.00)	(0.59)	(0.34)	(0.98)	(0.28)	(0.41)	(0.52)	(0.82)
14	−1.06**	0.21	−0.35	−1.35**	−1.93**	−2.13**	5.94	−0.21	−0.20	−0.05	−0.15	−0.29	−0.70	0.44
	(3.60)	(0.69)	(1.30)	(3.80)	(4.40)	(3.20)	(0.00)	(0.91)	(0.72)	(0.20)	(0.58)	(0.80)	(1.40)	(0.78)
16	−1.50**	0.15	−0.66	−1.96**	−2.68**	−2.83**	5.26	−0.51	−0.48	−0.28	−0.36	−0.70	−1.19	0.51
	(3.60)	(0.38)	(1.70)	(4.00)	(4.30)	(3.00)	(0.00)	(1.70)	(1.30)	(0.80)	(1.00)	(1.40)	(1.80)	(0.73)
R^2	0.14	0.12	0.13	0.14	0.14	0.13	–	0.10	0.08	0.09	0.09	0.09	0.08	–

[a]The slope coefficients for education were calculated as $\hat{\beta}_1 + 2\hat{\beta}_2 S$, where $\hat{\beta}_1$ and $\hat{\beta}_2$ are the coefficients of level of education and its square and S is the level of education reported in the first column. All regressions included an intercept and 31 additional explanatory variables; see data section for definitions. The absolute t values of the estimated slope coefficients are reported in parentheses; *, Significant at $P < 0.05$, **, significant at $P < 0.01$.

[b]F test for the equality of regression coefficients across the five quantiles; P values are reported in parentheses.

Table 4.5. Estimated effect of education on saturated fat intake.[a]

	Men							Women						
		Quantile							Quantile					
Years of education	OLS	0.10	0.25	0.50	0.75	0.90	*F* test[b]	OLS	0.10	0.25	0.50	0.75	0.90	*F* test[b]
10	−0.07	0.06	0.04	−0.03	−0.12	−0.31	0.76	0.11	0.08	0.07	0.09	0.05	0.17	0.37
	(0.83)	(0.71)	(0.50)	(0.37)	(1.00)	(1.30)	(0.55)	(1.90)	(1.30)	(0.92)	(1.40)	(0.56)	(1.30)	(0.83)
12	−0.25**	−0.03	−0.06	−0.25**	−0.39**	−0.52*	2.58	−0.03	−0.02	−0.02	−0.02	−0.09	−0.06	0.23
	(3.10)	(0.32)	(0.79)	(2.90)	(3.40)	(2.40)	(0.03)	(0.45)	(0.22)	(0.25)	(0.24)	(0.98)	(0.43)	(0.92)
14	−0.43**	−0.12	−0.17	−0.47**	−0.65**	−0.73*	3.24	−0.16*	−0.11	−0.10	−0.13	−0.23	−0.29	0.36
	(3.90)	(1.10)	(1.60)	(3.90)	(4.30)	(2.60)	(0.01)	(2.00)	(1.20)	(1.20)	(1.30)	(1.90)	(1.50)	(0.84)
16	−0.61**	−0.20	−0.27	−0.69**	−0.92**	−0.94*	3.08	−0.30*	−0.21	−0.19	−0.24	−0.38*	−0.51*	0.50
	(4.00)	(1.50)	(1.90)	(4.00)	(4.30)	(2.40)	(0.02)	(2.70)	(1.70)	(1.60)	(1.80)	(2.20)	(2.10)	(0.74)
R^2	0.14	0.12	0.13	0.13	0.13	0.13	–	0.09	0.07	0.08	0.08	0.08	0.07	–

[a]See footnote [a] in Table 4.4. The absolute *t* values of the estimated slope coefficients are reported in parentheses; *, significant at $P < 0.05$, **, significant at $P < 0.01$.
[b]*F* test for the equality of regression coefficients across the five quantiles; *P* values are reported in parentheses.

Table 4.6. Estimated effect of education on cholesterol intake.[a]

	Men							Women						
		Quantile							Quantile					
Years of education	OLS	0.10	0.25	0.50	0.75	0.90	*F* test[b]	OLS	0.10	0.25	0.50	0.75	0.90	*F* test[b]
10	−3.99**	0.11	−2.48*	−3.76**	−6.52**	−8.07**	3.14	−0.17	0.72	−0.57	−0.43	0.56	1.47	0.82
	(3.20)	(0.10)	(2.10)	(2.90)	(2.80)	(2.60)	(0.01)	(0.19)	(0.89)	(0.67)	(0.49)	(0.43)	(0.49)	(0.51)
12	−6.29**	−0.79	−3.08**	−5.17**	−10.35**	−11.45**	6.57	−1.88*	−0.54	−1.07	−1.59	−2.16	−2.51	0.38
	(5.70)	(0.81)	(3.10)	(4.40)	(5.20)	(4.60)	(0.00)	(2.20)	(0.69)	(1.30)	(1.80)	(1.50)	(1.10)	(0.82)
14	-8.58**	−1.68	−3.69**	−6.57**	−14.19**	−14.83**	6.30	−3.60**	−1.79	−1.57	−2.75*	−4.87*	-6.48*	0.96
	(6.00)	(1.30)	(3.00)	(4.30)	(5.60)	(4.20)	(0.00)	(3.10)	(1.80)	(1.60)	(2.30)	(2.30)	(2.20)	(0.43)
16	−10.89**	−2.57	−4.29*	−7.98**	−18.02**	−18.22**	4.77	−5.32**	−3.04	−2.08	−3.91*	−7.59**	-10.46*	1.40
	(5.30)	(1.50)	(2.50)	(3.80)	(5.00)	(3.40)	(0.00)	(3.30)	(2.20)	(1.50)	(2.30)	(2.70)	(2.40)	(0.23)
R^2	0.08	0.05	0.06	0.07	0.07	0.07	–	0.06	0.04	0.05	0.06	0.07	0.05	–

[a]See footnote [a] in Table 4.4. The absolute *t* values of the estimated slope coefficients are reported in parentheses; *, significant at $P < 0.05$, **, significant at $P < 0.01$.
[b]*F* test for the equality of regression coefficients across the five quantiles; *P* values are reported in parentheses.

Table 4.7. Estimated effect of education on fibre intake.[a]

	Men							Women						
		Quantile							Quantile					
Years of education	OLS	0.10	0.25	0.50	0.75	0.90	F test[b]	OLS	0.10	0.25	0.50	0.75	0.90	F test[b]
10	0.10	0.10	0.13*	0.25**	0.16	0.11	1.18	0.26**	0.20**	0.22**	0.33**	0.31**	0.28	1.60
	(1.60)	(1.60)	(2.30)	(3.70)	(1.60)	(0.70)	(0.32)	(6.10)	(5.10)	(6.10)	(6.40)	(5.20)	(1.90)	(0.17)
12	0.29**	0.18**	0.25**	0.38**	0.31**	0.42**	2.46	0.43**	0.27**	0.33**	0.48**	0.53**	0.47**	4.90
	(5.10)	(3.20)	(4.50)	(6.80)	(3.50)	(2.70)	(0.04)	(9.70)	(6.90)	(8.50)	(8.90)	(8.60)	(4.30)	(0.00)
14	0.47**	0.27**	0.37**	0.51**	0.45**	0.73**	2.27	0.59**	0.35**	0.43**	0.63**	0.76**	0.66**	5.70
	(6.60)	(3.50)	(5.30)	(6.80)	(3.80)	(3.40)	(0.06)	(10.00)	(6.60)	(8.20)	(8.40)	(8.80)	(4.90)	(0.00)
16	0.66**	0.35**	0.49**	0.63**	0.59**	1.04**	1.82	0.75**	0.42**	0.53**	0.79**	0.99**	0.86**	5.30
	(6.70)	(3.30)	(5.30)	(5.80)	(3.40)	(3.40)	(0.12)	(9.30)	(5.80)	(7.40)	(7.50)	(8.30)	(4.30)	(0.00)
R^2	0.08	0.07	0.07	0.07	0.07	0.07	–	0.11	0.09	0.11	0.11	0.11	0.10	–

[a]See footnote [a] in Table 4.4. The absolute t values of the estimated slope coefficients are reported in parentheses; *, significant at $P < 0.05$, **, significant at $P < 0.01$.
[b]F test for the equality of regression coefficients across the five quantiles; P values are reported in parentheses.

evidence on the effect of education on health and on health-related behaviours, such as smoking and excessive alcohol use (Grossman and Kaestner, 1997). When the overall results for men and women are compared, the results show a larger and more significant effect of education on men's diets than on women's diets for fat and saturated fat. On fibre intake, education has comparable impacts for men and women. One explanation for these findings is that, overall, men are at higher risk of overconsumption of fat and saturated fat compared with women, whereas for fibre the risk of underconsumption is similar.

Beyond confirming the beneficial effect of education on macronutrient intakes, our results paint a more detailed picture of the nature and degree of this effect. Examining men's intakes, the OLS estimates show the non-linear effect of education on total fat intake at the conditional mean, with larger declines in fat intake at higher levels of educational attainment. For a secondary-school graduate, an additional year of education lowers fat intake by 0.63 g. However, for a college graduate, the marginal decline in fat intake is more than double, at 1.5 g. This, however, is not the end of the story. Comparing the OLS and quantiles estimates shows why the OLS estimates may be undervaluing the benefit of education. As noted earlier, the OLS estimates give the 'average' effect of education over the whole distribution. However, the quantile estimates show that much of this effect is coming from the upper end of the fat-intake distribution. In Table 4.4, the significant quantile slope estimates are located in a 'box' bound by 12 or more years of education and 0.50 or upper quantiles. For the given level of education, all these slope estimates are higher than the OLS estimates. At 0.90 quantile, a part of the distribution well above the recommended fat-intake level, the marginal effect of education for a college graduate is 2.8 g, which is nearly 90% more than the effect estimated by OLS. Education has virtually no impact on fat intake at the lower end of the fat-intake distribution, as shown by the insignificant slope estimates at the 0.10 and 0.25 quantiles. This is not surprising since, at these quantiles, fat intake is substantially below the level of fat intake deemed to pose a health risk.

An important test of the usefulness of the quantile-regression method is whether the estimated slope coefficients are equal across all the quantiles or between specific quantiles. To conduct such equality tests, the joint variance–covariance matrix of the slope coefficients at the quantiles is required. This was obtained as a part of the bootstrap procedure by estimating quantile regressions separately at the five quantiles for each bootstrap sample. Once the joint variance–covariance matrix of the slope coefficients at the five quantiles is obtained, an F test for the equality of coefficients across quantiles can be performed.

While the quantile estimates show an increasing effect for education on fat intake at higher quantiles, are the estimates significantly different from each other? The *F* statistics in the final column for men in Table 4.4 show that the slope of education at the five quantiles are significantly different for 12 years or more of education at the 5% significance level. These results clearly illustrate the importance of examining the whole conditional distribution rather than focusing on just the conditional mean.

For men's saturated fat and cholesterol intakes (Tables 4.5 and 4.6), the results are similar to those for men's fat intake. For cholesterol, the effect of an increase in educational attainment appears even at 10 years of schooling and the effect increases significantly at upper quantiles. At 10 years of education, the 0.9 quantile estimate is twice the OLS estimate. The effects increase consistently across quantiles and years of education, reaching a high of an 18 mg reduction in cholesterol intake for a college graduate at the 0.90 quantile.

While education appears to have no influence on women's total fat intake, its effect on saturated fat and cholesterol intakes are in the expected direction. Saturated fat intake is significantly lower for college graduates at the 0.75 and 0.90 quantiles. For cholesterol, the significant effects are found in the 'box' bound by 14 or more years of schooling and above the median. Only the trend of the results for women is suggestive, since the *F* tests show insignificant differences across quantiles.

Nutritionally, dietary fibre is different from the other three macronutrients in two important respects. First, for fibre, the health risk stems from underconsumption rather than overconsumption. Secondly, as Tables 4.1 and 4.2 illustrate, for most age/sex groups, the observed intake distribution does not meet the recommended intake level. For example, for men between 21 and 50 years of age, the recommended intake for a 2900-calorie diet is about 33 g of dietary fibre. As Table 4.1 shows, this level is unmet even at the 90th percentile, where the intake is about 30 g. These two factors may help explain the estimated effect of education on fibre intake reported in Table 4.7. If the intake behaviour of those with higher levels of education is driven by a higher risk perception given their intake, then we would expect education to have a greater impact at the lower end of the fibre-intake distribution compared with the upper end. The results, however, show the effect of education rising rather than falling across quantiles. This could be because the entire distribution of fibre intake is below the healthful level and poses some level of health risk. Since the health risk from fibre has not received the same level of attention given to health risks from fat, saturated fat and cholesterol, fibre intake may be more an outcome of overall diet rather than specific dietary choices made by consumers to adjust the intake of fibre. Nevertheless, it is notable that education does have a significant effect on fibre intake at the lower quantiles.

A better view of the effects of educational attainment on the conditional distribution of intakes can be obtained by calculating the predicted intakes at specific levels of education, holding other explanatory variables constant at their sample means. Such predicted conditional intakes at the 0.10 and 0.90 quantiles for men and women are depicted in Figs 4.1 and 4.2. Predicted intakes at the 0.10 quantile are plotted off the left vertical axis and predicted intakes at the 0.90 quantile are plotted off the right vertical axis. To make the relative change in predicted intakes at the two quantiles comparable, the range of intake plotted on both the vertical axes is the same. The graphs show that intakes are much more responsive to the level of education at the 0.90 quantile compared with the 0.10 quantile. For men, the predicted intakes of fat, saturated fat and cholesterol are relatively flat at the 0.10 quantile. However, the intakes decrease rapidly at the 0.90 quantile with increasing level of education. Fibre intake remains flat at both quantiles up to about 11 years of education and increases thereafter. For women, the predicted cholesterol intake at the 0.90 quantile shows the most response, although fat and saturated-fat intakes also decline at the highest levels of education. The pattern of women's predicted fibre intake is broadly similar to that of men's.

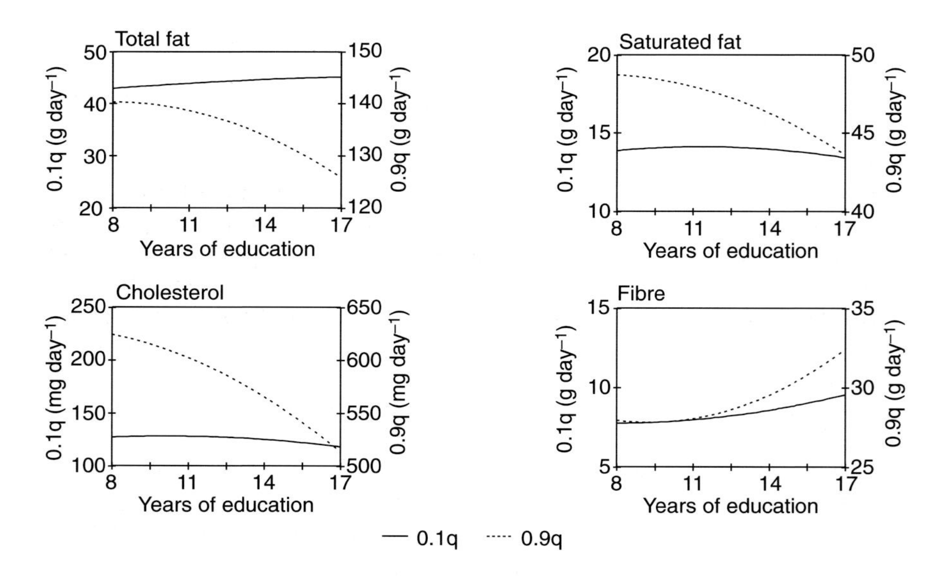

Fig. 4.1. Predicted intakes at 0.1 and 0.9 quantiles for men by years of educational attainment.

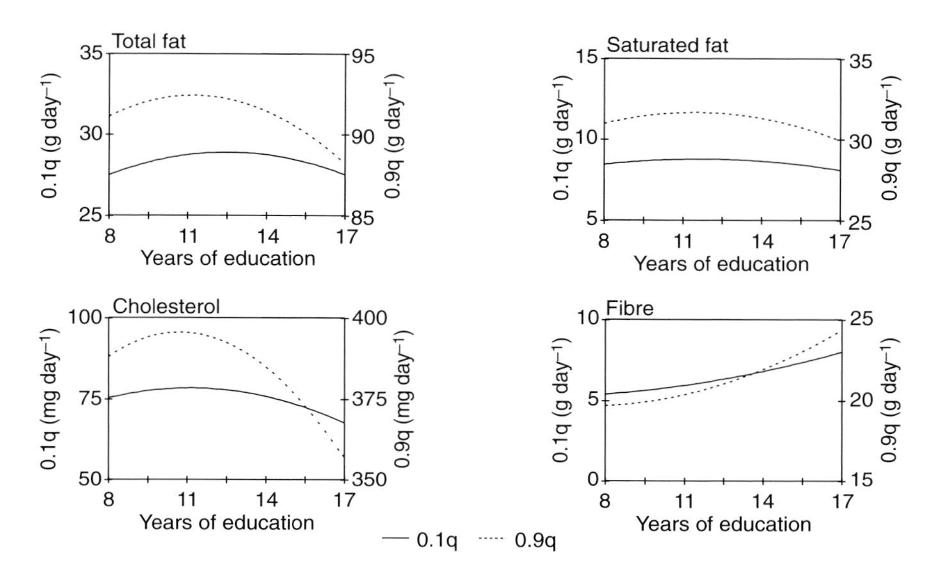

Fig. 4.2. Predicted intakes at 0.1 and 0.9 quantiles for women by years of educational attainment.

Conclusions

Our results clearly imply that the 'beneficial' effect of education on dietary behaviour may have been underestimated, due to the previous focus on conditional mean estimates. Such estimates provide a limited characterization of the conditional intake distribution by not telling us anything about the location of differences in marginal effects of a given characteristic along the entire distribution. Knowing the differential size and location of marginal effects is important for outcomes such as dietary intakes because the risk of dietary excess or inadequacy is greater at the tails than at the mean. For total fat, saturated fat and cholesterol intakes of men and to some degree for the saturated fat and cholesterol intakes of women, our results show that the effects of education are much larger at the upper end of the distribution, which are above the healthful levels, compared with the mean or lower parts of the distribution, which are within recommended levels. For these nutrients, the estimated effects of education reveal a double impact: intakes responding at greater levels to education at higher quantiles.

These results have important implications for future studies evaluating the dietary impact of many nutrition-related policy interventions, such as food-assistance programmes and food-labelling regulations. For such studies to fully uncover the extent and nature of the behavioural impact, it is essential to look beyond the conditional mean to parts of the dietary intake distribution where the risk of inadequacy or excess is greater.

The reduced-form nutrient-demand function we estimated gives the 'net' or 'total' effect of education on macronutrient intake at the conditional mean and conditional quantiles. Are these effects solely due to higher allocative efficiency – that is, the ability of those with more education to acquire, process and use information to produce health better than those with less education? Previous studies that have used educational attainment to capture informational effects on dietary behaviour hold this view (Ippolito and Mathios, 1990; Bhattacharya and Currie, 2001). If this view is accepted, then our results seem to suggest that those with more education have benefited most from the rapidly growing information linking diet and health made available, at least partly, through various nutrition-information initiatives undertaken by federal and private agencies. In this case, new efforts, beyond information programmes, that specifically target lower-educated groups may be necessary to achieve nutritional improvements for the overall population.

Notes

[1]Consumer's budget constraint is another factor, although, in general, this is outside the scope of public policy.

[2]If the intakes are long-run intakes, such as intakes measured using food-frequency questionnaires, anthropometric measures may themselves be determined by the intake levels. Such reverse causality is less of a concern with short-run intake data.

[3]The squared terms of these variables were statistically significant in at least a few of the regressions. For consistency across the estimated intake equations, we included the squared terms in all regressions.

[4]These quantiles, 0.10, 0.25, 0.50, 0.75 and 0.90, were chosen to be representative of the distribution and are consistent with quantiles used in the previous literature.

References

Adelaja, A., Nayga, R.M. Jr and Lauderbach, T. (1997) Income and racial differentials in selected nutrient intakes. *American Journal of Agricultural Economics* 79, 1452–1460.

Ashenfelter, O. and Krueger, A. (1994) Estimates of the economic return to schooling from a new sample of twins. *American Economic Review* 84, 1157–1174.

Behrman, J.R. and Stacey, N. (eds) (1997) *The Social Benefits of Education*. University of Michigan Press, Ann Arbor, Michigan, 266 pp.

Bhattacharya, J. and Currie, J. (2001) Youths at nutritional risk: malnourished or misnourished? In: Gruber, J. (ed.) *Risky Behaviour among Youth: An Economic Analysis*. University of Chicago Press, Chicago, Illinois, pp. 483–521.

Buchinsky, M. (1995) Estimating the asymptotic covariance matrix for quantile regression models: a Monte Carlo study. *Journal of Econometrics* 68, 303–338.

Buchinsky, M. (1998) Recent advances in quantile regression models: a practical guideline for empirical research. *Journal of Human Resources* 33, 88–126.

Deaton, A. (1997) *The Analysis of Household Surveys: a Microeconometric Approach to Development Policy*. Johns Hopkins University Press, Baltimore, Maryland, 479 pp.

Frazao, E. (1999) *The American Diet: Health and Economic Consequences*. Agricultural Information Bulletin No. 711, Economic Research Service, US Department of Agriculture, Washington, DC, 473 pp.

Fuchs, V.R. (1979) Economics, health and post-industrial society. *Milbank Memorial Fund Quarterly: Health and Society* 57, 153–182.

Gould, B.W. and Lin, H.C. (1994) Nutrition information and household dietary fat intake. *Journal of Agricultural and Resource Economics* 19, 349–365.

Grossman, M. (1972) On the concept of health capital and the demand for health. *Journal of Political Economy* 80, 223–255.

Grossman, M. and Kaestner, R. (1997) Effects of education on health. In: Behrman, J.R. and Stacey, N.G. (eds) *The Social Benefits of Education*. University of Michigan Press, Ann Arbor, Michigan, pp. 69–123.

Ippolito, P.M. and Mathios, A.D. (1990) Information, advertising, and health choices: a study of the cereal market. *Rand Journal of Economics* 21, 459–480.

Juster, F.T. (1975) *Education, Income, and Human Behaviour*. McGraw-Hill, New York, 438 pp.

Koenker, R. and Bassett, G. (1978) Regression quantiles. *Econometrica* 46, 33–50.

Levy, A.S., Fein, S.B. and Stephenson, M. (1993) Nutrition knowledge levels about dietary fats and cholesterol: 1983–1988. *Journal of Nutrition Education* 25, 60–66.

Lin, B.-H., Guthrie, J. and Frazao, E. (1999) *Away-from-home Foods Increasingly Important to Quality of American Diet*. Agricultural Information Bulletin No. 749, Economic Research Service, US Department of Agriculture, Washington, DC, 22 pp.

Mincer, J. (1974) *Schooling, Experience and Earnings*. National Bureau of Economic Research, New York, 167 pp.

National Research Council (1989) *Diet and Health: Implications for Reducing Chronic Disease Risk*. National Academy Press, Washington, DC, 749 pp.

Nayga, R.M. Jr (1994) Effects of socioeconomic and demographic factors on consumption of selected food nutrients. *Agricultural and Resource Economics Review* 23, 171–182.

Nayga, R.M. Jr (2000a) Schooling, health knowledge, and obesity. *Applied Economics* 32, 815–822.

Nayga, R.M. Jr (2000b) A note on schooling and obesity. *Journal of Health and Social Policy* 12, 65–72.

Strauss, J. and Thomas, D. (1996) Human resources: empirical modelling of household and family decision. In: Srinivasan, T.N. and Behrman, J.R. (eds) *Handbook of Development Economics*, Vol. IIIA. North-Holland, Amsterdam, pp. 1883–2023.

Thomas, P.R. (1991) *Improving America's Diet and Health*. National Academy Press, Washington, DC, 239 pp.

Tippett, K.S. and Cypel, Y.S. (1997) *Design and Operation: the Continuing Survey of Food Intakes by Individuals and the Diet and Health Knowledge Survey, 1994–1996*. NFS Report No. 96–1, Agricultural Research Service, US Department of Agriculture, Beltsville, Maryland, 150 pp.

Variyam, J.N., Blaylock, J.R. and Smallwood, D.M. (1998) Reduced-rank models for nutrition knowledge assessment. *Biometrics* 54, 1654–1661.

5 Assessing the Importance of Health Information in Relation to Dietary Intakes in the USA

SUNG-YONG KIM, RODOLFO NAYGA, JR AND ORAL CAPPS, JR

Texas A&M University, College Station, Texas, USA

Introduction

Many Americans are not meeting dietary recommendations. The US Department of Agriculture (USDA) revealed that only about 12% of Americans are eating healthfully (Bowman *et al.*, 1998). In addition, four of the top ten causes of death in the USA – heart disease, cancer, stroke and diabetes – are associated with poor diets (US Department of Health and Human Services, 1988). Diet-related health conditions cost society an estimated $250 billion annually in medical costs and lost productivity (Frazao, 1995).

During the last two decades product labelling has become a popular policy tool, particularly with respect to the provision of nutrition and health information (Padberg, 1992). It culminated in the passage of the Nutrition Labeling and Education Act (NLEA) in 1990.[1] The NLEA instituted sweeping changes to replace the voluntary system of labelling established by the Food and Drug Administration (FDA) in 1973. It requires mandatory nutrition labelling for almost all packaged food and strict regulation of nutrient content and health claims. In addition, it also requires a new format for the nutrition information panel called 'Nutrition Facts', standardization of serving sizes and strict regulation of use of descriptors and explicit health messages (Caswell, 1992; Ippolito and Mathios, 1993).

Nutrition labelling as a policy tool is appropriate and particularly valuable in that consumers have no way to evaluate the nutritional value of their food products (Padberg, 1992; Caswell and Mojduszka, 1996). Health information provided through food labelling may reduce consumer uncertainty (Zarkin and Anderson, 1992) or transform credence attributes of food products into search attributes (Caswell and

Mojduszka, 1996). Most of all, the passage of the NLEA has been expected to provide improvements in diet quality by helping consumers make healthier food choices.

The impact of providing health or nutritional information on consumer purchasing behaviour has been reported in the literature. Previous research, however, was limited to mainly examining the effectiveness of voluntary disclosure by firms in the form of producer advertising (Brucks *et al.*, 1984; Levy and Stokes, 1987; Ippolito and Mathios, 1990), the influence of informational labelling on the quality of food products (Caswell and Mojduszka, 1996; Mathios, 1998) or the impact of nutrient-content information use on individual nutrient intake (Kim *et al.*, 2000). Little information is available concerning the effectiveness of food-label use in improving overall diet quality of Americans. The purpose of this study is to assess the overall dietary impact of health information as represented by food-label use.

In evaluating the effectiveness of label use in relation to diet quality, label-use effects can be estimated by comparing the observed diets of label users and non-users. This comparison is appropriate only if label users and non-users are identical in all respects except label use. However, label users and non-users usually differ in observed socio-economic and demographic characteristics, which in turn induce differences in unobserved preferences for dietary patterns and food-label use. Attributing observed differences in dietary intake to label use alone ignores these other differences. This situation may subsequently lead to biased estimates of label-use effectiveness. The bias has been called a 'self-selection bias' because the simple comparison ignores how individuals self-select into label use. To avoid the bias, it is necessary to specify a model describing how consumer diet and label-use decisions are simultaneously induced by the observable and unobservable characteristics of an individual. Thus, an endogenous switching regression model is used to control for this self-selectivity bias.

The healthy eating index (HEI) developed by the USDA is used as a measure of diet quality in evaluating the effectiveness of label use. The responses to food-label use are expected to vary across individuals and across the types of information on the label. Thus, five types of nutritional-label information – nutritional panels, serving sizes, nutrient-content claims, list of ingredients and health claims – are examined to determine which type of label information provides the most improvement, if any, in diet quality.

The Model

Let the label-use decision be a dichotomous choice resulting from maximizing an individual's utility, which is a function of the consumption of food, non-food and health (Variyam *et al.*, 1996, 1998). The expected utility of label use, I_1* is compared with the utility of non-use, I_0*; the label is used by consumers if $I_1^* > I_0^*$; non-use of the label occurs if $I_1^* \leq I_0^*$. Define label use on the food products for the *i*th consumer as follows: $I_i = 1$ if $I_1^* > I_0^*$, and $I_i = 0$ if $I_1^* \leq I_0^*$. Then, the decision as to whether or not to use the label can be estimated by:

$$I_i = \gamma' \mathbf{Z}_i + \mu_i \tag{5.1}$$

where $\mathbf{Z}_i$ is a vector of characteristics that affect label use (Guthrie *et al.*, 1995; Nayga, 1996; Nayga *et al.*, 1998), γ is a vector of parameters and μ_i is an error term. Equation 5.1 is a probit specification for label use.

In theory, individuals not using food labels may be different in their food-consumption behaviour from those using food labels. Following Variyam *et al.* (1998), an individual diet can be transformed into a measure of quality expressed by the HEI. Label use, then, may affect an individual's diet quality. Define HEI_i as the observed diet quality of the *i*th consumer, and define HEI_{i1} and HEI_{i0} as the observed diet qualities of the label user i ($I_i = 1$) and non-label user i ($I_i = 0$), respectively. Then diet quality is specified for label users and non-users by:

$$HEI_{i1} = \beta_1' \mathbf{X}_i + \varepsilon_{i1} \tag{5.2}$$

and:

$$HEI_{i0} = \beta_0' \mathbf{X}_i + \varepsilon_{i0} \tag{5.3}$$

where $\mathbf{X}_i$ is a vector of the *i*th consumer's observed characteristics that affect diet quality (Variyam *et al.*, 1998), $\boldsymbol{\beta}_0$ and $\boldsymbol{\beta}_1$ are vectors of parameters and $\boldsymbol{\varepsilon}_{i1}$ and $\boldsymbol{\varepsilon}_{i0}$ are error-term vectors.

Estimation Procedure

Assume a trivariate normal distribution among the error terms of Equations 5.1, 5.2 and 5.3, with mean vector zero and the covariance matrix given below

$$\operatorname{cov}\left(\varepsilon_{i1}, \varepsilon_{i0}, \mu_i\right) = \begin{bmatrix} \sigma_1^2 & \sigma_{10} & \sigma_{1\mu} \\ \sigma_{10} & \sigma_0^2 & \sigma_{0\mu} \\ \sigma_{1\mu} & \sigma_{0\mu} & 1 \end{bmatrix}$$

Since the choice of using labels or not is endogenous, the error terms in Equations 5.2 and 5.3, conditional on the sample selection criterion, have non-zero expected values – that is:

$$E[\varepsilon_{i1} \mid I_i = 1] = -\sigma_{1\mu} \frac{\phi(\gamma' \mathbf{Z}_i)}{\Phi(\gamma' \mathbf{Z}_i)} \tag{5.4}$$

and:

$$E[\varepsilon_{i0} \mid I_i = 0] = \sigma_{0\mu} \frac{\phi(\gamma'\mathbf{Z}_i)}{1-\Phi(\gamma'\mathbf{Z}_i)} \tag{5.5}$$

Let us define, for convenience:

$$W_{1i} = \frac{\phi(\gamma'\mathbf{Z}_i)}{\Phi(\gamma'\mathbf{Z}_i)} \text{ and } W_{0i} = \frac{\phi(\gamma'\mathbf{Z}_i)}{1-\Phi(\gamma'\mathbf{Z}_i)}$$

Then we can write Equations 5.2 and 5.3 as:

$$HEI_{i1} = \beta_1'\mathbf{X}_i - \sigma_{1\mu}W_{1i} + \zeta_{1i} \text{ for } I_i = 1 \tag{5.6}$$

$$HEI_{i0} = \beta_0'\mathbf{X}_i + \sigma_{0\mu}W_{0i} + \zeta_{0i} \text{ for } I_i = 0 \tag{5.7}$$

where ζ_{1i} and ζ_{0i} are the new residuals, with zero conditional means:

$$\zeta_{1i} = \varepsilon_{1i} + \sigma_{1\mu}W_{1i} \quad \text{and} \quad \zeta_{0i} = \varepsilon_{0i} + \sigma_{0\mu}W_{0i}$$

Thus ordinary least squares (OLS) estimates of β_1' and β_0' in Equations 5.2 and 5.3 are biased.

Two procedures have been used in the estimation of endogenous switching regression models – Heckman's two-step estimation and maximum-likelihood estimation techniques (Lee, 1978; Willis and Rosen, 1979; Maddala, 1983; Gould and Lin, 1994). Monte Carlo experiment studies show that the two-step estimator performs poorly when there is a high degree of multicollinearity between $\mathbf{Z}_i$ and explanatory variables $\mathbf{X}_i$ (Nelson, 1984; Hartman, 1991; Nawata, 1994). Hence, the full-information maximum likelihood (FIML) is used to estimate Equations 5.1, 5.2 and 5.3. FIML estimation provides consistent and asymptotically efficient estimates (Lee and Trost, 1978). FIML parameter estimates, using two-step estimates as initial values can be obtained from the following log-likelihood function:

$$\ln L_i(\beta_1, \beta_0, \sigma_1, \sigma_0, \rho_{1\mu}, \rho_{0\mu}) = \sum_{i=1}^{I} \left\{ \begin{array}{l} I_i \left[\ln \frac{1}{\sigma_1} \phi\left(\frac{HEI_{i1} - \beta_1'\mathbf{X}_i}{\sigma_{i1}} \right) + \ln\Phi(\eta_{i1}) \right] + (1 - I_i) \\ \left[\ln \frac{1}{\sigma_0} \phi\left(\frac{HEI_{i0} - \beta_0'\mathbf{X}_i}{\sigma_{i0}} \right) + \ln\left(1 - \Phi(\eta_{i0})\right) \right] \end{array} \right\} \tag{5.8}$$

where:

$$\eta_{i1} = \frac{\left[\gamma'\mathbf{Z}_i - \frac{\rho_{1\mu}}{\sigma_1}(HEI_{i1} - \beta_1'\mathbf{X}_i) \right]}{\sqrt{1-\rho_{1\mu}^2}}$$

and:

$$\eta_{i0} = \frac{\left[\gamma'\mathbf{Z}_i - \frac{\rho_{0\mu}}{\sigma_0}(HEI_{i0} - \beta_0'\mathbf{X}_i)\right]}{\sqrt{1-\rho_{0\mu}^2}}$$

with $\rho_{1\mu}$, $\rho_{0\mu}$ the correlation coefficients of ε_{i1}, μ_i and ε_{i0}, μ_i, respectively.

Suppose there is a variable that appears both in $\mathbf{X}_i$ and $\mathbf{Z}_i$, say the kth element of these vectors. The conditional effect on those who actually use the label is given by:

$$\frac{\partial E(HEI_{i1}|I_i = 1)}{\partial \mathbf{X}_{ik}} = \beta_{1k} - \gamma_k \sigma_{1\mu} W_{1i}\left[\gamma'\mathbf{Z}_i + W_{1i}\right] \tag{5.9}$$

Similarly, the conditional effect on those who do not use the label is given by:

$$\frac{\partial E(HEI_{i0}|I_i = 0)}{\partial \mathbf{X}_{ik}} = \beta_{0k} + \gamma_k \sigma_{0\mu} W_{0i}\left[\gamma'\mathbf{Z}_i + W_{0i}\right] \tag{5.10}$$

Equations 5.9 and 5.10 decompose the effect of a change in $\mathbf{X}_{ik}$ into two parts. The first part is the direct effect on HEI_{i1} (HEI_{i0}). The second part captures an indirect effect that appears as a result of correlation between the unobservable components of HEI_{i1} (HEI_{i0}) and I_i (Poirier and Ruud, 1981; Maddala, 1983). If the kth element is not shared in $\mathbf{X}_i$ and $\mathbf{Z}_i$, β_{1k} and β_{0k} alone provide the correct marginal effects.

Healthy Eating Index

The *HEI* developed by the USDA Center for Nutrition Policy and Promotion provides an overall picture of the type and quantity of foods people eat, their compliance with specific dietary recommendations and the variety in their diet (Bowman *et al.*, 1998). Thus, the *HEI* is a summary measure of overall diet quality. The *HEI* has ten components, each representing different aspects of a healthful diet. Components one to five measure the degree to which a person's diet conforms to USDA's Food Guide Pyramid serving recommendations for the five major food groups – grains, vegetables, fruits, milk and meat. Components six and seven measure total fat and saturated fat consumption as a percentage of total food-energy intakes. Components eight and nine measure total cholesterol and sodium intakes, and component ten examines variety in a person's diet.

Data used to calculate the *HEI* are from the USDA's 1994–1996 Continuing Survey of Food Intakes by Individuals (CSFII), a nationally representative survey containing information on people's consumption of foods and nutrients. Each component has a possible range of 0 to 10. The maximum overall score, then, is 100. High component scores indicate intakes close to the recommended ranges or amounts; low component scores indicate less compliance with the recommended ranges or amounts. The mean *HEI* score for the USA as a whole is 63.6 for 1994, 63.5 for 1995 and 63.8 for 1996.

Data

Besides the 1994–1996 CSFII data for the *HEI* variable, the 1994–1996 Diet and Health Knowledge Survey (DHKS) data, the companion data set of the CSFII, are also used in this study. The DHKS includes detailed information about the individual's socio-economic background and questions on label usage. The empirical work uses DHKS respondent files, providing a sample size of 5343.

The dependent variables include the *HEI* and a binary label-use variable (Table 5.1). Label users' average diet quality is about 4 or 5 points higher in *HEI* than that of non-label users. The analysis is conducted by the type of information contained on food labels. The five types of information that are presented on the food label include: (i) the list of ingredients; (ii) the short phrases on the label, such as 'low-fat', 'light' or 'good source of fibre' (nutrient-content claims); (iii) the nutrition panels that tell the amount of calories, protein, fat and so on in a serving of the food; (iv) the information about the size of the serving; and (v) the statement on the label that describes health benefits of nutrients or foods (health claims). About 76.8% of the sample used the list of ingredients, 74.3% used the nutrient-content claims, 75.6% used the nutrition panel, 67.9% used the serving size and 68.2% used the health claims. Binary variables (1 = use; 0 = not use) capture the decision to use each type of information on the food labels.

Table 5.1. Comparison of *HEI* and label (health information) use by food-label information. (From: USDA, 1994–1996 CSFII and DHKS.)

	Healthy eating index		
Food label information	Label users	Non-label users	% of sample using label
List of ingredients	64.013	59.175	0.7679
Nutrient-content claims	64.014	59.635	0.7434
Nutritional panel	64.149	58.992	0.7559
Serving size	64.217	60.091	0.6785
Health claims	64.221	60.040	0.6818

The names, definitions and means of explanatory variables in the probit label-use and *HEI* models are given in Table 5.2. Independent variables consist of personal or household characteristics, sociodemographic factors, participation in government programmes, such as the food-stamp programme, and knowledge about the linkage between diet and diseases. Personal or household characteristics include age, gender, level of education, ethnicity, race, employment status and special diet status. Other sociodemographic factors include region, urbanization, household size and income. The diet–health-awareness variable is a dummy variable, which is constructed similarly to that reported by Variyam *et al.* (1996). The variable is given a value of one if respondents have heard about any health problems caused by being overweight or eating too much (or not eating enough) of a particular nutrient, such as fat, fibre, salt, calcium, cholesterol and sugar, while it is given a value of zero otherwise.[2] Two variables are included in the probit label-use model as explanatory variables but do not appear in the *HEI* equations. One of the variables is *Time*, which reflects time associated with reading the food label and it has a value of one if a respondent disagrees with the statement 'Reading the label takes more time than I can spare' and zero otherwise. The other is *Shopper*, which indicates whether a respondent is the main shopper of the household or not.

Empirical Results

Joint normality test

The joint normality assumption between the probit label-use error term and *HEI* equation error terms plays a key role in the estimation of an endogenous switching regression model. The normal selection-bias adjustment has been known to be sensitive to departures from normality (Goldberger, 1983; Pagan and Vella, 1989). The test of joint normality hinges on the procedure developed by Pagan and Vella (1989). It is like a regression specification error test (RESET) in which the predictions from the probit selection Equation 5.1 are powered up, although they are weighted by a function of W_{1i} (for label user) and W_{0i} (for non-label user).

The joint normality test results are reported in Table 5.3. Columns (1), (2) and (3) contain the absolute value of the t statistics for the null hypothesis that the coefficient is equal to zero. Column (4) contains the significance level of the χ^2 value for the null hypothesis that the parameters corresponding to columns (1), (2) and (3) are jointly zero. The results do not reflect any sign of model misspecification for both groups in all types of label information.

Table 5.2. Description, mean and standard deviation of explanatory variables. (From: USDA, 1994–1996 CSFII and DHKS.)

Explanatory variables	Description	Mean	Standard deviation
Income	Household income ($10,000)	3.5158	2.6416
Household size	Number of household members	2.5841	1.4538
Age	Age of respondent (in years)	50.8559	17.1703
Male	Respondent is male (1 = yes; 0 = no)	0.5025	[a]
Black	Respondent is black (1 = yes; 0 = no)	0.1159	
Others	Respondent is other non-white race (1 = yes; 0 = no)	0.0625	
Employed	Respondent is employed (1 = yes; 0 = no)	0.5826	
Non-Hispanic	Respondent is non-Hispanic (1 = yes; 0 = no)	0.9225	
City	Respondent resides in the central city (1 = yes; 0 = no)	0.2944	
Non-metropolitan	Respondent resides in a non-metropolitan area (1 = yes; 0 = no)	0.2667	
College	Respondent has at least some college education (1 = yes; 0 = no)	0.4434	
Food-stamp participation	Respondent participates in the food-stamp programme (1 = yes; 0 = no)	0.0779	
Special diet	Respondent has special diet (1 = yes; 0 = no)	0.1741	
Diet–health awareness	Respondent has heard about any health problem caused by being overweight or eating too much (or not eating enough) of a particular nutrient such as fat, fibre, salt, calcium, cholesterol and sugar (1 = yes; 0 = no)	0.9843	
Vegetarian	Respondent is a vegetarian (1 = yes; 0 = no)	0.0299	
Time	Respondent disagrees with the statement: 'Reading food labels takes more time than I can spare' (1 = yes; 0 = no)	0.4428	
Shopper	Respondent is a major shopper (1 = yes; 0 = no)	0.6597	
Northeast[b]	Respondent resides in the northeast (1 = yes; 0 = no)	0.1918	
West	Respondent resides in the west (1 = yes; 0 = no)	0.2023	
Midwest	Respondent resides in the Midwest (1 = yes; 0 = no)	0.2517	

[a] Blanks indicate that standard deviations cannot be computed for the dummy variables.
[b] South region is excluded in the model.

Table 5.3. Results for the joint normality test.[a]

Types of information/group	Pred* W (1)	Pred^{2}* W (2)	Pred^{3}* W (3)	P value (4)
List of ingredients				
User	0.443	1.307	1.469	0.120
Non-user	0.166	0.011	0.333	0.304
Nutrient content claims				
User	0.845	2.172	1.179	0.123
Non-user	0.660	0.673	0.621	0.821
Nutritional panels				
User	0.754	1.811	0.734	0.154
Non-user	0.764	0.676	0.559	0.843
Serving sizes				
User	1.124	1.206	0.310	0.430
Non-user	0.709	0.862	1.022	0.693
Health claims				
User	1.493	1.653	0.984	0.331
Non-user	0.547	0.668	0.794	0.818

[a] 'Pred' indicates the predicted value of $\gamma'\mathbf{Z}_i$. For food-label users, W is equal to $\phi(\gamma'\mathbf{Z}_i)/\Phi(\gamma'\mathbf{Z}_i)$ while, for food-label non-users, W is equal to $\phi(\gamma'\mathbf{Z}_i)/[1-\Phi(\gamma'\mathbf{Z}_i)]$. Columns (1), (2) and (3) contain the absolute value of the t statistic for the null hypothesis that the coefficient is equal to zero. Column (4) contains the significance level of the χ^2 value for the null hypothesis that the parameters corresponding to columns (1), (2) and (3) are jointly zero.

Probit food-label-use model

Parameter estimates of the probit food-label-use model for each of the five types of label information are presented in Table 5.4. Based on the results, the probability of using label information, except on health claims, increases with income, while the probability of using label information on nutritional panels, serving sizes and health claims decreases with age. Consistent with Nayga's (1996) finding, males are less likely to use the five types of label information than females. Results also indicate that education is significantly positively related to label use. Thus people with at least some college education are more likely to use the five types of label information than those who do not have a college education. This finding is consistent with those of Guthrie *et al.* (1995). Urbanization and regional differences are also evident in the results. Specifically, individuals who reside in non-metropolitan areas are less likely to use labels than those who reside in suburban areas.

Table 5.4. Parameter estimates of probit food-label-use equations.[a]

	List of ingredients	Nutrient-content claims	Nutritional panels	Serving sizes	Health claims
Constant	–0.2970	–0.1108	–0.1131	–0.1071	–0.1855
	(–1.355)	(–0.512)	(–0.516)	(–0.505)	(–0.884)
Income	0.0349***	0.0268***	0.0360***	0.0369***	0.0133
	(3.526)	(2.919)	(3.600)	(4.168)	(1.588)
Household size	–0.0234	–0.0018	–0.0106	–0.0193	0.0092
	(–1.454)	(–0.119)	(–0.646)	(–1.315)	(0.633)
Age	0.0003	–0.0012	–0.0045***	–0.0061***	–0.0033**
	(0.219)	(–0.838)	(–3.031)	(–4.571)	(–2.447)
Male	–0.4546***	–0.4464***	–0.5003***	–0.3779***	–0.2956***
	(–9.855)	(–10.304)	(–10.575)	(–9.094)	(–7.370)
Black	0.0204	–0.0141	–0.1207*	0.0086	0.0388
	(0.301)	(–0.216)	(–1.818)	(0.138)	(0.625)
Others	–0.0708	–0.2429***	–0.1541*	–0.2860***	–0.0637
	(–0.777)	(–2.746)	(–1.696)	(–3.497)	(–0.749)
Employed	0.1440***	0.1200**	0.1139**	0.0046	0.0489
	(2.731)	(2.407)	(2.132)	(0.096)	(1.054)
Non-Hispanic	–0.2071**	–0.2329***	–0.0522	–0.1609**	–0.1853**
	(–2.525)	(–2.785)	(–0.658)	(–2.154)	(–2.390)
City	–0.0652	–0.0226	–0.0215	–0.0153	–0.0712
	(–1.233)	(–0.456)	(–0.403)	(–0.322)	(–1.556)
Non-metro	–0.1867***	–0.2003***	–0.2561***	–0.1628***	–0.2214***
	(–3.690)	(–4.170)	(–5.100)	(–3.540)	(–4.867)
College	0.2873***	0.1941***	0.2533***	0.1259***	0.1768***
	(6.150)	(4.403)	(5.397)	(3.026)	(4.290)
Food-stamp	–0.1787***	–0.1625**	–0.2122***	–0.1545**	–0.1845**
participation	(–2.309)	(–2.153)	(–2.733)	(–2.126)	(–2.508)
Special diet	0.2673***	0.2775***	0.3618***	0.2917***	0.2323***
	(4.353)	(4.895)	(5.633)	(5.459)	(4.577)
Diet–health	0.7997***	0.7385***	0.6992***	0.7676***	0.7252***
awareness	(5.165)	(4.823)	(4.527)	(4.705)	(4.687)
Vegetarian	0.0808	–0.0232	–0.0746	–0.1522	0.0371
	(0.609)	(–0.200)	(–0.587)	(–1.363)	(0.331)
Time	0.8533***	0.6282***	0.9714***	0.6388***	0.5426***
	(18.826)	(15.454)	(20.745)	(16.779)	(15.037)
Shopper	0.2565***	0.2152***	0.2455***	0.2014***	0.1698***
	(5.435)	(4.986)	(4.977)	(4.731)	(4.294)

[a]The number in parentheses are *t* values. LIMDEP was used to carry out the FIML estimation of endogenous switching regression model. *, ** and *** indicate significance at the 10%, the 5%, and the 1% level.

Non-Hispanics are less likely to use label information, with the exception of nutritional panels, than Hispanics. Individuals who are on a special diet are more likely to use labels than individuals who are not on a special diet. Individuals who are more informed about the link between diet and health are also more likely to use nutritional labels. This result is consistent with the argument that poorly informed consumers tend to underestimate the marginal benefit of label use. Individuals who participate in the food-stamp programme are less likely to use label information than others. Time associated with reading the label is significant in all types of label information. Individuals who agreed with the statement that reading food labels takes more time than they can spare are less likely to use label information. Major food shoppers are more likely to use label information when shopping than others. The household's major food shopper, then, can potentially influence the quality of the diet of individual household members just from the types of foods he or she decides to purchase.

HEI models

Parameter estimates of the *HEI* equations for the different types of information on labels are exhibited in Table 5.5, which also contains the estimated standard deviation of the *HEI* equation error terms and the correlation coefficients between probit label use and *HEI* equations for label users and non-label users. For all types of label information, the estimated correlation coefficients are significant for label users. Conversely, these correlation coefficients are not significant for non-label users in most cases. For nutrient-content claims, self-selectivity occurs because the correlation coefficients are significant in the *HEI* equations for both label users and non-label users.

Results based on the estimates of β indicate that income is positively related to diet quality (i.e. *HEI*) regardless of label use. For non-users, there is a non-linear relationship between income and *HEI*. The age of label users and non-label users is positively related to *HEI* for all types of label information. Black label users and non-label users have *HEI*s that are about three or four points lower than the *HEI*s of white label users and non-label users, respectively. Label users of other races on all types of label information, except nutritional panels and health claims, however, have higher *HEI*s than white label users. For non-label users, except list of ingredients, other races also have higher *HEI*s than whites.

Male users of nutrient-content claims or serving sizes have higher *HEI*s than female label users. For non-label users, males have lower *HEI*s than females. Employed label users have lower *HEI*s than unemployed label users. The reason for this result is not clear. However, it is possible that the diet quality of employed label users is lower

Table 5.5. Parameter estimates of the HEI equations.[a]

	List of ingredients		Nutrient content claims		Nutritional panels		Serving sizes		Health claims	
	User	Non-user	User	Non-user	User	Non-user	User	Non-user	User	Non-user
Constant	57.302***	51.406***	58.855***	48.267***	55.357***	49.783***	57.948***	49.182***	62.939***	44.379***
	(20.129)[a]	(14.117)	(21.175)	(12.246)	(19.246)	(13.930)	(18.993)	(14.365)	(21.233)	(12.617)
Income	0.7518**	1.6644***	0.6283**	1.7901***	0.8803**	1.2408**	0.6977**	1.1558**	0.6869**	1.3486***
	(2.522)	(3.031)	(2.082)	(3.371)	(2.919)	(2.353)	(2.176)	(2.546)	(2.159)	(2.948)
Income^2	−0.0297	−0.1464***	−0.0251	−0.1394***	−0.0388	−0.1061**	−0.0405	−0.0759*	−0.0266	−0.0951**
	(−1.036)	(−2.649)	(−0.864)	(−2.678)	(−1.344)	(−1.991)	(−1.326)	(−1.675)	(−0.874)	(−2.104)
Household size	−0.1772	−0.4252	−0.2026	−0.4346	−0.3197*	−0.2828	−0.0340	−0.4582*	−0.2998	−0.3572
	(−1.036)	(−1.559)	(−1.152)	(−1.586)	(−1.875)	(−1.078)	(−0.180)	(−1.917)	(−1.589)	(−1.419)
Age	0.1255***	0.1019***	0.1413***	0.0930***	0.1431***	0.0967***	0.172***	0.116***	0.1592***	0.1201***
	(7.836)	(3.988)	(8.527)	(3.567)	(8.932)	(3.729)	(9.442)	(5.016)	(8.699)	(5.149)
Male	0.2514	−1.8239*	1.0128**	−1.0230	−0.0487	−1.2731	0.9499*	−1.526*	0.4924	−0.3418
	(0.547)	(−1.850)	(2.124)	(−0.998)	(−0.105)	(−1.317)	(1.892)	(−1.828)	(0.986)	(−0.401)
Black	−3.4605***	−4.0191***	−3.2602***	−4.0911***	−3.2915***	−3.1650***	−3.374***	−3.942***	−3.5675***	−4.0307***
	(−4.972)	(−3.271)	(−4.443)	(−3.465)	(−4.779)	(−2.617)	(−4.384)	(−3.726)	(−4.500)	(−3.662)
Others	2.2623**	1.5976	2.9629***	3.2107*	1.4947	4.3512***	3.0255***	3.255***	1.4042	3.8294**
	(2.208)	(0.941)	(2.753)	(1.927)	(1.467)	(2.626)	(2.642)	(2.290)	(1.220)	(2.566)
Employed	−2.0633***	−2.2929**	−2.1387***	−2.3788***	−2.0228***	−2.0456**	−1.591***	−1.758**	−2.0126***	−1.8256**
	(−3.971)	(−2.417)	(−3.911)	(−2.632)	(−3.923)	(−2.251)	(−2.746)	(−2.174)	(−3.391)	(−2.291)
Non-Hispanic	−1.4814*	−1.81	−1.4703	−0.3137	−2.4437***	−0.8115	−2.0275**	−0.6352	−2.1782**	1.092
	(−1.647)	(−1.117)	(−1.564)	(−0.191)	(−2.728)	(−0.517)	(−2.086)	(−0.448)	(−2.137)	(0.774)
City	0.3589	1.9139*	0.2116	2.2608**	0.0725	2.1209**	−0.1154	2.093**	0.5824	1.7153**
	(0.706)	(1.945)	(0.395)	(2.389)	(0.145)	(2.235)	(−0.205)	(2.569)	(1.009)	(2.013)

Non-metro	−1.1004**	−2.1055**	−1.0920*	−0.8895	−1.2469**	−1.3106	−1.1604*	−1.3211*	−0.9218	−0.4958
	(−2.007)	(−2.284)	(−1.901)	(−0.979)	(−2.254)	(−1.485)	(−1.907)	(−1.715)	(−1.493)	(−0.589)
College	1.4125***	2.9144***	1.4946***	2.2085**	1.9037***	2.5078***	1.6834***	2.937***	1.0597**	2.4432***
	(2.934)	(2.993)	(2.997)	(2.452)	(3.989)	(2.758)	(3.252)	(3.861)	(1.977)	(3.014)
Food stamp	−3.7429***	−1.5581	−3.5118***	−1.7178	−3.3053***	−2.6352**	−2.728***	−3.189***	−2.2011**	−3.105**
program	(−4.145)	(−1.169)	(−3.718)	(−1.320)	(−3.617)	(−2.072)	(−2.737)	(−2.719)	(−2.149)	(−2.548)
Special diet	3.0666***	4.4666***	2.4095***	4.6411***	2.7823***	5.3438***	2.2726***	3.789***	2.194***	3.6419***
	(5.588)	(3.534)	(4.170)	(3.655)	(5.128)	(4.186)	(3.669)	(3.700)	(3.501)	(3.400)
Midwest	1.1496**	3.8206***	1.5343***	2.5246***	1.2795***	3.108***	1.3559***	2.782***	1.7458***	1.816**
	(2.188)	(3.829)	(2.914)	(2.604)	(2.399)	(3.257)	(2.431)	(3.340)	(3.196)	(2.165)
West	2.0994***	1.8478*	2.2607***	1.4804	2.0287***	2.1046**	2.0329***	2.1643**	2.2728***	1.6647*
	(3.554)	(1.755)	(3.801)	(1.452)	(3.425)	(2.017)	(3.257)	(2.384)	(3.697)	(1.821)
σ_1, σ_0[b]	13.568***	12.588***	14.28***	13.027***	13.186***	12.602***	14.534***	12.654***	15.122***	13.356***
	(56.658)	(42.487)	(52.948)	(31.682)	(60.766)	(42.621)	(46.386)	(41.813)	(48.169)	(25.906)
$\rho_{1\mu}, \rho_{0\mu}$[c]	−0.571***	0.0615	−0.7224***	0.2458*	−0.4286***	0.1074	−0.707***	0.1751	−0.7975***	0.3637***
	(−10.422)	(0.528)	(−19.48)	(1.915)	(−6.424)	(1.000)	(−17.790)	(1.492)	(−28.846)	(2.973)
N[d]	4103	1240	3972	1371	4039	1304	3625	1718	3643	1700

[a] The numbers in parentheses are *t* values. *, ** and *** indicate significance at the 10%, the 5 % and the 1% level, respectively. LIMDEP was used to carry out the FIML estimation of endogenous switching regression model.

[b] σ_1, σ_0 indicates the standard deviation of error terms of *HEI* equations for label users and non-label users, respectively.

[c] $\rho_{1\mu}$, $\rho_{0\mu}$ indicates the correlation coefficients between probit label use and conditional *HEI* equation for label users and non-label users, respectively.

[d] N indicates the number of observations in the label user and non-label user groups.

because they do not have as much time as the unemployed to spend on food shopping to make the more appropriate decisions regarding the quality of foods they need to buy.

Non-label users from central cities have *HEI*s that are about two points higher than non-label users from suburban areas. On the other hand, label users (except health claims) from non-metropolitan areas have *HEI*s that are one point lower than label users from suburban areas. Regionally, label and non-label users from the South have lower *HEI*s than those from other regions.

Interestingly, results indicate that people with at least some college education have higher *HEI*s than people with no college education, regardless of label use and label information type. The effect of education may operate through several vehicles. First, it may improve the efficiency of the production process directly. For example, well-educated consumers may better understand the label information and the effect of food on health. They will, therefore, be more able to adapt their diet behaviour in a more healthy way. Secondly, education may provide better access to information. Well-educated consumers may be better aware of effective methods to improve their diets, or they may face lower costs of gathering information. Thirdly, education may be associated with a preference for healthier diets.

Non-Hispanic label users, except in the use of nutrient-content claims, have *HEI*s that are one or two point(s) lower than Hispanic label users. More importantly, food-stamp participants who are label users have *HEI*s that are almost three points lower than non-food-stamp participants who are label users. This result implies that the food-stamp programme does not improve the diet quality of participants to the level of non-participants, despite the use of the labels. This finding is consistent with that of Butler and Raymond (1996). They observed that, controlling for participation in the food-stamp programme, nutrition intake is negatively affected by food-stamp income for a sample of elderly people. As expected, those who are on a special diet have higher *HEI*s than those who are not on a special diet.

Food-label use and diet-quality improvements

For each label user with characteristics **X** and **Z**, we can compare the expected outcome *HEI* when using the label and the expected outcome in the absence of the food label. For a label user with characteristics **X** and **Z**, the expected value of *HEI* is:

$$\mathrm{E}(HEI_1 \mid I = 1) = \boldsymbol{\beta}_1{}'\mathbf{X} + \sigma_1\rho_{1\mu}\frac{\phi(\gamma'\mathbf{Z})}{\Phi(\gamma'\mathbf{Z})} \tag{5.11}$$

Similarly, the expected value of *HEI* for label users in the absence of food labels is:

$$E(HEI_0 \mid I = 1) = \boldsymbol{\beta_0}'\mathbf{X} + \sigma_0\rho_{0\mu}\frac{\phi(\gamma'\mathbf{Z})}{\Phi(\gamma'\mathbf{Z})} \quad (5.12)$$

Thus, for a label user with characteristics **X** and **Z**, the expected change in diet quality due to food-label use is:

$$E(HEI_1 \mid I = 1) - E(HEI_0 \mid I = 1) = (\boldsymbol{\beta_1} - \boldsymbol{\beta_0})'\mathbf{X} + (\sigma_1\rho_{1\mu} - \sigma_0\rho_{0\mu})\frac{\phi(\gamma'\mathbf{Z})}{\Phi(\gamma'\mathbf{Z})} \quad (5.13)$$

The expected *HEI* before using label information and after using it, and the net change in *HEI* are reported in Table 5.6. The changes in diet quality due to label use are estimated for each of the five types of label information. Consumer label use increases the average *HEI* by a range of 3.5 to 6.1 points, depending on the type of label information use. Improvement in the diet is highest when consumers use the health-claim information on the label.[3]

Concluding Comments

Concerns about the effect of diet on health resulted in the legislation of the NLEA in 1990. To assess the effect of health-related information (i.e. consumer label use) on diet quality, endogenous switching regression techniques are employed to control for self-selectivity in the label-use decision and diet intakes. The results show that label use does indeed have a positive effect in improving diet quality. Improvements in diet quality, as measured by the *HEI*, range from 3.5 points (list of ingredients) to 6.1 points (health claims), depending on the type of label information. Of

Table 5.6. The effect of health information (food-label use) on the healthy eating index.

	Healthy eating index					
	Before using	After using	Net change (= B − A)[a]			
Food label information	food label (A)[a]	food label (B)[a]	Mean	Std dev.	Min.	Max.
List of ingredients	60.523	64.024	3.500	2.589	−6.252	11.994
Nutrient-content claims	58.649	64.047	5.398	2.694	−5.423	13.640
Nutritional panels	59.644	64.152	4.509	2.534	−6.375	12.199
Serving sizes	60.089	64.260	4.171	2.473	−6.683	10.884
Health claims	58.161	64.299	6.138	3.066	−6.742	16.243

[a] Descriptive statistics of the net change in the *HEI* across all individuals.

interest in the results as well is the negative relationship between diet quality and food-stamp participation because it raises questions about the role of the food-stamp programme in improving the diets of participants.

The key findings in this study are of great importance in terms of public policy because of the benefits that improved diets can provide society in general in terms of lives saved and reduction of health-care costs. For instance, McNutt (1992) estimated that health care savings from improved and better diets could amount to $3.6–$21 billion. Zarkin *et al.* (1993) also estimated that the number of discounted life-years that could be gained in the USA during the first 20 years after the implementation of the NLEA ranges from about 40,000 to a high of 1.2 million. While no information is currently available concerning the effect of *HEI* improvements on life expectancy, USDA estimated that improved dietary patterns could save $43 billion in medical care costs and lost productivity resulting from disability associated with coronary heart disease, cancer, stroke and diabetes each year, and prevent over 119,900 premature deaths among individuals 55–84 years of age, valued at $28 billion year^{-1} (Frazao, 1995). On the other hand, the FDA estimated that the food labelling required by NLEA would cost the food industry $1.4–$2.3 billion and the government $163 million over the next 20 years. These estimates, however, are contingent upon the presumption that consumers' diets are improved by their use of food labels. While our study does not address the effects of the NLEA *per se*, our analysis supports the assumption that use of food labels can affect consumers' diets.

There are several areas of future research. One area relates to the ongoing debate on a specific type of health information – health claims (both in the USA and in the EU). For example, it would be interesting to know if some people will use health claims without looking at other health-information aspects of the product. Our study also cannot say anything about whether the switch to mandatory labelling under the NLEA was an improvement in the *HEI*. Data from pre- and post-NLEA would be needed for this task.

While this chapter relates to the USA, future studies could also conduct a similar type of analysis for the EU. It would then be interesting to compare the findings between the USA and the EU.

Notes

[1] In addition to NLEA, the Dietary Supplement Health and Education Act (DSHEA) was passed in 1994, which defined dietary supplements, created a mechanism for dealing with safety issues and regulated health claims and labelling of dietary supplements.

[2] This variable may be a potential endogenous variable in the probit label-use or *HEI* equations, as discussed in Variyam *et al.* (1996). However, there are no good instruments in the data set that can serve as a proxy of this variable to

circumvent the potential endogeneity. Nakamura and Nakamura (1998) oppose an 'always instrumentation' policy for endogenous explanatory variables because in cross-sectional data sets there is usually little real evidence that the instruments that are used are exogenous themselves, and it encourages applied researchers to limit the variables they include in their models so as to avoid this difficult instrument variable (IV) problem.

[3] When self-selectivity is not taken into account in the estimation, the net changes in *HEI* are quite different from the estimates presented here when self-selectivity is accounted for. Estimates range from 4.18 for health claims to 5.15 for nutritional panels.

References

Bowman, S.A., Lino, M., Gerrior, S.A. and Basiotis, P.P. (1998) *The Healthy Eating Index: 1994–1996.* CNPP-5, Center for Nutrition Policy and Promotion, US Department of Agriculture, Washington, DC, 19 pp.

Brucks, M., Mitchell, A.A. and Staelin, R. (1984) The effect of nutrition information disclosure in advertising: an information processing approach. *Journal of Public Policy and Marketing* 3, 1–27.

Butler, J.S. and Raymond, J.E. (1996) The effect of the food stamp programme on nutrient intakes. *Economic Inquiry* 34, 781–798.

Caswell, J.A. (1992) Current information levels on food labels. *American Journal of Agricultural Economics* 74, 1196–1201.

Caswell, J.A. and Mojduszka, E.M. (1996) Using informational labelling to influence the market for quality in food products. *American Journal of Agricultural Economics* 78, 1248–1253.

Frazao, E. (1995) *The American Diet: Health and Economic Consequences.* Information Bulletin No. 711. ERS, US Department of Agriculture, Washington, DC.

Goldberger, A.S. (1983) Abnormal selection bias. In: Karlin, S., Amemiya, T. and Goodman, L. (eds) *Studies in Econometrics, Time Series and Multivariate Statistics*. Academic Press, New York, 570 pp.

Gould, B.W. and Lin, H.C. (1994) Nutrition information and household dietary fat intake. *Journal of Agricultural and Resource Economics* 19, 349–365.

Guthrie, J., Fox, J., Cleveland, L. and Welsh, S. (1995) Who uses nutrition labelling and what effects does label use have on diet quality? *Journal of Nutrition Education* 27, 153–172.

Hartman, R.S. (1991) A Monte Carlo analysis of alternative estimators in models involving selectivity. *Journal of Business and Economic Statistics* 9, 41–49.

Ippolito, P.M. and Mathios, A.D. (1990) Information, advertising and health choice: a study of the cereal market. *Rand Journal of Economics* 21, 459–480.

Ippolito, P.M. and Mathios, A.D. (1993) New food labelling regulations and the flow of nutrition information to consumers. *Journal of Public Policy and Marketing* 12, 188–205.

Kim, S., Nayga, R.M. Jr and Capps, O. Jr (2000) The effect of food-label use on nutrient intakes: an endogenous switching regression analysis. *Journal of Agricultural and Resource Economics* 25, 215–231.

Lee, L. (1978) Unionism and wage rates: a simultaneous equation model with qualitative and limited dependent variables. *International Economic Review* 19, 415–433.

Lee, L.F. and Trost, R.P. (1978) Estimation of some limited dependent variable models with application to housing demand. *Journal of Econometrics* 8, 357–382.

Levy, A.S. and Stokes, R. (1987) Effects of a health promotion campaign on sales of ready-to-eat cereals. *Public Health Reports* 102, 398–403.

McNutt, K. (1992) $3.6 to 21 billion benefit from new labelling regulations. *Nutrition Today* 27, 39–43.

Maddala, G.S. (1983) *Limited Dependent and Qualitative Variables in Econometrics*. Cambridge University Press, New York, 401 pp.

Mathios, A.D. (1998) The importance of nutrition labelling and health claim regulation on product choice: an analysis of the cooking oils market. *Agricultural and Resource Economics Review* 27, 159–168.

Nakamura, A. and Nakamura, M. (1998) Model specification and endogeneity. *Journal of Econometrics* 83, 213–237.

Nawata, K. (1994) Estimation of sample selection bias models by the maximum likelihood estimator and Heckman's two-step estimator. *Economics Letters* 45, 33–40.

Nayga, R.M. Jr (1996) Determinants of consumers' use of nutritional information on food packages. *Journal of Agricultural and Applied Economics* 28, 303–312.

Nayga, R.M. Jr, Lipinski, D. and Savur, N. (1998) Consumers' use of nutritional labels while food shopping and at home. *Journal of Consumer Affairs* 32, 106–120.

Nelson, F.D. (1984) Efficiency of the two-step estimator for models with endogenous sample selection. *Journal of Econometrics* 24, 181–186.

Padberg, D.I. (1992) Nutritional labelling as a policy instrument. *American Journal of Agricultural Economics* 74, 1208–1212.

Pagan, A. and Vella, F. (1989) Diagnostic tests for models based on individual data: a survey. *Journal of Applied Econometrics* 4, 29–59.

Poirier, D.J. and Ruud, P.A. (1981) On the appropriateness of endogenous switching. *Journal of Econometrics* 16, 249–256.

US Department of Health and Human Services (1988) *The Surgeon General's Report on Nutrition and Health*. Public Health Service, Washington, DC, 727 pp.

Variyam, J.N., Blaylock, J. and Smallwood, D. (1996) A probit latent variable model of nutrient information and dietary fibre intake. *American Journal of Agricultural Economics* 78, 629–639.

Variyam, J.N., Blaylock, J. and Smallwood, D. (1998) *USDA's Healthy Eating Index and Nutrition Information*. TB-1866, ERS, US Department of Agriculture, Washington, DC, 21 pp.

Willis, R.J. and Rosen, S. (1979) Education and self-selection. *Journal of Political Economy* 87, 7–36.

Zarkin, G.A. and Anderson, D.W. (1992) Consumer and producer responses to nutrition label changes. *American Journal of Agricultural Economics* 74, 1202–1207.

Zarkin, G.A., Dean, N., Mauskopf, J.A. and Williams, R. (1993) Potential health benefits of nutrition label changes. *American Journal of Public Health* 83, 717–724.

A Sample Selection Model with Endogenous Health Knowledge: Egg Consumption in the USA

6

KAMHON KAN[1] AND STEVEN T. YEN[2]

[1]*Academia Sinica, Taipei, Taiwan;*
[2]*University of Tennessee, Knoxville, Tennessee, USA*

Introduction

Per capita consumption of eggs in the USA has declined from 335 eggs in 1960 to an all-time low of 235.4 eggs in 1995, and has since then increased steadily to 258.7 eggs in 2000 (Fig. 6.1). Identification of causes for changes in egg consumption is of great importance for egg producers and processors. The roles of health knowledge and other determinants in relation to egg consumption are also of interest for nutrition educators and policy makers concerned about improving the dietary outcomes of the consumer.

Eggs contain much more dietary cholesterol than most other foods. Previous studies suggest that diffusion of information on the linkage between cholesterol and arterial disease has been a major cause of decline in egg consumption in recent decades (Putler, 1987; Stillman, 1987; Brown and Schrader, 1990). Brown and Schrader (1990) constructed a cholesterol-information index to estimate the effect of cholesterol information on egg consumption. Their index suggests that consumers' cholesterol awareness has increased since the 1960s. Analysis by Brown and Schrader (1990) based on time-series data suggests that information on the link between cholesterol and heart disease has reduced per capita shell-egg consumption. This cholesterol-information index has also been used to examine the impact of cholesterol information on consumer demand for pork, beef, poultry, fish, fats and oils (Capps and Schmitz, 1991; Yen and Chern, 1992; Chern *et al.*, 1995).

Time-series measures of health-information and health-risk variables may be poor proxies for individual-specific health concerns. Individuals

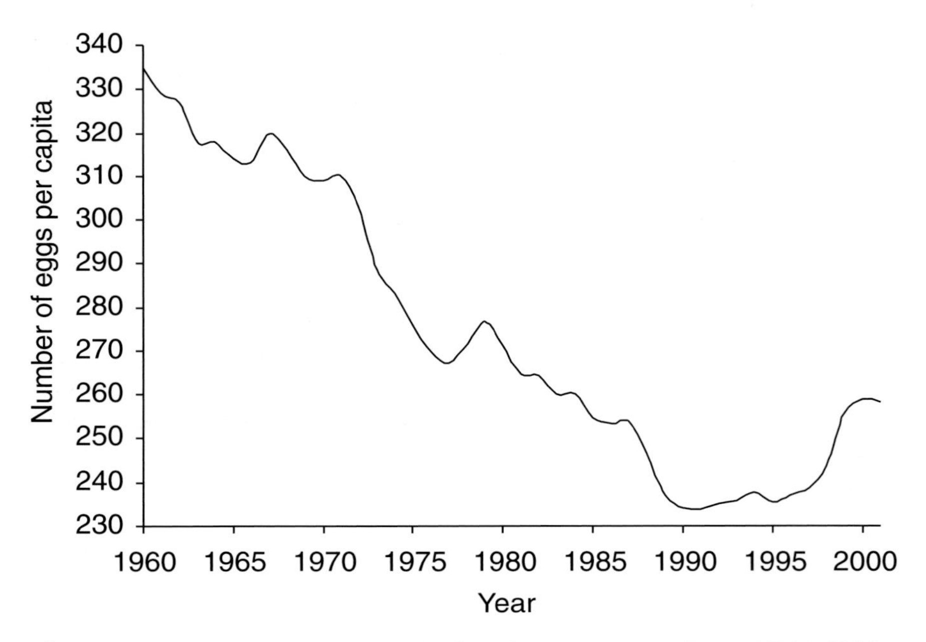

Fig. 6.1. Per capita egg consumption in the USA, 1960–2001. (From: USDA-ESCS, 1978; Putnam and Allshouse, 1999; and USDA-ERS, 2000.)

may differ in exposure to media sources and in cognitive skills to process health and diet information (Jensen *et al.*, 1992; Lin and Milon, 1993). Although household and individual survey data have been used in modelling consumer demand for food products, except for Yen *et al.* (1996), time series findings about the effects of cholesterol information on egg consumption have not been corroborated by microsurvey data.

This study aims to assess the effects of cholesterol information, health knowledge and demographic variables on egg consumption, using microdata from the 1994–1996 Continuing Survey of Food Intakes by Individuals (CSFII). The following sections present the econometric model, describe the data source and sample, present the empirical estimates and summarize major findings.

Health Information, Health Knowledge and Consumption

There are four dependent variables: health information (I_i), health knowledge (K_i), whether an individual consumes egg (P_i) and the quantity of egg consumed (C_i). These four variables are modelled as a recursive system such that P_i and C_i are explained by K_i, while K_i is explained by I_i.[1] In what follows, vectors of explanatory variables x_i^I, x_i^K,

x_i^P, x_i^C, are used to explain information, knowledge, participation and consumption, respectively, with corresponding parameter vectors β^I, β^K, β^P, β^C and random errors ε_i^I, ε_i^K, υ_i^P, υ_i^C. Health information is a binary variable indicating whether an individual has heard about the health effect of cholesterol ($I_i = 1$) or not ($I_i = 0$) and is characterized by a binary response model:

$$I_i = \begin{cases} 1 & \text{if } \beta^I x_i^I + \varepsilon_i^I > 0 \\ 0 & \text{otherwise} \end{cases} \tag{6.1}$$

where random error ε_i^I is distributed as $N(0, 1)$.

Health knowledge (K_i) is a categorical variable, measuring how important an individual feels it is to choose a diet low in cholesterol. The variable ranges from 1 (not at all important) to 4 (very important) and is characterized by an ordered polychotomous response model (McKelvey and Zavoina, 1975):

$$\begin{aligned} K_i^* &= \beta^K x_i^K + \alpha^K I_i + \varepsilon_i^K \\ K_i &= j \text{ if } \mu_{j-1} < K_i^* \le \mu_j,\ j = 1, 2, 3, 4 \end{aligned} \tag{6.2}$$

where α^K is a scalar parameter, ε_i^K is distributed as $N(0, 1)$, and threshold parameters μ_j are normalized such that $\mu_0 = -\infty$, $\mu_1 = 0$ and $\mu_4 = \infty$ for identification. Realizations of health knowledge level K_i are determined by its latent counterpart K_i^*.

Consumption is modelled using a sample selection model (Heckman, 1979); the participation component is modelled as a probit based on the binary outcomes $P_i \in \{0, 1\}$:

$$P_i = \begin{cases} 1 & \text{if } \beta^P x_i^P + \alpha^P K_i^* + \upsilon_i^P > 0 \\ 0 & \text{otherwise} \end{cases} \tag{6.3}$$

where α^P is a scalar parameter and υ_i^P is distributed as $N(0, 1)$: that is, if the ith individual is observed to consume a positive quantity of eggs, then $P_i = 1$. The second component of consumption concerns the actual quantity (C_i) consumed, such that:

$$C_i = \begin{cases} \beta^C x_i^C + \alpha^C K_i^* + \upsilon_i^C & \text{if } P_i = 1 \\ \text{unobserved} & \text{otherwise} \end{cases} \tag{6.4}$$

where α^C is a scalar parameter and $\upsilon_i^C \sim N(0, \sigma_C^2)$. To sum up, the random-error vector $\{\varepsilon_i^I, \varepsilon_i^K, \upsilon_i^P, \upsilon_i^C\}$ is normally distributed with zero mean vector and covariance matrix:

$$\begin{bmatrix} 1 & \sigma_{KI} & 0 & 0 \\ & 1 & \sigma_{KP} & \sigma_{KC} \\ & & 1 & \sigma_{PC} \\ & & & \sigma_C^2 \end{bmatrix}$$

It is noted that, since K_i^* is unobserved and K_i is not a good proxy for it, we use $\hat{K}_i^* = \beta^K x_i^K + \alpha^K I_i$, instead to explain participation and consumption. Therefore, the models for participation and consumption become:

$$P_i = \begin{cases} 1 & \text{if } \beta^P x_i^P + \alpha^P \hat{K}_i^* + \varepsilon_i^P > 0 \\ 0 & \text{otherwise} \end{cases} \tag{6.5}$$

$$C_i = \begin{cases} \beta^C x_i^C + \alpha^C \hat{K}_i^* + \varepsilon_i^C & \text{if } P_i = 1 \\ \text{unobserved} & \text{otherwise} \end{cases} \tag{6.6}$$

where composite errors $\varepsilon_i^P = \upsilon_i^P + \alpha^P \varepsilon_i^K$ and $\varepsilon_i^C = \upsilon_i^C + \alpha^C \varepsilon_i^K$, the terms $\alpha^P \varepsilon_i^K$ and $\alpha^C \varepsilon_i^K$ resulting from the use of $\hat{K}_i^*$ in lieu of K_i^*. Accordingly, the composite error vector $\{\varepsilon_i^I, \varepsilon_i^K, \varepsilon_i^P, \varepsilon_i^C\}$ is distributed as multivariate normal with zero mean vector and covariance matrix:

$$\Omega = \begin{bmatrix} 1 & \sigma_{KI} & \alpha^P \sigma_{KI} & \alpha^C \sigma_{KI} \\ & 1 & \sigma_{KP} + \alpha^P & \sigma_{KC} + \alpha^C \\ & & 1 + 2\alpha^P \sigma_{KP} + (\alpha^P)^2 & \sigma_{PC} + \alpha^C \sigma_{KP} + \alpha^P \sigma_{KC} + \alpha^P \alpha^C \\ & & & \sigma_C^2 + 2\alpha^C \sigma_{KC} + (\alpha^P)^2 \end{bmatrix} = \begin{bmatrix} \Omega_{11} & \Omega_{12} \\ & \Omega_{22} \end{bmatrix} \tag{6.7}$$

partitioned such that Ω_{11} is 3×3, Ω_{12} is 3×1 and Ω_{22} is 1×1. The four equations are estimated jointly with maximum likelihood. To construct the sample likelihood function, we first introduce the conditional and marginal distributions of the error terms. The conditional distribution of $\{\varepsilon_i^I, \varepsilon_i^K, \varepsilon_i^P, | \varepsilon_i^C\}$ is trivariate normal with mean vector and covariance matrix, respectively:

$$\xi_{1\cdot 2} = \Omega_{12} \Omega_{22}^{-1} \varepsilon_i^C \tag{6.8}$$

$$\Omega_{11\cdot 2} = \Omega_{11} - \Omega_{12} \Omega_{22}^{-1} \Omega_{12}' \tag{6.9}$$

whereas the marginal distribution of $\{\varepsilon_i^I, \varepsilon_i^K, \varepsilon_i^P\}$ is trivariate normal with zero mean vector and covariance matrix Ω_{11} (Kotz *et al.*, 2000). Then the conditional probabilities for a participant are (for $j = 1, 2, \ldots, 4$ and $\ell = 0, 1$)

$$\begin{aligned} &\Pr(P_i = 1, K_i = j, I_i = \ell \mid \varepsilon_i^C) \\ &= \Psi\left[(2\ell - 1)(\beta^I x_i^I + \xi_{1\cdot 2(1)}), \mu_j - \beta^K x_i^K - \alpha^K I_i - \xi_{1\cdot 2(2)}, \beta^P x_i^P + \alpha^P \hat{K}_i^* + \xi_{1\cdot 2(3)}; Z_i' \Omega_{11\cdot 2} Z_i\right] \\ &- \Psi\left[(2\ell - 1)(\beta^I x_i^I + \xi_{1\cdot 2(1)}), \mu_{j-1} - \beta^K x_i^K - \alpha^K I_i - \xi_{1\cdot 2(2)}, \beta^P x_i^P + \alpha^P \hat{K}_i^* + \xi_{1\cdot 2(3)}; Z_i' \Omega_{11\cdot 2} Z_i\right] \end{aligned} \tag{6.10}$$

where $\varepsilon_i^C = C_i - (\beta^C x_i^C + \alpha^C \hat{K}_i^*)$, $Z_i = \text{diag}\{(2\ell - 1), 1, 1\}$, $\xi_{1\cdot 2(j)}$, $j = 1, 2, 3$ are elements of the conditional mean vector $\xi_{1\cdot 2}$ defined in Equation 6.8, and $\psi(\cdot, \cdot, \cdot; \cdot)$ is the trivariate normal cumulative density function (CDF) with the last element being the covariance matrix. In Equation 6.10, the definition of the probability of being in the jth knowledge category using threshold parameters μ_j and μ_{j-1} is similar to the case of the univariate-order probit model (McKelvey and Zavoina, 1975), and multiplication of $(\beta^I x_i^I + \xi_{1\cdot 2(1)})$ by $(2\ell - 1)$ and $\Omega_{11\cdot 2}$ by Z_i

accommodates the necessary sign changes in the integration limit and covariance matrix while evaluating the trivariate normal probabilities as lower-tailed CDFs. Likewise, using the marginal distribution of $\{\varepsilon_i^I, \varepsilon_i^K, \varepsilon_i^P\}$ with zero mean and covariance matrix Ω_{11}, the probabilities for a non-participant are (for $j = 1, 2, \ldots, 4$ and $\ell = 0, 1$):

$$\begin{aligned}&\Pr(P_i = 0, K_i = j, I_i = \ell)\\&= \Psi\left[(2\ell-1)\beta^I x_i^I,\ \mu_j - \beta^K x_i^K - \alpha^K I_i, -(\beta^P x_i^P + \alpha^P \hat{K}_i^*);\ W_i'\Omega_{11}W_i\right]\\&-\Psi\left[(2\ell-1)\beta^I x_i^I,\ \mu_{j-1} - \beta^K x_i^K - \alpha^K I_i, -(\beta^P x_i^P + \alpha^P \hat{K}_i^*);\ W_i'\Omega_{11}W_i\right]\end{aligned} \tag{6.11}$$

where $W_i = \text{diag}\{(2\ell - 1), 1, -1\}$, again with $(2\ell - 1)$ and W_i accommodating sign changes in $\beta^I x_i^I$ and Ω_{11}. Define a dichotomous index d_{ij} such that $d_{ij} = 1$ if $K_i = j$ and zero otherwise. Then, using components in Equations 6.10 and 6.11, the sample likelihood function is:

$$\begin{aligned}L = &\prod_{P_i=1, I_i=1} \prod_{j=1}^{4} \left[\Pr(P_i = 1, K_i = j, I_i = 1 \mid \varepsilon_i^C)\ \phi\left(\varepsilon_i^C ; \sigma_C^2\right)\right]^{d_{ij}}\\&\times \prod_{P_i=1, I_i=0} \prod_{j=1}^{4} \left[\Pr(P_i = 1, K_i = j, I_i = 0 \mid \varepsilon_i^C)\ \phi\left(\varepsilon_i^C ; \sigma_C^2\right)\right]^{d_{ij}}\\&\times \prod_{P_i=0, I_i=1} \prod_{j=1}^{4} \left[\Pr(P_i = 0, K_i = j, I_i = 1)\right]^{d_{ij}}\\&\times \prod_{P_i=0, I_i=0} \prod_{j=1}^{4} \left[\Pr(P_i = 0, K_i = j, I_i = 0)\right]^{d_{ij}}\end{aligned} \tag{6.12}$$

where $\phi(\varepsilon_i^C; \sigma_C^2)$ is the univariate normal probability density function of ε_i^C with mean zero and variance σ_C^2.

To calculate the marginal effects of explanatory variables, we present relevant probability and mean expressions. The unconditional probabilities for a participant are:

$$\Pr(P_i = 1, K_i = j, I_i = \ell) = \int_{-\infty}^{\infty} \Pr(P_i = 1, K_i = j, I_i = \ell \mid \varepsilon_i^C)\, \phi(\varepsilon_i^C)\, d\varepsilon_i^C \tag{6.13}$$

for $j = 1, 2, \ldots, 4$ and $\ell = 0, 1$, where the conditional probability $\Pr(P_i = 1, K_i = j, I_i = \ell \mid \varepsilon_i^C)$ in the integrand is defined in Equation 6.10. Using the unconditional probabilities in Equations 6.11 and 6.13, the probability of having information on the health effects of cholesterol is:

$$\Pr(I_i = 1) = \sum_{j=1}^{4} \sum_{\ell=0}^{1} \Pr(P_i = \ell,\ K_i = j,\ I_i = 1) \tag{6.14}$$

the probability of being in the jth knowledge category is:

$$\Pr(K_i = j) = \sum_{h=0}^{1}\sum_{\ell=0}^{1} \Pr(P_i = h, K_i = j, I_i = \ell),\ j = 1, 2, \cdots, 4 \qquad (6.15)$$

and the probability of participation is:

$$\Pr(P_i = 1) = \sum_{j=1}^{4}\sum_{\ell=0}^{1} \Pr(P_i = 1, K_i = j, I_i = \ell) \qquad (6.16)$$

The conditional mean of consumption is:

$$E(C_i \mid P_i = 1, K_i, I_i) = \beta^C x_i^C + \alpha^C \hat{K}_i^* + \rho_{PC}\left[\frac{\partial \Pr(P_i = 1)}{\partial(\beta^P x_i^P + \alpha^P \hat{K}_i^*)}\right]\left[\frac{1}{\Pr(P_i = 1)}\right] \qquad (6.17)$$

where ρ_{PC} is the correlation between ε_i^P and ε_i^C, $\Pr(P_i = 1)$ is defined in Equation 6.16 and the last term in brackets is the multivariate counterpart of the inverse Mills' ratio. Using Equations 6.16 and 6.17, the unconditional mean of consumption is:

$$E(C_i \mid P_i, K_i, I_i) = \Pr(P_i = 1)\, E(C_i \mid P_i = 1, K_i, I_i) \qquad (6.18)$$

We calculate the marginal effects of explanatory variables by differentiating Equations 6.14, 6.15, 6.16 and 6.18.

Data and Variable Definitions

Data used in this study are drawn from the 1994–1996 CSFII and the follow-up Diet and Health Knowledge Survey (DHKS) (USDA-ARS, 2000). Each year the CSFII collects data on individual's food and nutrient intakes and demographic information. The DHKS, administered to designated main meal planners or preparers, also includes information on each individual's attitudes and knowledge about diet and health.

Quantity is defined as the 2-day average of egg consumption (C_i), including egg and mixtures consisting mainly of egg (e.g. omelette, egg salad). Participation (P_i) is a binary realization of C_i, namely $P_i = 1$ if $C_i > 0$ and $P_i = 0$ if $C_i = 0$. In addition, health information (I_i) is a dummy variable which equals 1 if the individual has heard about the health effect of cholesterol and 0 otherwise. Finally, health knowledge (K_i) is a categorical variable ranging from 1 (not at all important) to 4 (very important), reflecting the importance of choosing a low-cholesterol diet.

The explanatory variables include income, body-mass index, education, age, household size and dummy variables indicating race (black), ethnicity (Hispanic), self-evaluated health status, vegetarianism, gender (male), urbanization (suburban), region (South, West) and whether the individual had been diagnosed as having high blood pressure, heart disease/problem or high blood cholesterol.

The final sample includes 5093 adult individuals age 18 and over with complete records in the 1994–1996 CSFII and DHKS. The definitions and sample statistics of all variables are reported in Table 6.1. About 89% of the whole sample were aware of the link between

Table 6.1. Definitions of variables and sample statistics: egg consumption by individuals in the USA.[a] (Compiled from the Continuing Survey of Food Intakes by Individuals, 1994–1996.)

Variable	Definition	Mean	SD
Egg	Egg consumption (g day^{-1})	20.04 (58.27)	36.68 (40.05)
Cholesterol information	Heard about health problem caused by eating too much cholesterol (yes = 1; no = 0)	0.89 (0.87)	0.32 (0.33)
Health knowledge	How important to choose a diet low in cholesterol (not at all important = 1; not too important = 2; somewhat important = 3; very important = 4)	3.45 (3.42)	0.77 (0.79)
Income	Per capita income in thousands of US dollars	16.44	13.88
Body-mass index	Quetelet's body-mass index = (weight in kg)/(height in m)^2	26.58	5.42
Education	Years of formal education	12.80	2.95
Age	Age in years	50.41	16.95
Size	Household size	2.58	1.44
Dummy variables (yes = 1, no = 0)			
Black	Individual is black	0.11	
Hispanic	Individual is Hispanic	0.03	
Health	Self-evaluated health status fair or better	0.84	
Vegetarian	Individual is vegetarian	0.03	
Male	Individual is male	0.51	
High blood pressure	Individual is diagnosed with high blood pressure	0.26	
Heart problem	Individual is diagnosed with a heart problem/disease	0.11	
High blood cholesterol	Individual is diagnosed with high blood cholesterol	0.17	
Suburban	Resides in suburban area	0.45	
South	Resides in the South	0.35	
West	Resides in the West	0.20	

[a] Sample size is 5093. Figures in parentheses are results calculated from a consuming subsample of 1752 individuals.

cholesterol and health problems, an increase from 82% during 1989–1991 (Yen *et al.*, 1996). This proportion is only slightly lower among the consumers (87%). The sample mean for health knowledge is about 3.46 (out of 4) for the full sample, which is comparable to 3.43 for the consuming sample.

Estimation Results

Maximum-likelihood estimation is carried out by maximizing the (logarithm of the) likelihood function (Equation 6.12), programmed in Gauss, using numerical optimization routine 'maxlik'. Results are presented in Table 6.2. Correlation between the error terms of the participation and health-knowledge equations is significant at the 1% level, whereas that between the participation and consumption equations is significant at the 10% level. Correlation between the cholesterol information and health-knowledge equations, while not significant at the 10% level, has a P value of 0.13. Thus, estimation of these equations in a system is justified.

Significant variables in the cholesterol-information equation are (at the 10% level): education, suburban and high blood cholesterol, which all have a positive effect on information. In addition, age is marginally significant (negative; P value = 0.126). Thus, individuals with higher education may have better access to the media and therefore are more likely to be aware of the health problems caused by too much cholesterol intake, as are individuals who had been diagnosed as having high blood cholesterol than individuals who had not. The negative coefficient of age suggests that, as a person grows older, he/she is less likely to be aware of the cholesterol information. Surprisingly, men, who traditionally are less

Table 6.2. Maximum-likelihood estimation of sample selection model with endogenous health information and knowledge.[a]

Variable	Cholesterol information	Health knowledge	Participation	Consumption
Constant	0.872‡	2.359‡	0.444	28.968
	(6.000)	(7.549)	(0.955)	(1.464)
Cholesterol information		0.229		
		(0.703)		
Health knowledge			−0.205†	1.501
			(−1.735)	(0.325)
Income			−0.001	−0.092
			(−0.870)	(−1.118)
Body-mass index			0.001	0.026
			(0.378)	(1.345)

Table 6.2. *Continued.*

Variable	Cholesterol information	Health knowledge	Participation	Consumption
Education	0.043‡	−0.040‡	−0.020‡	−0.289
	(5.393)	(−6.125)	(−2.459)	(−0.698)
Age (× 10^{-3})	−2.169	5.435‡	1.408	−4.000
	(−1.529)	(5.156)	(1.031)	(−0.058)
Suburban	0.086†			
	(1.774)			
South	−0.008			
	(−0.155)			
West	0.011			
	(0.164)			
Size			0.015	2.503‡
			(0.963)	(3.108)
Black	−0.088	0.047	0.160‡	4.760
	(−1.295)	(0.840)	(2.584)	(1.457)
Hispanic	−0.025	0.063	0.164	12.870‡
	(−0.191)	(0.685)	(1.437)	(2.282)
Male	−0.025	−0.337‡		
	(−0.520)	(−9.809)		
Health		−0.269‡	−0.049	1.296
		(−6.010)	(−0.852)	(0.416)
Vegetarian	−0.026	0.127	−0.043	
	(−0.192)	(1.246)	(−0.394)	
High blood pressure	−0.006			
	(−0.098)			
Heart problem	0.023			
	(0.306)			
High blood cholesterol	0.158†	0.252‡		
	(1.907)	(5.126)		
μ_2		0.774‡		
		(21.399)		
μ_3		1.666‡		
		(36.539)		
σ_C^2				18.297‡
				(7687.389)
Correlation (ρ_{ij})				
Health knowledge	0.123			
	(1.506)			
Participation		−0.178‡		
		(−3.142)		
Consumption		−0.185	0.377†	
		(−1.107)	(1.722)	
Log-likelihood	−14812.903			

[a] Asymptotic *t* ratios in parentheses.
Daggers ‡ and † denote significance at the 5 and 10% levels, respectively.

conscious of health risk, do not appear to be less aware of the health problems caused by too much cholesterol than women.

As to the health-knowledge equation, both threshold parameters are positive and significant at the 1% level of significance. Negative threshold parameters would have indicated misspecification of the model. Not surprisingly, individuals who are older recognize the importance of a low-cholesterol diet more than others, as do those who have been diagnosed as having high blood cholesterol. Those who believe themselves to be healthy downplay the importance of a low-cholesterol diet. Men, who are found to be less aware of the cholesterol information, also tend not to rank cholesterol as important. For no obvious reason, education decreases the awareness of the importance of a low-cholesterol diet. Common knowledge would seem to suggest otherwise, as the educated would have more and better access to health-related publications. Perhaps it is because the educated are more health-conscious and believe they have better and other ways to deal with health risks and therefore tend to downplay the importance of a low-cholesterol diet.

Turning now to results of the participation equation in Table 6.2, health knowledge and education have negative effects on participation (i.e. decision to consume eggs), while blacks and Hispanics are more likely to participate in egg consumption than others. Finally, results for the consumption equation suggest that, conditional on participation, Hispanics and those who come from a larger household consume more eggs than others. Interestingly, income does not play a significant role in the level of egg consumption.

As in the univariate ordered polychotomous response model, it is difficult to interpret the coefficients of the latent equation in the ordered probit (Greene, 2000). The marginal effects of education and age (the two significant and continuous variables in the knowledge equation) on probabilities are reported in Table 6.3. The results suggest that education increases the probability of being in the lower-knowledge category ($K = 1$) and decreases the probability of being in the higher-knowledge category ($K = 4$). The marginal effects of age are opposite; that is, age decreases the probability of being in the lower-knowledge category ($K = 1$) and increases the probability of being in the higher-knowledge category ($K = 4$).

Table 6.3. Marginal effects of continuous variables on probabilities of knowledge categories.

	Knowledge category			
Variable	$K = 1$	$K = 2$	$K = 3$	$K = 4$
Education	0.021	0.054	0.077	−0.138
Age	−0.003	−0.007	−0.010	0.020

As in other limited dependent-variable models, the effects of variables can be explored further by calculating marginal effects. The marginal effects of cholesterol-information probability, participation probability and consumption level with respect to continuous explanatory variables are reported in Table 6.4. Overall, these marginal effects are rather small (in absolute values). For instance, the marginal effects of education suggest that when year (level) of education increases by 1 year, all else being equal, the probability of receiving cholesterol information increases by 0.08, the probability of consuming eggs decreases by 0.06 and the amount of eggs consumed decreases by about 0.36 g day^{-1}. Negative effects of education and very small effects of income are also reported by Yen *et al.* (1996), based on an earlier sample of the CSFII. The marginal effects of other variables can be interpreted in the same manner.

Table 6.4. Marginal effects of continuous variables on probability of access to information, probability of participation and level of consumption.

Variable	Cholesterol information	Participation	Consumption
Health knowledge		−0.074	−2.962
Income ($\times 10^{-4}$)		−2.000	−3.000
Body-mass index ($\times 10^{-4}$)		−3.000	6.000
Education	0.076	−0.058	−0.355
Age	−0.004	0.005	0.0002
Size		−0.002	0.646

Concluding Remarks

Interest in the roles of cholesterol information and health-risk knowledge in food consumption has continued to grow. We investigate egg consumption by individuals in the USA using a recursive system in which cholesterol information affects health knowledge, and health knowledge in turn affects participation and consumption. Correlations are accommodated among these equations and our results confirm these correlations. We find that health knowledge plays a negative role in participation but not in level of egg consumption. Demographic variables, such as education, age and household size, also play significant roles in determining egg consumption. In concluding this chapter we note that results for this current analysis are somewhat sparsely significant, with quite a number of variables insignificant in determining cholesterol information, health knowledge and the probability and level of egg consumption. Future studies might consider applying the proposed econometric framework to consumption of other foods, such as red meat, fats and oils, and raw fish/meat, in which the roles of health information and health-risk knowledge can be further investigated.

Note

[1] Results of our preliminary analysis suggest that I_i has little explanatory power in explaining P_i and C_i.

References

Brown, D.J. and Schrader, L.F. (1990) Cholesterol information and shell egg consumption. *American Journal of Agricultural Economics* 72, 548–555.

Capps, O. Jr and Schmitz, J.D. (1991) A recognition of health and nutrition factors in food demand analysis. *Western Journal of Agricultural Economics* 16, 21–35.

Chern, W.S., Loehman, E.T. and Yen, S.T. (1995) Information, health risk beliefs, and the demand for fats and oils. *Review of Economics and Statistics* 77, 555–564.

Greene, W.H. (2000) *Econometric Analysis*, 4th edn. Prentice Hall, Upper Saddle River, New Jersey, 1004 pp.

Heckman, J.J. (1979) Sample selection bias as a specification error. *Econometrica* 47, 153–161.

Jensen, H., Kesavan, T. and Johnson, S. (1992) Measuring the impact of health awareness on food demand. *Review of Agricultural Economics* 14, 299–312.

Kotz, S., Balakrishnan, N. and Johnson, N.L. (2000) *Continuous Multivariate Distributions*, Vol. 1: *Models and Applications*, 2nd edn. John Wiley & Sons, New York, 752 pp.

Lin, C.T. and Milon, J.W. (1993) Attribute and safety perceptions in a double-hurdle model of shellfish consumption. *American Journal of Agricultural Economics* 75, 724–729.

McKelvey, R.D. and Zavoina, W. (1975) A statistical model for the analysis of ordinal level dependent variables. *Journal of Mathematical Sociology* 4, 103–120.

Putler, D.S. (1987) *The Effect of Health Information on Shell Egg Consumption.* Department of Agricultural and Resource Economics Working Paper No. 448, California Agricultural Experiment Station, University of California, Berkeley, California.

Putnam, J.J. and Allshouse, J.E. (1999) *Food Consumption, Prices, and Expenditures, 1970–1997.* Statistical Bulletin No. 965, Economic Research Service, US Department of Agriculture, Washington, DC.

Stillman, R.P. (1987) *A Quarterly Forecasting Model of the US Egg Sector.* Tech. Bull. No. 1729, Economic Research Service, US Department of Agriculture, Washington, DC.

US Department of Agriculture (USDA), Agricultural Research Service (ARS) (2000) *Continuing Survey of Food Intakes by Individuals 1994–1996, 1998.* CD-ROM, USDA, Washington, DC.

US Department of Agriculture (USDA), Economic Research Service (ERS) (2000) *Livestock, Dairy and Poultry Situation and Outlook.* LDP-M-77, USDA, Washington, DC. Available at http://usda.mannlib.cornell.edu/reports/erssor/livestock/ldp-mbb/2000/ldp-m77.pdf

US Department of Agriculture (USDA), Economics, Statistics and Cooperatives Service (ESCS) (1978) *Food consumption, prices, expenditures: supplement for 1976 to Agricultural Economics Report No. 138.* USDA, Washington, DC.

Yen, S.T. and Chern, W.S. (1992) Flexible demand systems with serially correlated errors: fat and oil consumption in the United States. *American Journal of Agricultural Economics* 74, 689–697.

Yen, S.T., Jensen, H.H. and Wang, Q. (1996) Cholesterol information and egg consumption in the US: a nonnormal and heteroscedastic double-hurdle model. *European Review of Agricultural Economics* 23, 343–356.

Health Information and Food Demand in Eastern and Western Germany

7

SUSANNE WILDNER[1] AND
STEPHAN VON CRAMON-TAUBADEL[2]
[1]*University of Kiel, Kiel, Germany;*
[2]*University of Göttingen, Göttingen, Germany*

Introduction

Food demand evolved in different environments in the German Democratic Republic (GDR), or East Germany (EG), and the Federal Republic of Germany (FRG), or West Germany (WG), prior to 1989. The main characteristic of the GDR's economic system was central planning. This was accompanied by low real incomes relative to the FRG, strong subsidization of basic foods and very high prices for 'luxury' foods. Nevertheless, consumption structures in the GDR and the FRG differed less than might be expected (Donat, 1996: 2), which suggests that a common cultural heritage was an important determinant of the demand for food.

In both the GDR and the FRG, consumption of food was subject to structural change over time. Two main trends were characteristic of this change. First, the consumption of high-value vegetable products increased. Secondly, the consumption of coffee, cocoa and other 'luxury' products also increased (Donat, 1996: 8). After monetary union in 1990, diets changed quite rapidly in the former GDR and in the course of 1990/91 an obvious convergence towards food-consumption structures in WG could be observed.

Changes in food-consumption structures in the former GDR since reunification could be due to several factors, including price and income changes, changing preferences and the abrupt changes in the availability of some foods, such as bananas, that accompanied reunification.

The main aim of this work is to develop a demand model for EG and WG that can be used to test the hypothesis that there is a link between the dissemination of health information and the demand for specific types of food. The analysis focuses on meat, as meat accounts for an important share of food consumption in EG and WG and meat may be expected to be especially responsive to health information. Health campaigns geared to reducing the incidence of heart disease often stress the benefits of reducing the intake of animal fats.

After a brief review of German food-demand studies, differences in meat- and fish-consumption structures between EG and WG are presented. Next, an almost ideal demand system (AIDS) model of food demand is specified and the integration of health-information indices into this model is discussed. This AIDS model is estimated for EG and WG using different health-information indices. The results of this estimation and their implications are discussed and we close with conclusions.

A Review of German Food-demand Studies

Table 7.1 presents an overview of German food-demand studies that have been published since 1980. Most studies to date have been based on straightforward single-equation methods. Estimations based on demand systems that are consistent with consumer theory have been comparatively rare, with the most recent such study appearing in 1995 (Recke, 1995). The study by Sienknecht (1986) is interesting in that it represents a concerted attempt to compare the results of applying different types of demand system to the same data set. All of the studies in Table 7.1 have been published in German; we are not aware of any specific studies of German food demand that have been published in English. Several demand systems, including the AIDS, the linear expenditure system (LES) and the quadratic expenditure system (QES), have been estimated.

It is beyond the scope of this chapter to present a detailed comparison of the empirical estimates generated by the studies summarized in the table. Wildner (2001: 25–27) compares the estimated income elasticities of demand for aggregate food from those studies that provide such estimates and finds that they are remarkably consistent, given the variety of underlying methods, data sources and sample periods employed; with few exceptions, these elasticities all range between 0.25 and 0.6, pointing to the expected income-inelastic demand for food.

Food Consumption in Eastern and Western Germany

The analysis in this chapter is based on data from the *Einnahmen und Ausgaben ausgewählter Haushalte* collected by the Federal Statistical Office

Table 7.1. An overview of German food-demand studies since 1980. (From: Wildner, 2001: Table 2-1.)

Reference	Sample period[a]	Method	Products
Wildner (2001)	1966/1–1997/12	LA/AIDS	5 food groups and 5 types of meat
Röder (1998)	1985–1989	Single equation	24 food products
Eckert (1998)	1994–1996	Single equation	Beef
Hoff and Claes (1997)	1986–1996	Single equation	Beef
Giere *et al.* (1997)	1970–1994	Single equation	Butter
Recke (1995)	1968–1991	Rotterdam model	5 types of meat
Hagner (1994)	1992/1–1993/10	Single equation	Muesli
Grings (1993)	1964–1985	LA/AIDS	7 food groups
Pawlik (1993)	1975–1989	Single equation	Frozen food
Schons (1993)	1965–1988	Single equation	13 food groups and types of meat
Henning and Michalek (1992)	1970–1985	3-stage LES-LES-AIDS	Food, beverages and tobacco, and non-food products
Michalek and Keyzer (1992)	1970–1985	2-stage LES-AIDS	Food, beverages and tobacco, and non-food products
Grings (1991)	1964–1985	LA/AIDS	5 food groups
Appel *et al.* (1989)	1965–1984	Single equation	11 food groups and types of meat
Sienknecht (1986)	1963–1980	GDS, AIDS, QES, GTL, NGLES	3 groups of durable goods and 4 groups of non-durable goods
Sommer (1985)	1965–1981	Single equation	Fish and fish products
Schmidt (1984)	1973–1982	Single equation	Butter
Höh (1984)	1970/71–1982/83	Single equation	Processed potato products
Filip and Wöhlken (1984)	1978	Single equation	Food groups and individual food products
Ryll (1984)	1961–1980	Single equation	Fish
Rettig (1983)	1950–1980	LES	8 groups of non-durable goods
Hansen (1983)	1960–1979	AIDS	6 groups of non-durable goods
Filip (1983)	1954–1980	Single equation	11 food groups
Wöhlken and Filip (1982)	1954–1980	Single equation	10 groups of non-durable goods
deHaen *et al.* (1982)	1970–1976	LES	Food and beverages as well as individual food groups
Wöhlken *et al.* (1981)	1960–1977	Single equation	11 food groups and types of meat

Table 7.1. *Continued.*

Reference	Sample period[a]	Method	Products
Eschenbach (1981)	1973	Single equation	Food groups and individual food products
Funkte (1981)	1973	Single equation	Alcoholic beverages
Murty and Pradesh (1981)	1960–1976	QES	Food and beverages, tobacco, and other non-durable goods
Weindlmaier and Neubüser (1980)	1960–1978	Single equation	Beer
Hasenkamp (1980)	1954–1967	LES	Non-durable goods and disaggregated product groups
Merz (1980)	1969	LES, ELES, FELES	Durable goods, non-durable goods and savings
Meyer (1980a)	1962/63 and 1973	Single equation	Various milk products
Meyer (1980b)	1962/63 and 1973	Single equation	5 groups of oils and fats
Graser and Sagmeister (1980)	1952/53–1976/77	Single equation	Sugar

[a] In some studies, shorter sample periods than reported were used for specific products or household types.

GDS, generalized demand system; GTL, generalized translog system; NGLES, non-separable generalization of the linear expenditure system; ELES, extended linear expenditure system; FELES, functionalized linear expenditure system.

Table 7.2. Annual per capita consumption of major food items (kg) in Germany. (From: Statistisches Bundesamt, *Statistisches Jahrbuch für die Bundesrepublik Deutschland*, various issues.)

Food item	1980[a]	1985[a]	1990[a]	1995[a]	1998[a, b]
Meat	100.5	100.6	102.0	92.0	93.3
Beef	24.7	23.1	22.2	16.6	15.1
Pork	58.2	60.1	60.1	54.9	56.0
Poultry	9.9	9.7	11.7	13.4	15.1
Other meat	7.7	8.3	8.2	7.1	7.1
Fish	11.2	11.9	13.9	14.1	14.0
Eggs	17.2	17.0	15.2	13.7	13.7
Cheese	13.7	15.8	17.3	19.8	20.4
Fluid milk	84.5	87.7	91.5	91.0	88.3
Fresh vegetables	73.4	72.5	82.3	81.6	86.9
Fresh fruits	88.8	85.0	89.3	92.9	93.6
Grains and cereals	68.3	74.0	74.1	72.3	76.2
Sugar	36.9	35.6	36.6	33.1	33.0
Fats and oils	25.8	25.7	26.0	28.4	29.4

[a] 1980 and 1985 data refer to Germany's Western Länder (the Federal Republic of Germany prior to reunification). Later data refer to reunified Germany as a whole. Data for plant products are on a crop-year basis (i.e. 1980 refers to 1979/80) while data for animal products are on a calendar-year basis.
[b] 1998 for animal products, as 1999 not available. Data for plant products are from 1998/99.

Germany (various issues). These data are available as a monthly time series covering the consumption of three different standard household types. Results for household type III from 1991 to 1998 are used in the following:[1] household type III consists of a married couple with two children and at least one child younger than 15 years of age. The household type has one main earner but the marital partner can earn as well. The main earner's gross monthly income from professional, and not self-employed, work is between 6500 and 8800 DM in WG and between 5500 and 7400 DM in EG. The household's gross income may not exceed the income of the main earner by more than 40%.

Annual per capita consumption of major food items is presented in Table 7.2. The data for 1980 and 1985 refer to WG while later data refer to reunified Germany as a whole. Beef consumption has been declining, while consumption of poultry has increased. The importance of meat and processed meat in both EG and WG food budgets was considerable prior to reunification and remains so today. The share of meat and processed meat in food expenditure was 28% in EG and 24% in WG in 1991 and fell to 25 and 21%, respectively, by 1998.

According to Grings (1993), per capita pork consumption was greater in the GDR than in the FRG between 1970 and 1988, while the opposite was true for fish and fish products. Until the beginning of the 1980s, per capita consumption of beef and veal and poultry was also greater in the FRG. However, this changed in the early 1980s (Grings, 1993: 25). Generally, prices of meat and fish tended to be slightly lower in the GDR than in the FRG between 1970 and 1988. However, poultry prices were significantly higher (Grings, 1993: 37).

Immediately following reunification, households in EG continued to consume more pork and sausage and processed-meat products than households in WG, while the opposite was true for beef (Anon., 1992: 14). These differences in consumption resulted from differences in price structures, but also from differences in preferences (Halk, 1992: 24) that are still prevalent (Grunert, 2000: 209).

Since reunification, incomes and price structures in EG and WG have been converging (Anon., 1999: 2). At the same time, food-consumption structures in EG and WG have been converging. Consumption of meat and sausage has been declining in both EG and WG, while the consumption of fruits, vegetable, dairy products and fish has been increasing (Knötzsch, 1992: 158). This trend predates reunification in WG and in EG, although it was repressed in EG by trade restrictions and central planning and the resulting shortages (Halk, 1992: 24).

Changes in poultry consumption are similar in EG and WG over the period 1991–1998, while consumption of beef and veal, and fish and fish products differs more between the two regions. Consumers eat more poultry, beef, veal, fish and fish products in WG than in EG. In both EG and WG, budget shares for beef and veal have been declining in recent years, while those of pork, poultry and fish show a slight but steady increase. Budget shares for sausages and processed meat have remained constant in EG and decreased slightly in WG. Calculations show that the differences between budget shares in EG and WG have remained constant for sausages and processed meat and have fallen for beef and veal and fish over the period 1991–1998. For pork and poultry, however, the differences between budget shares in EG and WG have tended to increase.

Note, however, that these aggregate differences disguise considerable variability between individuals and regions. The demand for meat and fish was mainly contingent on supply in the GDR, and there were significant differences in the consumption of these goods between individuals (Ulbricht, 1996: 225). Furthermore, regional consumption patterns were and remain distinct; differences between regions within EG or WG (in WG, for example, between Bavaria and Schleswig-Holstein) can exceed those between EG and WG (between, for example, Mecklenburg-Vorpommern in EG and Schleswig-Holstein in WG) for some foods.

In summary, food-consumption structures in EG and WG have been converging. Nevertheless, characteristic differences in food consumption based on different habits and preferences remain. Common trends in food consumption, such as growing health consciousness regarding the diet (Donat, 1996: 7), can be observed in EG and WG before as well as after reunification. We suspect that any differences in revealed health consciousness between EG and WG will be largely due to different socio-economic environments as opposed to differences in the availability of health information.

An AIDS Model for Meat in Eastern and Western Germany

The basic model

Preferences are assumed to be weakly separable between meat, on the one hand, and other foods and non-durable goods, on the other. Hence, it is assumed that within the subset 'meat' the consumption of different types of meat is independent of changes in the prices of other food products. Meat is divided into beef and veal, pork, poultry, fish and fish products, and sausages and processed meat.

The AIDS model is chosen because of its desirable features and the ease of estimation using Stone's price index (Deaton and Muellbauer, 1980: 312). The so-called linear approximate AIDS (LA/AIDS) is specified as follows:

$$w_{it} = \alpha_i^* + \sum_j^n \gamma_{ij} \log p_{jt} + \beta_i (\log x_t - \sum_j^n w_j^0 \log p_{jt}) + u_{it} \quad (7.1)$$

where w_{it} is the expenditure share of the ith good in period t, p_{jt} is the price of the jth good, x_t is total expenditure, w_j^0 is the average expenditure share of the jth good and u_{it} is the stochastic error term in the ith equation.

Homogeneity, symmetry and negativity all follow from the budget constraint and utility maximization and can be tested (see below). The problems concerned with testing for adding up are discussed in Edgerton *et al.* (1996: 245 ff.). Deaton and Muellbauer (1980) presented the AIDS and stated that the adding-up constraint will be automatically satisfied by these estimates.

Several means of incorporating dynamic adjustments into the AIDS model are proposed in the literature (see, for example, Edgerton *et al.*, 1996; Moschini and Moro, 1996: 248; Larivière *et al.*, 2000). In a study of the demand for food in Norway, Rickertsen (1998: 95) finds that incorporating dynamic elements in the AIDS model improves the consistency between theory and observed data. Our results support this finding. Under the assumption that preferences have changed smoothly over the observed period, a trend variable is introduced. F tests confirm that there is a significant difference between the models with and without the trend in both EG and WG. Since we use monthly data and meat consumption is known to be seasonal, monthly dummy variables are also introduced.

Misspecification tests are applied and indicate that autocorrelation (first- and 12th-order) plays an important role. To correct this autocorrelation, the Bewley (1979) transformation is used.[2] The final specification for EG and WG is:

$$w_{it} = \alpha_i^* + \tau_i TT_t + \sum_j^{n-1} z_{ij} \Delta w_{jt} + \sum_j^{n-1} s_{ij} \Delta w_{jt-11} + \sum_{d=1}^{11} \delta_{id} D_{dt} + \sum_j^{n} \gamma_{ij} \log p_{jt} + \beta_i (\log x_t - \sum_j^{n} w_j^0 \log p_{jt}) + u_{it} \qquad (7.2)$$

where TT is a trend variable and D a monthly dummy variable. Since we apply the Bewley transformation, the independence of the explanatory variables and the stochastic error does not hold and ordinary least squares (OLS) estimates will be biased. Hence, in estimating Equation 7.2, w_{jt-1} and w_{jt-12} are used as instruments for Δw_{jt} and Δw_{jt-11} (Wickens and Breusch, 1988), respectively. Moreover, since instruments are used, three-stage least squares (3SLS) rather than seemingly unrelated regression (SUR) is applied (Koutsoyiannis, 1983: 270 ff.). Anderson and Blundell (1982) show that parameter estimates can be obtained by arbitrarily deleting an equation. In the following, the equation for sausage and processed meat is deleted. The lagged expenditure share of this group is also not taken into consideration in the Bewley transformation. As a result, the parameters associated with these lagged expenditure shares are not identified (Anderson and Blundell, 1982: 1564).

The LA/AIDS in Equation 7.2 was estimated with homogeneity and symmetry imposed. At the 5% level, homogeneity is only rejected for poultry in EG; symmetry is not rejected at the 5% level in either EG or WG (likelihood-ratio test).

The task of checking curvature properties is exacerbated by the fact that there are geometric and algebraic definitions of concavity and convexity. Many of the geometric definitions are quite intuitive. Their drawback is that techniques do not exist to check these geometric properties of a function directly; rather, the algebraic equivalent (usually expressed in terms of the signs of the minors, eigenvalues or Cholesky values of Hessians or bordered Hessians) of a geometric definition must be checked. The problem is that those definitions that are intuitive from a geometric point of view often have no algebraic equivalent (e.g. strict concavity and strict quasi-concavity), and those definitions that are simple in algebraic terms often have complicated and non-intuitive geometric equivalents (e.g. strong concavity and strong quasi-concavity). The assumption of a strongly quasi-concave utility function is consistent with the postulates of ordinal neoclassical consumer theory and with 'standard' indifference curves. Everywhere diminishing marginal utility (the Marshallian cardinal approach to consumer theory) requires the assumption of strongly concave utility (Morey, 1986: 214). Calculating the signs of the eigenvalues of the Hessian (or bordered Hessian, as appropriate) is one of the simplest methods of checking local curvature properties (Morey, 1986: 217). In the present case, the necessary and sufficient conditions for strong quasi-concavity are accepted at the mean data values and for more than 50% of the individual observations.

In addition to the test for autocorrelation mentioned above, other misspecification tests are applied. These include the Jarque–Bera test for normality, a test for heteroscedasticity, an autoregressive conditional heteroscelasticity (ARCH) test developed by Engle and Ramsey's regression specification error test (RESET). The latter is used to test whether unknown variables have been omitted from a regression specification, and it can also help identify a misspecified functional form. Table 7.3 shows the *P* values for these equation-by-equation misspecification tests.

The *P* values of the Jarque–Bera test for normality suggest that the null hypothesis that the residuals are normally distributed cannot be rejected for each equation for EG and WG. A Lagrange multiplier test indicates that homoscedasticity is only rejected for beef and veal in WG at the 5% level. The results of an ARCH(1) test do not suggest that the error variance is serially correlated, while the results of Ramsey's RESET (with $\hat{w}_i^2$, $\hat{w}_i^3$ as additional regressors) indicate that all equations for both EG and WG are well specified.

The incorporation of health information indices

The authors of this chapter have participated in a European Union (EU) research project entitled 'Nutrition, Health and the Demand for Food', with partners in Scotland, Norway, Spain, France and Germany. In the EU study, a global health-information index (GIA) (see Rickertsen and von Cramon-Taubadel, Chapter 3, this volume) was developed. The GIA is also included in Equation 7.2 above.

Two additional indices were developed for this study of food demand in EG and WG. The 'deutsches' health information index (DHII) and the restricted 'deutsches' health information index (D_rHII) are based on publications in English by German scientists and also on German-language publications. The same keywords are used for GIA and D_rHII. Because the D_rHII contains very few articles, the scope of DHII was enlarged by omitting the link to diet. The Medline database through the Lexis/Nexis services was not available for the German indices, so the usual Medline, which contains only medical journals, was used.

Table 7.3. *P* values for equation-by-equation misspecification tests.[a]

	West Germany	East Germany
Normality		
Beef and veal	0.998	0.998
Pork	0.997	0.998
Poultry	0.996	0.997
Fish and fish products	0.997	0.998
Heteroscedasticity		
Beef and veal	**0.026**	0.188
Pork	0.897	0.139
Poultry	0.384	0.711
Fish and fish products	0.990	0.101
ARCH(1)		
Beef and veal	0.924	0.627
Pork	0.053	0.387
Poultry	0.226	0.259
Fish and fish products	0.197	0.808
Ramsey's RESET		
Beef and veal	0.983	0.990
Pork	0.981	0.988
Poultry	0.985	0.986
Fish and fish products	0.988	0.992

[a] Significant values at the 5% level are printed in bold type. The equation for sausage and processed meat is not included because it was deleted from the estimation.

Since the GIA is only available to 1997, both German indices were also only created to 1997. Thus, the demand analysis incorporating health-information indices is based on the period 1991–1997. In this period, the GIA contains 743 articles, the DHII 76 and the D_rHII 17. Figure 7.1 illustrates the number of articles per month for each index, while Table 7.4 presents information on correlations between the different indices themselves and a trend.

While the two German indices are quite similar, the GIA differs from both considerably. None of the three indices is significantly correlated with a trend. This is reassuring because it means that the indices can be included in the AIDS model discussed above without the danger of collinearity between indices and trend.

Figure 7.2 presents information on the seasonal structure of the three indices. In constructing the GIA, articles that could not be assigned to a specific month (i.e. in journals that appear irregularly) were automatically assigned to January. Note that this convention contributes to a noticeable seasonality in the GIA. Since meat demand is

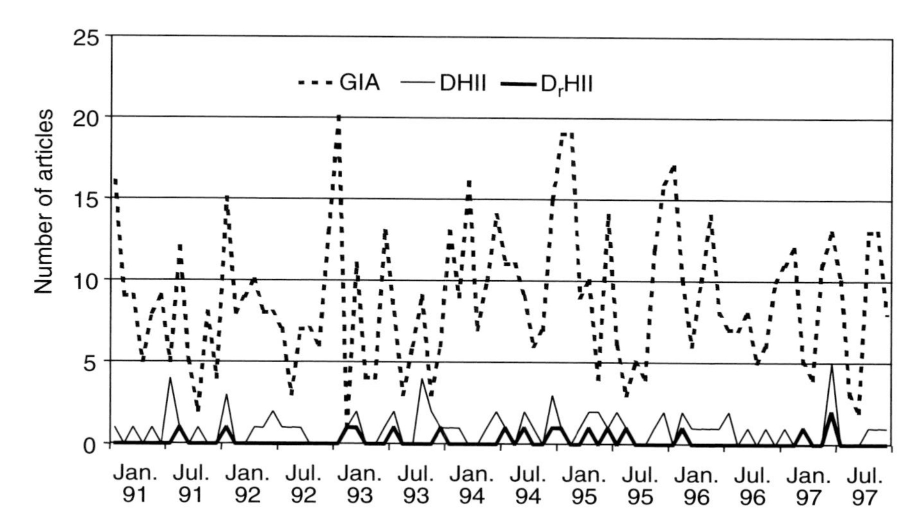

Fig. 7.1. The number of articles in the GIA, DHII and D$_r$HII, 1991–1997.

Table 7.4. Correlation between the different indices and a trend.

Index	GIA	DHII	D$_r$HII	Trend
GIA	1.000	0.153	0.160	0.073
DHII	0.153	1.000	0.478	0.003
D$_r$HII	0.160	0.478	1.000	0.067

known to be seasonal, the possibility that the health index could be spuriously correlated with demand must be considered when interpreting the results. In the case of the German indices, only a few articles could not be clearly assigned to a specific month, so these articles were simply omitted.

In order to include the health-information indices in the AIDS model, it is necessary to specify some form of lag structure. Health information can have two different types of impact on household behaviour: immediate or delayed. One way of approaching the problem of formulating an appropriate lag structure is to use the polynomial distributed lag (PDL). An F test was applied to investigate the difference between employing a PDL structure and simply including lags without structure for different lag lengths. For all equations and indices the test indicated no significant difference, so the unstructured approach was adopted.

The maximum length of the lag must be specified in advance. The Schwarz information criterion (Gujarati, 1999: 423) indicates that for

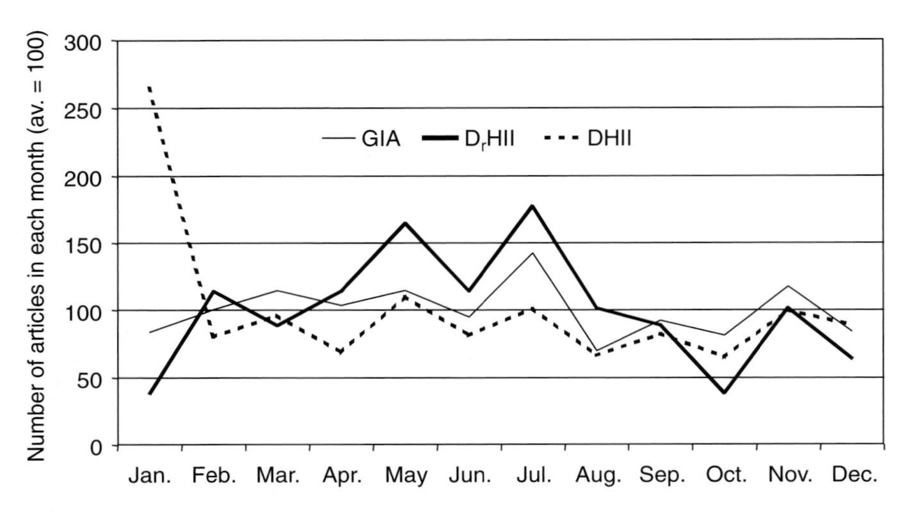

Fig. 7.2. The seasonal structure of the health-information indices.

the majority of the equations and indices only the contemporaneous value of the health-information index should be included. Capps and Schmitz (1991) also find that the contemporaneous effects of health information in general are larger than the lagged effects. Kinnucan *et al.* (1997) interpret similar findings to imply a relatively rapid decay of health-information effects. The final model incorporating health effects for EG and WG is therefore:

$$
\begin{aligned}
w_{it} &= \alpha_i^* + \tau_i TT_t + \sum_j^{n-1} z_{ij}\,\Delta w_{jt} + \sum_j^{n-1} s_{ij}\,\Delta w_{jt-11} + \sum_{d=1}^{11} \delta_{id} D_{dt} \\
&+ \sum_j^{n} \gamma_{ij} \log p_{jt} + \beta_i (\log x_t - \sum_j^{n} w_j^0 \log p_{jt}) + \eta_{it} HI_t + u_{it}
\end{aligned}
\tag{7.3}
$$

where HI is one of the health-information indices (GIA, DHII or D_rHII). Estimation with (Equation 7.3) and without (Equation 7.2) health information produced similar results as regards tests of homogeneity, symmetry and negativity, on the one hand, and specification tests (autocorrelation, normality of the residuals, etc.), on the other. Since the results of these tests have been outlined above under the heading 'The basic model' for the model without health information, we do not repeat this here. However, in this connection, note that the addition of health-information indices, in particular the GIA, did not result in any noticeable reduction in the autocorrelation mentioned above. This is reassuring as it suggests that any significance of the GIA variable is not spurious, i.e. not caused by seasonality in meat consumption, on the one hand, and in the GIA index, on the other (i.e. the convention of

Table 7.5. Marshallian expenditure and price elasticities in East Germany.[a]

Elasticity with respect to	Beef and veal	Pork	Poultry	Fish and fish products	Processed meat
Expenditure	**1.457**	**1.085**	**0.966**	**0.563**	**0.959**
	(5.807)	(8.530)	(4.361)	(2.036)	(21.981)
Price					
Beef and veal	−0.484	−0.143	−0.226	0.121	−**0.726**
	(−1.204)	(−1.035)	(−0.977)	(0.835)	(−2.067)
Pork	−0.017	−**0.813**	**0.149**	−0.043	−**0.362**
	(−0.485)	(−8.227)	(3.282)	(−1.754)	(−2.496)
Poultry	−0.153	**0.489**	−**1.135**	0.106	−0.273
	(−0.827)	(3.383)	(4.042)	(0.940)	(−0.870)
Fish and fish products	0.290	−0.236	0.299	−0.390	−0.526
	(0.984)	(−1.131)	(1.034)	(−1.751)	(−1.083)
Sausage and processed meat	0.005	−0.058	0.008	0.007	−**0.788**
	(0.115)	(−1.533)	(0.203)	(0.220)	(−11.893)
Mean values of the expenditure shares	0.051	0.200	0.064	0.025	0.660

[a] Calculated at mean values with *t* values in parentheses. Significant values at the 5% level are printed in bold type.

assigning to the month of January articles that could not be assigned to a specific month). However, this is only indirect evidence and future research could study the impact of either completely omitting the 'artificial' January articles or assigning them at random to different months.[3]

Results

Results without the use of health-information indices

Estimated parameters from Equation 7.2 are available from the authors on request. They do not differ in any major way from the estimated parameters of the model with health information (Equation 7.3) discussed below and reported in the Appendix. Estimated expenditure effects show that total meat expenditure is not a significant determinant of the demand for individual types of meat in EG and WG. However, the estimated expenditure elasticities in Tables 7.5 and 7.6 confirm that all types of meat are normal goods and that beef and veal and pork are luxuries in EG and WG.[4] Only three of 15 price coefficients in WG (five of 15 in EG) are statistically significant at the 5% level, indicating that not all expenditure shares are sensitive to cross-price effects. Considering the elasticities, all own-price elasticities are

Table 7.6. Marshallian expenditure and price elasticities in West Germany.[a]

Elasticity with respect to	Beef and veal	Pork	Poultry	Fish and fish products	Processed meat
Expenditure	**1.268**	**1.572**	**0.621**	0.466	**0.888**
	(2.668)	(4.570)	(2.112)	(1.287)	(7.837)
Price					
Beef and veal	−**1.149**	−0.290	−0.138	−0.014	0.295
	(−5.278)	(−1.624)	(−1.419)	(−0.211)	(0.938)
Pork	−**0.206**	−**0.696**	0.055	−0.126	−**0.799**
	(−2.097)	(−2.519)	(0.675)	(−1.906)	(−2.736)
Poultry	−0.090	0.266	−**0.966**	0.017	0.151
	(0.800)	(1.694)	(−5.344)	(0.166)	(0.527)
Fish and fish products	0.114	−0.275	0.048	−**0.747**	0.394
	(0.697)	(−1.154)	(0.239)	(−3.550)	(0.983)
Sausage and processed meat	**0.166**	−0.051	0.076	0.095	−**0.874**
	(2.041)	(−0.567)	(1.020)	(1.268)	(−7.866)
Mean values of the expenditure shares	0.103	0.171	0.0912	0.047	0.588

[a] Calculated at mean values with *t* values in parentheses. Significant values at the 5% level are printed in bold type.

negative, but not always significant at the 5% level in EG. Only some categories of meat, such as beef and veal in WG and poultry in EG, are price-elastic.

In both EG and WG, compensating for the income effects of price changes leads to (Hicksian) elasticities that are, as expected, smaller than uncompensated elasticities in absolute value.[5] Most uncompensated and compensated cross-price elasticities are not significant at the 5% level. Hicksian cross-price elasticities indicate that most meat products are substitutes for one another. Those Hicksian cross-price elasticities that do indicate complementarity are not significant at the 5% level. The Hicksian cross-price elasticities between pork and poultry in EG and WG suggest significant substitution.

The results of the AIDS estimation can also be used to illustrate how elasticities have changed over time. The strongest change is an increase in the expenditure elasticities of demand for beef and veal in EG and WG. The EG expenditure elasticity has increased from 1.3 to 1.6 and that of the WG from 1.2 to 1.4. It is perhaps surprising that beef and veal should become more 'luxurious' over time; certainly it is not the case that real incomes have fallen in EG and WG over the period in question. A possible explanation for this phenomenon could lie in the impact that the bovine spongiform encephalopathy (BSE) crises have had on the market for beef and veal. BSE has caused this market to shrink, and many consumers who continue to consume beef have reduced consumption and shifted to imports (e.g. Argentinean beef) or organically produced beef, both of which are considerably more expensive than 'mass-produced' EU beef.

Expenditure elasticities of demand for poultry in WG and fish and fish products in WG and EG also increase over time, while other expenditure elasticities remain more or less constant. The development of uncompensated own-price elasticities of beef and veal and pork differs between WG and EG. The own-price elasticity of beef and veal slightly increases in WG and decreases in EG, but most individual yearly own-price elasticities of beef and veal in EG are not significant at the 5% level (results available from the authors).

The consumption of beef and veal declined in EG over the 1990s, and it appears that consumers' preferences for pork are much stronger in EG than in WG. The uncompensated own-price elasticities of pork declined in absolute values in WG between 1991 and 1994. In these years there were many media reports of swine fever in WG (Mahlau, 1999: 212). EG was less affected by swine fever. Uncompensated own-price elasticities of demand for poultry, fish and fish products as well as sausages and processed meat follow slightly increasing trends in absolute values in WG.

Uncompensated own-price elasticities of demand for poultry and sausages and processed meat remain roughly constant in EG over the estimation period. The increases in the absolute value of the uncompensated own-price elasticity of demand for fish and fish products over time are stronger in EG than in WG. Cross-price elasticities are not subject to notable changes over time.

Results of the estimation with health-information indices

Appendix Tables A7.1 and A7.2 contain the estimated parameters for the AIDS model with health information. As the estimated price and expenditure elasticities already reported in Tables 7.5 and 7.6 are quite insensitive to the inclusion of health information, the corresponding elasticities from the estimation with health information are not presented here. Instead, we focus on health-information effects. Note that this insensitivity of price and income elasticities to the inclusion of health information is also reported by other authors. Yen and Chern (1992: 695) also find that including health information has little impact on estimated price and expenditure elasticities, as do Schmitz and Capps (1993). However, the latter do find that the inclusion of health information reduces the magnitude of the coefficients associated with lagged expenditure shares, a result that we cannot confirm.[6]

Table 7.7. The significance of health-information indices, *P* values.[a]

Region and product	GIA	DHII	D_rHII
West Germany			
Beef and veal	**0.027**	0.391	1.000
Pork	**0.000**	1.000	1.000
Poultry	0.270	0.196	1.000
Fish and fish products	**0.001**	1.000	0.631
Sausage and processed meat	0.598	0.560	0.368
East Germany			
Beef and veal	1.000	1.000	0.119
Pork	**0.002**	1.000	0.102
Poultry	**0.048**	**0.036**	0.328
Fish and fish products	**0.050**	1.000	1.000
Sausage and processed meat	0.131	0.619	0.129

[a] Significant values at the 5% level are printed in bold type.

Table 7.8. Elasticities of demand with respect to the global health-information index.

Region and product	Elasticity	*P* value
West Germany		
Beef and veal	0.044	0.171
Pork	−0.064	0.004
Poultry	0.007	0.763
Fish and fish products	0.066	0.005
Sausage and processed meat	0.004	0.598
East Germany		
Beef and veal	0.011	0.742
Pork	−0.034	0.038
Poultry	−0.026	0.346
Fish and fish products	0.056	0.089
Sausage and processed meat	0.010	0.131

An *F* test was applied to investigate differences between the models with and without health-information indices.[7] Table 7.7 presents the *P* values of these tests for all three indices. With one exception (poultry in EG), the two German indices have no significant impact on meat demand in either EG or WG. The GIA is significant at the 5% level in three of five equations in both EG and WG. Therefore, elasticities of demand for meat with respect to health information are computed for this index in Table 7.8.

Significant health-information effects appear to be greater in WG than in EG. This is plausible given the changes that have occurred in

EG since reunification. Prices and incomes have changed, shortages of goods have disappeared and consumers have other concerns and problems in EG than in WG.

Fish and fish products appear to have benefited at the expense of pork from the dissemination of cholesterol- and fat-related health information. Capps and Schmitz (1991), in comparison, find that poultry gained from increases in health information at the expense of pork. Other meats appear to have been unaffected by health information, although the significant coefficients associated with the trend variable in all equations suggest that other sources of structural change may be at work. These might include demographic changes, increased consumption of meals away from home, changing product characteristics or simply changing preferences. Thus, the diagnosis and prescription are not clear even though symptoms are apparent.

The estimated elasticities of demand with respect to the GIA appear to be reasonable. First, fish is regarded as a healthy food. Even fish with a high fat content contains high proportions of beneficial fatty acids (e.g. omega-3 fatty acids). So the positive effect of health information on fish and fish products is plausible. Secondly, the negative effects of health information on pork demand can result from the fact that pork is considered less healthy (for example, than poultry) due to its cholesterol content and the composition of its fatty acids. In addition, German consumers already consume a great deal of pork. Note that Kinnucan *et al.* (1997) find that poultry, rather than fish, benefits from health information, in their case largely at the expense of beef (rather than pork). This study is also consistent with ours in finding relatively significant trend effects, as discussed in the previous paragraph.

It is perhaps surprising that health-information indices have no impact on the demand for beef and veal according to our estimations. However, it may be that consumers' concerns with these products have focused more on BSE than on links to arteriosclerosis, etc. in recent years. The observed decline in consumption of beef and veal over the estimation period is probably mainly due to the BSE crisis and ensuing price increases.[8]

Conclusions

Following reunification differences in food consumption between EG and WG remain. Convergence can be observed for the consumption of some but not all food groups. Differences also exist regarding health-information effects; these appear to be less pronounced in EG than in WG.

While specific German health-information indices might be expected to produce more plausible results in a German food-demand system than the GIA, our results do not bear this out. Several explanations for this result might be proposed. First, the Medline database does not cover all German medical publications, and there are very few such publications in the first place (one of our German indexes only contains 17 articles). Secondly, it may simply be that German researchers are less active in the areas of diet, health and disease prevention.

Thirdly, as the scientific literature and mass media become increasingly 'global', it may not make sense to expect a distinct German transmission of health information. Certainly, the link between research, consumer awareness and, ultimately, consumer behaviour is an important future field of research. This research would have to be based on sociological and psychological surveys and experiments, tasks that are not simple and do not necessarily correspond to the comparative advantage of economists. Due to the importance of cohort effects that might be expected, detailed panel data could cast light on these questions.

Finally, of course, it may simply be the case that health information as defined here with reference to fat and heart disease is not that important and that other concerns, such as BSE, dioxin or hormones, play more important roles. Separating the impacts of these factors from changes in demographics and product characteristics, among other things, represents a major challenge for future research.

Acknowledgements

We gratefully acknowledge the helpful comments of two anonymous reviewers. The financial support from the Commission of the EU, contract FAIR–CT97–3373, is also appreciated.

Notes

[1] Household type I is a two-person household of welfare recipients or pensioners and is deemed too narrow to be of interest for this analysis. Household type II differs from type III in terms of its (lower) income and the employment of the main earner. Estimation with data from household types II and III leads to similar results and conclusions.

[2] Larivière *et al.* (2000), who also estimate a demand model with monthly data, incorporate dynamic adjustment by augmenting the standard LA/AIDS model with lagged quantities, a time trend and trigonometric variables to capture seasonal effects. We find that the use of the Bewley transformation, which introduces lagged expenditure shares, combined with a trend variable and seasonal dummy variables, produces plausible results that are broadly consistent with demand theory. The approach of Larivière *et al.* (2000) was not pursued, but certainly is an avenue for future research.

[3] We are grateful to an anonymous reviewer for pointing this out.
[4] All elasticities in this section and below are conditional elasticities calculated according to the formulae in Asche and Wessells (1997); see also Rickertsen and von Cramon-Taubadel (Chapter 3, this volume). Since the Bewley transformation is applied, estimated coefficients and elasticities can be interpreted as long run.
[5] Available from the authors on request.
[6] Schmitz and Capps (1993: 21) interpret these coefficients as measures of habit persistence. Our use of the Bewley transformation does not allow us to interpret the coefficients of lagged expenditure shares in this manner (Anderson and Blundell, 1982: 1564).
[7] In the case of the excluded *n*th equation, a Wald test was used.
[8] The inclusion of an index of BSE newspaper coverage in a similar model of meat demand (results available from the authors) does produce significant effects on beef demand. Interestingly, tests indicate that this BSE index, too, is best included as a contemporaneous variable, i.e. with no lagged terms.

References

Anderson, G. and Blundell, R. (1982) Estimation and hypothesis testing in dynamic singular equation systems. *Econometrica* 50, 1559–1571.

Anon. (1992) Trends und Tendenzen bei Fleisch und Wurst, Talfahrt des roten Sortiments dauert an. *Lebensmittel-Praxis* 9, 14–18.

Anon. (1999) Steine auf dem Weg des Fortschritts. In: *Institut der deutschen Wirtschaft* e.V. Köln (ed.). *iwd* 25(44), 2–3.

Appel, V., Ferber, P. and Rickli, T. (1989) *Vorschätzung des Nahrungsmittelverbrauchs in den Ländern der EG (12) im Zieljahr 1990/91.* No. 339, Landwirtschaftsverlag, Münster-Hiltrup, 140 pp.

Asche, F. and Wessells, C.R. (1997) On price indices in the almost ideal demand system. *American Journal of Agricultural Economics* 74, 1182–1185.

Bewley, R.A. (1979) The direct estimation of the equilibrium response in a linear dynamic model. *Economic Letters* 61, 357–361.

Capps, O. Jr and Schmitz, J.D. (1991) A recognition of health and nutrition factors in food demand analysis. *Western Journal of Agricultural Economics* 16, 21–35.

Deaton, A. and Muellbauer, J. (1980) An almost ideal demand system. *American Economic Review* 70, 312–326.

deHaen, H., Murty, K.N. and Tangermann, S. (1982) *Künftiger Nahrungsmittelverbrauch – Ergebnisse eines simultanen Nachfragesystems –. Die mittel- bis langfristige Verbrauchsentwicklung von Agrarprodukten in den Ländern der Europäischen Gemeinschaft.* No. 271, Landwirtschaftsverlag, Münster-Hiltrup, 98 pp.

Donat, P. (1996) Die Entwicklung des Ernährungsverhaltens der 'DDR-Bevölkerung' vor und nach der Wende. In: Kutsch, T. and Weggemann, S. (eds) *Ernährung in Deutschland nach der Wende: Veränderungen in Haushalt, Beruf und Gemeinschaftsverpflegung.* 14. und 15. Wissenschaftliche Arbeitstagung der Arbeitsgemeinschaft Ernährungsverhalten e.V. Bonner Studien zur Wirtschaftssoziologie, Vol. 3, Verlag M. Wehle, Witterschlick, Bonn, pp. 1–20.

Eckert, S. (1998) *Ökonomische Effekte von Lebensmittelskandalen: Das Beispiel BSE.* Arbeitsbericht No. 25, Institut für Agrarpolitik und Marktforschung, Justus-Liebig-University Giessen, Giessen, Germany, 114 pp.

Edgerton, D.L., Assarsson, B., Hummelmose, A., Laurila, I.P., Rickertsen, K. and Vale, P.H. (1996) *The Econometrics of Demand Systems. With Applications to Food Demand in the Nordic Countries.* Kluwer Academic Publishers, Dordrecht, 290 pp.

Eschenbach, J. (1981) Lebensmittelnachfrage und privater Verbrauch ausgewählter Haushalte in der Bundesrepublik Deutschland – eine Auswertung der Einkommens- und Verbrauchsstichprobe 1973. Dissertation, Justus-Liebig-University of Giessen, Giessen, Germany, 234 pp.

Federal Statistical Office Germany (Statistisches Bundesamt) (various issues) *Einnahmen und Ausgaben ausgewählter Haushalte. Wirtschaftsrechnungen.* Fachserie 15. Reihe 1. Metzler-Poeschel-Verlag, Stuttgart (time series can be purchased online at: http://www-zr.statistik-bund.de/zeitreih/home.htm).

Filip, J. (1983) Lebensmittelnachfrage ausgewählter privater Haushalte in der Bundesrepublik Deutschland. Eine Auswertung der laufenden Wirtschaftsrechnungen für drei Haushaltstypen. Dissertation, Justus-Liebig-University of Giessen, Giessen, Germany, 413 pp.

Filip, J. and Wöhlken, E. (1984) *Nachfrage nach Lebensmitteln in privaten Haushalten. Eine Auswertung der Einkommens- und Verbrauchsstichprobe 1978.* Vol. I: *Analyseergebnisse für Nahrungsmittel tierischer Herkunft und Nahrungsfette.* No. 304, 244 pp. Vol. II: *Analyseergebnisse für Nahrungsmittel pflanzlicher Herkunft und Getränke*, No. 305, Landwirtschaftsverlag, Münster-Hiltrup, 241 pp.

Funkte, H. (1981) Nachfrage nach alkoholischen Getränken in der BR Deutschland. Auswertung der Einkommens- und Verbrauchsstichprobe 1973. *Agrarwirtschaft* 30(5), 142–148.

Giere, A., Herrmann, R. and Böcher, K. (1997) Ernährungsinformationen und Nahrungsmittelkonsum: Theoretische Überlegungen und empirische Analyse am Beispiel des deutschen Buttermarktes. *Agrarwirtschaft* 46(8/9), 283–293.

Graser, S. and Sagmeister, L. (1980) Prognose des Verbrauchs an Zucker (Saccharose) in der BR Deutschland unter Berücksichtigung der Marktentwicklung bei Saccharosesubstitution. *Agrarwirtschaft* 29(2), 29–36.

Grings, M. (1991) Modelluntersuchung über den Nahrungsmittelverbrauch in der DDR. In: Schmitz, P.M. and Weindlmaier (eds) *Land- und Ernährungswirtschaft im europäischen Binnenmarkt und in der internationalen Arbeitsteilung.* Schriften der Gesellschaft für Wirtschafts- und Sozialwissenschaften des Landbaues e.V., 27, Landwirtschaftsverlag, Münster-Hiltrup, pp. 129–136.

Grings, M. (1993) Die Nachfrage nach Nahrungsmitteln in Ost- und Westdeutschland. Vergleichende Analyse auf der Grundlage eines ökonometrischen Modells. Habilitation Thesis, Georg-August University Göttingen, Germany, 208 pp.

Grunert, R. (2000) Bei Annäherung weiterhin Besonderheiten der Konsumstruktur in Ostdeutschland. *Wirtschaft im Wandel* 6(7), 204–209.

Gujarati, D.N. (1999) *Essentials of Econometrics*, 2nd edn. McGraw-Hill, New York, 559 pp.

Hagner, C. (1994) *Nachfrage nach Produkten des Ökologischen Landbaus – Stand der Literatur und Ergebnisse einer ökonometrischen Nachfrageanalyse für Müsliprodukte.* Agrarökonomische Diskussionsbeiträge No. 23, Justus-Liebig-University of Giessen, Giessen, Germany, 45 pp.

Halk, K. (1992) Anforderungen der Verbraucher an die Nahrungsmittelqualität – Ein Vergleich zwischen Ost- und Westdeutschland. *Ifo-Schnelldienst* 45(29), 20–27.

Hansen, G. (1983) *Nachfrage nach nichtdauerhaften Gütern in der Bundrepublik 1960–1979.* Institut für Statistik und Ökonometrie, No. 17, Christian-Albrechts-University, Kiel, 24 pp.

Hasenkamp, G. (1980) A demand system analysis of disaggregated consumption. In: Gollnick, H. and Scherf, H. (eds) *Studien zur angewandten Wirtschaftsforschung und Statistik aus dem Institut für Statistik und Ökonometrie der Universität Hamburg.* No. 10. Vandenhoeck und Ruprecht, Göttingen, 214 pp.

Henning, C. and Michalek, J. (1992) Innovatives Konsumverhalten für Nahrungsmittel? Ableitung und Schätzung eines auf Nahrungsmittel fokussierten kompletten Nachfragesystems unter Berücksichtigung von zeitlichen Präferenzänderungen. *Agrarwirtschaft* 41(11), 330–342.

Hoff, K. and Claes, R. (1997) Der Einfluß von Skandalen und Gemeinschaftswerbung auf die Nachfrage nach Rindfleisch. *Agrarwirtschaft* 46(10), 332–344.

Höh, H. (1984) Nachfrage nach Kartoffeledelerzeugnissen in der BR Deutschland. *Agrarwirtschaft* 33(8), 248–253.

Kinnucan, H.W., Xiao, H., Hsia, C.-J. and Jackson, J.D. (1997) Effects of health information and generic advertising on US meat demand. *American Journal of Agricultural Economics* 79, 13–23.

Knötzsch, P. (1992) Das Ernährungsverhalten der neuen Bundesbürger ändert sich. *AID-Verbraucherdienst* 37(8), 157–163.

Koutsoyiannis, A. (1983) *Theory of Econometrics. An Introductory Exposition of Econometric Methods.* Macmillan Press, London, 689 pp.

Larivière, É., Larue, B. and Chalfant, J. (2000) Modeling the demand for alcoholic beverages and advertising specifications. *Agricultural Economics* 22, 147–162.

Mahlau, G. (1999) Das Image der Landwirtschaft. Ein Vergleich zwischen Medienberichterstattung, Bevölkerungsmeinung und Realität. In: Kutsch, T. (ed.) *Bonner Studien der Wirtschaftssoziologie,* Vol. 11. Verlag M. Wehle, Witterschlick, Bonn, no page numbers.

Merz, J. (1980) *Die Ausgaben privater Haushalte. Ein mikroökonomisches Modell für die Bundesrepublik.* Sonderforschungsbereich 3 of the Universities of Frankfurt and Mannheim 'Mikroanalytische Grundlagen der Gesellschaftspolitik', Schriftenreihe 5, Campus Verlag, Frankfurt, 252 pp.

Meyer, H. (1980a) Nachfrage nach Molkereiprodukten in der BR Deutschland. *Agrarwirtschaft* 29(4), 105–116.

Meyer, H. (1980b) Querschnittsanalyse der Nahrungsfettnachfrage in der BR Deutschland. Eine Auswertung der Einkommens- und Verbrauchsstichproben von 1962/63 und 1973. *Agrarwirtschaft* 29(6), 182–187.

Michalek, J. and Keyzer, M.A. (1992) Estimation of a two-stage LES-AIDS consumer demand system for eight EC countries. *European Review of Agricultural Economics* 19, 137–163.

Morey, E.R. (1986) An introduction to checking, testing, and imposing curvature properties: the true function and the estimated function. *Canadian Journal of Economics* 19, 207–235.

Moschini, G. and Moro, D. (1996) Structural change and demand analysis: a cursory review. *European Review of Agricultural Economics* 23, 239–261.

Murty, K.N. and Pradesh, A. (1981) Analysis of food consumption in the Federal Republic of Germany. *Empirical Economics* 6, 75–86.

Pawlik, H. (1993) Die Nachfrage nach Tiefkühlkost – Struktur, Bestimmungsgründe und Perspektiven. In: Stamer, H. (ed.) *Agrarmarkt-Studien.* No. 39, Paul Parey Verlag, Berlin, 197 pp.

Recke, G. (1995) *Sind Nachfragetheorie und Empirie unvereinbar? Ein Beitrag zum Test auf Homogenität und auf Symmetrie.* Wissenschaftsverlag Vauk, Kiel, Germany, 68 pp.

Rettig, R. (1983) Ein vollständiges Nachfragesystem für den privaten Verbrauch. *Zeitschrift für Wirtschafts- und Sozialwissenschaften* 103(3), 205–224.

Rickertsen, K. (1998) The demand for food and beverages in Norway. *Agricultural Economics* 18, 89–100.

Röder, C. (1998) *Determinanten der Nachfrage nach Nahrungsmitteln und Ernährungsqualität in Deutschland. Eine ökonometrische Analyse auf der Grundlage der Nationalen Verzehrsstudie.* Sonderheft 161, Agrarwirtschaft, Frankfurt-on-Main, 317 pp.

Ryll, E. (1984) Bestimmungsgründe und Elastizitäten der Nachfrage nach Fisch und Fischwaren in der Bundesrepublik Deutschland. *Berichte über Landwirtschaft* 62(2), 208–221.

Schmidt, E. (1984) *Analyse der Entwicklung der Nachfrage nach Butter in der BR Deutschland unter Verwendung ökoskopischer und demoskopischer Informationen.* Arbeitsbericht 84/1, Institut für landwirtschaftliche Marktforschung der Bundesanstalt für Landwirtschaft, Braunschweig-Völkenrode, 124 pp.

Schmitz, J.D. and Capps, O. Jr (1993) *A Complete Systems Analysis of Nutritional Awareness on Food Demand.* Texas A&M University System, College Station, Texas, 38 pp.

Schons, H.-P. (1993) *Vorschätzung des Nahrungsmittelverbrauchs in den Mitgliedsländern der EG (12) und ausgewählten Drittländern für die Zieljahre 1995 und 2000.* No. 421, Landwirtschaftsverlag, Münster-Hiltrup, 202 pp.

Sienknecht, H.-P. (1986) *Probleme der Konstruktion und Überprüfung ökonometrischer Modelle der Konsumgüternachfrage.* Schriften zur angewandten Ökonometrie No. 17, Haag & Herchen Verlag, Frankfurt-on-Main, 165 pp.

Sommer, U. (1985) Quantitative Analyse der Nachfrage nach Fisch und Fischwaren in der Bundesrepublik Deutschland. *Berichte über Landwirtschaft* 63(1), 82–101.

Ulbricht, G. (1996) Analyse des Ernährungsverhaltens in der DDR bzw. in den Neuen Bundesländern auf der Grundlage von Haushaltsbudgeterhebungen. In: Kutsch, T. and Weggemann, S. (eds) *Ernährung in Deutschland nach der Wende: Veränderungen in Haushalt, Beruf und Gemeinschaftsverpflegung.* 14. und 15. Wissenschaftliche Arbeitstagung der Arbeitsgemeinschaft Ernährungsverhalten e.V. Bonner Studien zur Wirtschaftssoziologie, Vol. 3, Verlag M. Wehle, Witterschlick, Bonn, pp. 219–236.

Weindlmaier, H. and Neubüser, E. (1980) *Entwicklungstendenzen am Biermarkt in der BR Deutschland.* No. 1, Institut für landwirtschaftliche Betriebslehre, Friedrich-Wilhelms University Bonn, Bonn, 43 pp.

Wickens, M.R. and Breusch, T.S. (1988) Dynamic specification, the long-run and estimation of transformed regression models. *Economic Journal* 98, 189–205.

Wildner, S. (2001) *Die Nachfrage nach Nahrungsmitteln in Deutschland unter besonderer Berücksichtigung von Gesundheitsinformationen.* Sonderheft 169, Agrarwirtschaft, Frankfurt-on-Main, 194 pp.

Wöhlken, E. and Filip, J. (1982) Einkommensverwendung zu Konsumzwecken in ausgewählten privaten Haushalten der Bundesrepublik Deutschland. *GfK – Jahrbuch der Absatz- und Verbrauchsforschung* 28(2), 169–197.

Wöhlken, E., Filip, J., Quinckardt, F. and Salamon, P. (1981) *Nahrungsmittelverbrauch im Mehrländervergleich. Analyse von Niveau und Entwicklungstendenz für OECD- bzw. EG-Länder.* No. 249, Landwirtschaftsverlag, Münster-Hiltrup, 297 pp.

Yen, S.T. and Chern, W.S. (1992) Flexible demand systems with serially correlated errors: fat and oil consumption in the United States. *American Journal of Agricultural Economics* 74, 689–697.

Appendix

Table A7.1. Estimated parameters of the AIDS model with the GIA in East Germany 1991–1998.

		Beef and veal		Pork		Poultry		Fish and fish products		Sausage and processed meat	
Parameter	Variable	Parameter	t value	Parameter	t value	Parameter	t value	Parameter	t value	Parameter	t value
α	Constant	0.038	(2.657)	0.194	(6.201)	0.139	(9.768)	0.056	(8.652)	0.573	(18.804)
γ	Beef and veal	0.029	(1.325)	0.008	(1.103)	−0.017	(−1.504)	0.004	(0.553)	−0.023	(−1.284)
	Pork	0.008	(1.103)	0.016	(0.987)	0.041	(5.063)	−0.007	(−1.558)	−0.057	(−2.982)
	Poultry	−0.017	(−1.504)	0.041	(5.063)	−0.027	(−1.575)	0.003	(0.474)	0.0001	(0.008)
	Fish and fish products	0.004	(0.553)	−0.007	(−1.558)	0.003	(0.474)	0.010	(2.009)	−0.010	(−1.199)
	Sausage and processed meat	−0.023	(−1.284)	−0.057	(−2.982)	0.0001	(0.008)	−0.010	(−1.199)	0.090	(2.646)
β	Expenditure	0.034	(2.269)	−0.021	(−0.764)	−0.011	(−0.713)	−0.018	(−2.690)	0.016	(0.471)
δ	Dummy	0.064	(2.461)	−0.077	(−1.402)	−0.096	(−3.914)	−0.043	(−3.665)	0.152	(2.502)
		0.032	(2.288)	−0.022	(−0.773)	−0.086	(−6.311)	−0.032	(−5.149)	0.108	(3.359)
		0.032	(2.392)	−0.023	(−0.883)	−0.084	(−6.721)	−0.027	(−4.857)	0.103	(3.452)
		0.034	(2.290)	−0.012	(−0.405)	−0.092	(−6.693)	−0.031	(−4.521)	0.101	(2.987)
		0.027	(1.965)	0.022	(0.727)	−0.099	(−7.164)	−0.037	(−5.544)	0.087	(2.553)
		0.019	(1.449)	0.011	(0.405)	−0.096	(−7.490)	−0.040	(−7.017)	0.105	(3.370)
		0.022	(1.547)	0.013	(0.434)	−0.103	(−7.584)	−0.038	(−6.027)	0.106	(3.291)
		0.027	(1.976)	0.008	(0.288)	−0.098	(−7.497)	−0.035	(−5.860)	0.098	(3.050)
		0.030	(2.284)	−0.023	(−0.848)	−0.088	(−7.062)	−0.035	(−6.266)	0.116	(3.884)
		0.035	(2.791)	−0.024	(−0.922)	−0.084	(−6.828)	−0.032	(−5.691)	0.105	(3.481)
		0.026	(2.414)	−0.018	(−0.772)	−0.069	(−6.241)	−0.030	(−6.004)	0.090	(3.369)
z	Δw	−0.102	(−0.823)	−0.023	(−0.100)	0.086	(0.727)	0.059	(0.973)	−0.019	(−0.069)
		−0.020	(−0.375)	−0.017	(−0.156)	0.054	(1.070)	−0.029	(−0.907)	0.011	(0.078)
		0.384	(2.576)	−0.811	(−2.391)	−0.114	(−0.800)	−0.179	(−2.619)	0.719	(1.932)
		0.219	(1.403)	0.354	(0.869)	−0.095	(−0.538)	0.085	(0.918)	−0.562	(−1.069)
s	Δw_{-11}	0.166	(2.035)	0.164	(1.149)	−0.095	(−0.079)	0.110	(2.842)	−0.433	(−2.448)
		−0.021	(−0.535)	0.178	(2.435)	−0.015	(−0.408)	0.036	(2.102)	−0.178	(−2.006)
		−0.094	(−0.727)	−0.263	(−1.044)	0.095	(0.663)	−0.012	(−0.186)	0.273	(0.806)
		−0.141	(−0.889)	−0.509	(−1.480)	0.045	(0.263)	0.041	(0.400)	0.564	(1.250)
τ	Trend	−0.0004	(−9.988)	0.0005	(7.157)	0.0001	(3.577)	0.00005	(2.849)	−0.0003	(−2.909)
η	Health information	0.00005	(0.330)	−0.0007	(−2.072)	−0.0002	(−0.942)	0.0001	(1.701)	0.0007	(1.509)
	R^2		0.892		0.903		0.971		0.925		0.874

Table A7.2. Estimated parameters of the AIDS model with the GIA in West Germany 1991–1998.

		Beef and veal		Pork		Poultry		Fish and fish products		Sausage and processed meat	
Parameter	Variable	Parameter	*t* value	Parameter	*t* value	Parameter	*t* value	Parameter	*t* value	Parameter	*t* value
α	Constant	0.180	(10.383)	0.135	(7.734)	0.110	(13.568)	0.048	(7.256)	0.527	(25.411)
γ	Beef and veal	−0.006	(−0.278)	0.002	(0.119)	−0.002	(−0.225)	−0.008	(−1.320)	0.014	(0.685)
	Pork	0.002	(0.119)	0.123	(3.715)	0.026	(2.068)	−0.025	(−2.370)	−0.126	(−3.172)
	Poultry	−0.002	(−0.225)	0.026	(2.068)	−0.002	(−0.143)	−0.006	(−0.744)	−0.015	(−0.779)
	Fish and fish products	−0.008	(−1.320)	−0.025	(−2.370)	−0.006	(−0.744)	0.026	(2.868)	0.013	(0.833)
	Sausage and processed meat	0.014	(0.685)	−0.126	(−3.172)	−0.015	(−0.779)	0.013	(0.833)	0.114	(1.898)
β	Expenditure	0.014	(0.338)	0.089	(1.950)	−0.016	(−0.738)	−0.023	(−1.631)	−0.064	(−1.237)
δ	Dummy	−0.033	(−1.423)	0.010	(0.468)	−0.032	(−3.091)	−0.014	(−1.552)	0.069	(2.596)
		−0.050	(−3.112)	0.018	(1.138)	−0.032	(−4.379)	−0.003	(−0.536)	0.066	(3.580)
		−0.046	(−3.307)	0.007	(0.408)	−0.036	(−5.192)	−0.001	(−0.200)	0.076	(4.140)
		−0.060	(−3.809)	0.021	(1.532)	−0.037	(−5.379)	0.001	(0.254)	0.074	(4.335)
		−0.073	(−5.309)	0.032	(2.434)	−0.039	(−6.520)	−0.009	(−1.780)	0.090	(5.372)
		−0.080	(−5.206)	0.044	(2.972)	−0.043	(−5.861)	−0.018	(−3.045)	0.096	(5.310)
		−0.084	(−4.857)	0.061	(3.419)	−0.050	(−6.173)	−0.016	(−2.472)	0.089	(4.095)
		−0.064	(−3.848)	0.046	(2.660)	−0.049	(−6.789)	−0.011	(−1.823)	0.077	(3.931)
		−0.049	(−2.993)	0.011	(0.645)	−0.038	(−5.419)	−0.008	(−1.256)	0.085	(4.255)
		−0.041	(−3.153)	0.008	(0.604)	−0.035	(−5.593)	−0.012	(−2.129)	0.079	(4.759)
		−0.021	(−1.876)	0.007	(0.540)	−0.024	(−4.779)	−0.011	(−2.358)	0.049	(3.160)
z	Δw	0.149	(1.400)	−0.130	(−1.292)	0.030	(0.693)	0.069	(2.117)	−0.118	(−0.978)
		0.335	(2.292)	−0.264	(−1.423)	−0.104	(−1.455)	−0.005	(−0.089)	0.039	(0.206)
		0.174	(0.607)	−0.420	(−1.556)	−0.091	(−0.720)	−0.240	(−2.421)	0.578	(1.789)
		0.151	(0.920)	−0.161	(−0.641)	−0.021	(−0.213)	0.057	(0.741)	−0.027	(−0.100)
s	Δw_{-11}	0.156	(1.254)	−0.100	(−0.850)	−0.027	(−0.382)	−0.015	(−0.382)	−0.014	(−0.110)
		−0.075	(−0.595)	0.065	(0.390)	0.007	(0.098)	0.038	(0.736)	−0.035	(−0.199)
		0.410	(1.682)	−0.040	(−0.175)	0.195	(1.421)	−0.136	(−1.519)	−0.428	(−1.572)
		0.168	(0.836)	−0.183	(−0.888)	0.175	(2.022)	0.355	(5.744)	−0.515	(−2.030)
τ	Trend	−0.0007	(−8.523)	0.0002	(2.063)	0.0002	(5.784)	0.0001	(4.254)	0.0001	(1.024)
η	Health information	0.0005	(1.371)	−0.001	(−2.864)	0.00006	(0.302)	0.0003	(2.799)	0.0002	(0.527)
	R^2		0.916		0.819		0.910		0.872		0.881

Health Information and Food Demand in France

8

VÉRONIQUE NICHÈLE

Institut National de la Recherche Agronomique (INRA), Ivry-sur-Seine, France

Introduction

During the last four decades of the 20th century, several epidemiological and clinical studies have shown the link between diet and the most prevalent diseases in industrialized countries. Particularly, it was gradually admitted that the consumption of dietary cholesterol and saturated fats increases the risk of coronary disease. Nutritional recommendations emanating from health professionals reflected the progress of medical research. In France, the health status related to cardiovascular diseases has improved considerably since the beginning of the 1980s. As for mortality rates from coronary disease, France, together with Japan, remains the best ranked in the world. However, this favourable situation is very similar to the one prevailing in all southern European countries, so much so that the idea of a specific French paradox does not seem to hold any longer. Actually, cardiovascular diseases are still the primary cause of mortality in France, accounting for 32 out of every 100 deaths. The direct and indirect costs of treating coronary disease are estimated to be around 30 billion francs for the year 1986.[1] All things considered, these figures imply that improvements are still necessary. Although the influence of dietary factors on the occurrence of cardiovascular diseases is not clearly quantified, it is of interest to investigate how consumers react to nutritional information related to these diseases. Such a study requires the adoption of a method to measure information about fat and cholesterol received by consumers.

Previous studies have addressed the role of health information in food demand. Brown and Schrader (1990) proposed a cholesterol-information index discussed in Rickertsen and von Cramon-Taubadel (Chapter 3, this volume). Brown and Schrader (1990) used the index to analyse the US demand for shell eggs and, as expected, a negative influence is found. Several studies used Brown and Schrader's (1990) index. For example, investigations have been made on cholesterol-information effects on the demand for beef, pork, poultry and fish (Capps and Schmitz, 1991) and fats and oils (Chern *et al.*, 1995). Some papers have also focused on the influence of this index in combination with advertising (Chang and Kinnucan, 1991; Kinnucan *et al.*, 1997). As discussed by Rickertsen and von Cramon-Taubadel (Chapter 3, this volume), Brown and Schrader's (1990) index is often criticized because, by construction, it is similar to a time trend. Chern and Zuo (1995) proposed a new health-information index solving most of the problems raised by the cholesterol index and tested the index on the demand for fresh milk. Chern (Chapter 2, this volume) also included this index in a demand system with ten food groups and confirmed that it performs well. With the exception of Chang and Kinnucan's (1991) study using Canadian data, all the empirical analyses mentioned above are run on US data. Most of the results show that health information is a significant determinant of the US demand for food.

To our knowledge, only one study has hitherto investigated the impact of health information on European food demand. Rickertsen and von Cramon-Taubadel (Chapter 3, this volume) assembled and put in perspective the results of a European research programme for France, Germany, Norway, Scotland and Spain.[2] The results obtained do not allow the identification of an overall European pattern and suggest that country-specific models are required. The present study estimates a French model, which differs from the French model discussed in Rickertsen and von Cramon-Taubadel (Chapter 3, this volume), mainly with respect to the demand system and the data used. In this chapter, the quadratic almost ideal demand system (QUAIDS) is estimated. This specification is more encompassing than the standard almost ideal demand system (AIDS) and, clearly, than its linear approximate version. To fulfil the commodity specification of the common model, the European study used annual time-series data from the National Accounts.[3] The present study is based on pooled microdata from the National Food Survey (NFS). Information collected in this survey enhances the empirical analysis in several respects. First, food products are aggregated into 15 categories, which are partly selected according to high and low fat content. Secondly, sociodemographic variables are introduced to take into account heterogeneity in household behaviour. Thirdly, the variability of the data and the use of unit values permit the estimation of more reasonable price effects than the quite unexpected effects for France presented in the European paper. Finally, the temporal structure of the data warrants the inclusion of a monthly health index. The empirical findings provide evidence that the dissemination of fat and cholesterol information should be considered when French food-demand systems are estimated.

There are relatively few studies of French food demand and an additional objective is to fill this gap. With the exception of the reports based on the French NFS (Bertrand, 1993) and conventional summary statistics on household consumption derived from the National Accounts (Cases, 1999), to our knowledge, only two econometric studies investigate French food demand. First, Fulponi (1989) presents income and price elasticities calculated from a linear almost ideal demand system (LAIDS) estimated over the period 1959–1985. Elasticities are evaluated at different points of time to examine possible changes in the patterns of French food-demand behaviour. Secondly, Combris (1992) describes the evolution of food consumption between 1949 and 1988 and tests for structural change. Both articles use data from the French National Accounts and both focus on the demand for meat.

Trends in French Household Food Consumption

The figures reported in Table 8.1 are from the French National Accounts (ANS, 1991/92, 1999, 2000) and show the trends in food consumption over the period 1980–1998. The data are constructed from food balance sheets corrected with information from household food-consumption surveys and the food distribution sector.

The trends in food consumption have been stable since the early 1990s, after a period of some fluctuations. Substitutions within the groups explain most of the development in consumed quantities.

From 1980 to 1998 and in spite of some fluctuations, the consumption of meats as a group has increased by more than 3 kg per capita. However, the composition of meat consumption has changed. As expected, a continuous decline is observed for beef (−4.35 kg). This decline takes place to the benefit of all other meats, particularly poultry (+3.57 kg), other meats and meat products (+2.61 kg) and, to a lesser extent, pork (+1.52 kg). Within the group of other meats and meat products, French households have consumed more meat products and canned meat (+5.79 kg) but less veal (−1.55 kg).

Although the data for fresh fish are incomplete, a 5 kg increase in the per capita consumption in the fish and seafood group is observed over the first 10 years (1980–1990). The same holds, over the entire period, for canned fish and the group of frozen fish and shellfish (respectively, +2.44 and +1.89 kg).

Consumption of eggs has remained fairly stable at around 15 kg per capita. Milk suffered several fluctuations from 1980 to 1998, resulting in a decrease of 5 l per capita. Conversely, the per capita consumption of

Table 8.1. Annual per capita consumption in France (kg of food and l of milk). (From: ANS, 1991/92, 1999, 2000.)

Food group	1980	1985	1990	1995	1997	1998
Meat	84.54	86.47	87.47	86.66	86.76	87.89
Beef	19.25	18.80	17.58	16.56	15.42	14.90
Pork	9.25	9.52	9.96	10.06	10.11	10.77
Poultry	24.78	25.75	26.83	26.79	28.17	28.35
Other meats and meat products	31.26	32.40	33.10	33.25	33.06	33.87
Veal	5.43	5.18	4.49	4.13	3.95	3.88
Meat products	16.68	18.25	19.84	21.19	21.56	22.47
Other meats	9.15	8.97	8.77	7.93	7.55	7.52
Fish, shellfish and molluscs	18.08	19.75	24.76	[a]		
Fresh	12.85	13.78	14.75			
Fish and shellfish, frozen	2.48	2.77	4.20	4.22	4.33	4.37
Fish, canned	2.75	3.20	5.81	4.84	4.69	5.19
Eggs	14.25	14.97	14.34	15.49	14.89	14.85
Cheese and cream	26.32	24.62	26.58	29.98	30.76	32.09
Fluid milk	74.03	77.22	68.16	69.52	68.07	68.96
Yoghurt and dairy desserts	11.75	16.82	22.20	25.57	26.90	28.29
Vegetables and fruits	176.02	187.49	189.23	189.62	187.37	187.56
Fresh vegetables	88.41	91.05	88.34	91.32	89.84	91.33
Fresh fruits	59.40	64.03	62.38	58.74	57.72	55.81
Processed vegetables	21.41	24.77	29.97	31.61	31.94	32.44
Processed fruits	6.80	7.64	8.54	7.95	7.87	7.98
Potatoes	89.01	80.65	62.42	69.93	69.79	65.78
Grains and cereals	94.48	91.37	90.75	88.25	88.84	88.21
Bread	70.64	66.45	63.37	59.89	60.02	59.87
Pasta	6.20	6.30	6.75	7.30	7.57	7.39
Biscuits	5.91	7.19	8.78	9.80	10.32	10.30
Other cereals	11.73	11.43	11.85	11.26	10.93	10.65
Sugar	14.98	12.38	10.02	8.92	8.20	7.71
Fats and oils	22.57	22.31	22.14	23.34	23.35	21.90
Butter	9.42	8.55	8.43	8.05	7.83	7.65
Oils	10.84	11.32	11.37	13.17	13.20	11.99
Margarine	2.31	2.44	2.34	2.12	2.32	2.26
Other foods	12.44	11.91	12.73	14.11	14.90	14.64

[a] Blanks indicate that data are not available.

yoghurt and dairy desserts has almost doubled over the 1980–1990 period and it continues to rise but at a slower rate. The consumption of cheese has shown a growing trend since 1985 (+5.77 kg per capita).

Potato consumption has declined by 23.23 kg per capita. Per capita consumption of vegetables and fruits increased by 11.54 kg. Consumption of fresh vegetables varies slightly, but shows an increase of 2.92 kg over the 18-year period. The consumption of processed vegetables has grown by more than 10 kg between 1980 and 1995 and later at a somewhat slower rate. After 1985 there is a downward trend in the consumption of fresh fruits (−3.59 kg). Processed fruit consumption fluctuates but has increased by 1.18 kg since 1980.

Cereal products show a decline of 6.27 kg in per capita consumption. Bread consumption has decreased by a considerable 10.77 kg. On the other hand, biscuits and, to a lesser extent, pasta show a growing consumption (respectively, +4.39 and +1.19 kg).

The consumption of fats and oils has decreased somewhat. In 1998, French households consumed 1.77 kg per capita less of butter than in 1980. The consumption of oils increased by 1.15 kg over the same period.

The consumption of sugar has decreased considerably, by 7.27 kg per capita, but we need to remember that sugar in processed fruits, biscuits and other foods (chocolate, sweets, etc.) is likely to have offset this decrease in direct consumption.

Model

The QUAIDS introduced by Banks *et al.* (1997) is used to describe household consumption behaviour. The QUAIDS possesses many of the attractive properties of the popular AIDS, while allowing for more non-linear Engel curve behaviour. The budget share, w_i, for good i takes the following form:

$$w_i = \alpha_i + \sum_{j=1}^{N} \gamma_{ij} \ln p_j + \beta_i \ln\left(\frac{x}{a(p)}\right) + \frac{\lambda_i}{b(p)}\left(\ln\left(\frac{x}{a(p)}\right)\right)^2 + u_i \qquad (8.1)$$

where p_i is the price of good i, x is the total expenditure on the N goods, α_i, γ_{ij}, β_i, and λ_i are parameters to be estimated, u_i is an error term and $a(p)$ and $b(p)$ are the following non-linear price aggregators:

$$\ln a(p) = \alpha_0 + \sum_{i=1}^{N} \alpha_i \ln p_i + \frac{1}{2}\sum_{i=1}^{N}\sum_{j=1}^{N} \gamma_{ij} \ln p_i \ln p_j \qquad (8.2)$$

$$b(p) = \prod_{i=1}^{N} p_i^{\beta_i} \qquad (8.3)$$

The effect of health information is introduced through a modification of the constant terms. Specifically, the preference parameters, α_i, are assumed to vary linearly with the health information as follows:

$$\alpha_i = \alpha_{i0} + \sum_{h=1}^{H} \omega_{ih} I_{t-h} + \sum_{k=1}^{K} \alpha_{ik} Z_k \qquad (8.4)$$

where I_{t-h} is the lagged health-information index (to be discussed later). In order to take into account the heterogeneity in behaviour, individual characteristics denoted as Z_k are also included in the constant term.

Economic theory places restrictions upon the value of parameters. Adding up requires:

$$\sum_i \alpha_{i0} = 1, \quad \sum_i \omega_{ih} = 0\,\forall h, \quad \sum_i \alpha_{ik} = 0\,\forall k,$$
$$\sum_i \gamma_{ij} = 0\,\forall j, \quad \sum_i \beta_i = 0 \quad \text{and} \quad \sum_i \lambda_i = 0 \tag{8.5}$$

Homogeneity of degree zero of Marshallian demands in prices and total expenditure and symmetry of the Slutsky matrix require:

$$\sum_j \gamma_{ij} = 0 \quad \forall i \quad \text{and} \quad \gamma_{ij} = \gamma_{ji} \tag{8.6}$$

The concavity of the cost function implies that the Slutsky matrix is negative semidefinite. Consequently, its diagonal elements and, therefore, compensated own-price elasticities must be non-positive.

Uncompensated price elasticities, e_{ij}, are computed from the estimated parameters as follows:

$$e_{ij} = -\delta_{ij} + \frac{1}{w_i}\left(\begin{array}{l} \gamma_{ij} - \left(\beta_i + \dfrac{2\lambda_i}{b(p)}\ln\left(\dfrac{x}{a(p)}\right)\right)\left(\alpha_j + \displaystyle\sum_{k=1}^{N}\gamma_{jk}\ln p_k\right) \\ -\dfrac{\lambda_i\,\beta_j}{b(p)}\left(\ln\left(\dfrac{x}{a(p)}\right)\right)^2 \end{array} \right) \tag{8.7}$$

where δ_{ij} is the Kronecker delta, which equals 1 if $i = \gamma_j$ (i.e. for own-price elasticities) and 0 otherwise. Compensated price elasticities, e^*_{ij}, are deduced from uncompensated price elasticities with Slutsky's formula as follows:

$$e^*_{ij} = e_{ij} + w_j e_i \tag{8.8}$$

where e_i is the expenditure elasticity given by:

$$e_i = 1 + \frac{1}{w_i}\left(\beta_i + \frac{2\lambda_i}{b(p)}\ln\left(\frac{x}{a(p)}\right)\right) \tag{8.9}$$

Health-information elasticity η_i can be calculated as:

$$\eta_i = \sum_{h=1}^{H} \omega_{ih}\frac{I_{t-h}}{w_i} \tag{8.10}$$

Table 8.2. Fat contents in grams and cholesterol contents in milligrams of the 15 food groups.[a] (Calculations based on information on nutrient values taken from the REGAL Database (Favier *et al.*, 1995), the French equivalent of the USDA's Nutrient Database.)

		Fatty acids			
Food group	Lipids	Saturated	Mono-unsaturated	Poly-unsaturated	Cholesterol
Beef	8.6	3.6	4.0	0.3	63.0
Pork	14.4	5.5	6.5	1.3	73.4
Other meats	9.4	4.1	3.5	0.8	111.2
Meat products	21.4	8.1	9.6	2.3	72.7
Poultry	4.4	1.5	1.6	0.9	69.1
Eggs	10.5	3.1	4.2	1.3	380.0
Fish	4.2	0.9	1.3	1.3	57.5
Milk	2.0	1.2	0.6	0.1	8.1
Cheese	23.6	14.8	6.9	0.7	75.1
Yoghurt and dairy desserts	1.9	1.1	0.6	0.1	15.1
Butter and animal fats	74.5	45.8	22.4	1.9	226.1
Oils	93.2	20.6	30.1	37.9	0.1
Vegetables	0.3	0.0	0.0	0.1	0.0
Fruits	0.8	0.1	0.2	0.2	0.0
Grain products	1.2	0.2	0.2	0.5	1.4

[a] The base is 100 g of the edible portion of each food.

Data

The data were obtained from the French NFS over the period 1978–1991 (Bertrand, 1993). This was a stratified survey of 10,000 households and the response rate was about 67.5%.[4] The sampling was completed by a time sampling: the year was divided into eight periods of 6 weeks each. No survey was done during the first 2 weeks of August and the last 2 weeks of December.[5] Expenditure and quantity data were obtained from a 7-day diary completed by one member of the household who was the main food planner. The survey also contained detailed information on a wide variety of household and personal characteristics. According to the availability of the survey, the data were pooled for the years 1978 to 1983, 1985, 1987, 1989 and 1991, providing a sample of 63,978 observations for estimation. The purchases of 15 food categories are studied: beef, pork, other meats, meat products, poultry, eggs, fish, milk, cheese, yoghurt and dairy desserts, butter and animal fats, oils, vegetables, fruits and grain products.[6] Conventional classification was used in the construction of the 15 food groups, but fat and cholesterol contents helped to refine them.[7] Table 8.2 gives the average fat and cholesterol contents of each group. The contents are in the edible portion of 100 g of food. Some of the groups, such as oils, butter, meat products, eggs, cheese, other meats and, to a lesser extent, beef and pork, have a high content of fats and cholesterol.

Since prices were not recorded in the survey, unit values are used. First, for each commodity group and each household with positive expenditure, unit values were computed by dividing expenditure by quantity purchased. Secondly, these unit values were averaged in clusters. Following Deaton (1987), a cluster is defined as a particular temporal and geographical area. The clusters were constructed by concatenating four variables: year, month, size of residence area and region. Finally, the average unit values so obtained were imputed to each household according to its own cluster. The use of unit values to approximate market prices in demand systems raises the problem of quality effects. Two different households facing the same prices may exhibit different unit values because their shares of the varieties comprising an aggregated group are different. The unit value of a food group reflects its average market price and consumers' choices of food quality. In a series of papers, Deaton (1987, 1988, 1990) developed a methodology that allowed estimation of quantity and quality effects separately. More recently, Deaton's methodology has been criticized by Crawford *et al.* (1996), who found an inconsistency in the procedure. They proposed an alternative form for the unit-values system, which, together with an AIDS specification, does not contradict the restrictions imposed in Deaton's methodology.[8] In a detailed review of Deaton's approach, Huang and Lin (2000) pointed out that the main problem of applying this approach lies with the lack of a guarantee to obtain accurate estimates of price responses.[9] Finally, Nichèle and Robin (1999) noticed that Deaton (1988, 1990) and followers (Deaton and Grimard, 1991; Ayadi *et al.*, 1995; Crawford *et al.*, 1996) adopt an *ad hoc* specification for the unit-values equations associated with the linearized version of the AIDS. Moreover, they proposed a structural model that allows them to estimate a non-linear AIDS specification for budget shares. This approach has been empirically tested on the series of the French NFS over the period 1981–1991. We used the QUAIDS specification for 17 food and beverage groups and obtained rather weak quality effects.[10] These results suggest that food-quality effects are not very important in this data set. Given the laboriousness of the procedure, no attempt was made to implement it here.

A monthly version of the adjusted global index (GIA) discussed by Rickertsen and von Cramon-Taubadel (Chapter 3, this volume) is used.[11] Including a measure of health information in a demand system involves specifying a lag distribution. Once information on health is established, it can have two distinct types of impact on household behaviour: an immediate and a delayed impact. It is further recognized that the impact of an article declines over time. Several attempts have been made to investigate the lag structure of the health index. Brown and Schrader (1990) used an Almon lag procedure. This approach produced unclear conclusions concerning the pattern of decay of the influence of health information. In constructing a new monthly index, Chern and Zuo (1995) used a third-degree distributed lag function with the specific assumption of 24 lagged periods and a maximum impact on the second month. More recently, some tests with the GIA and a polynomial distributed lag (PDL) procedure on German monthly time series concluded that the imposition of a PDL structure on the health index is not necessary. Wildner and von Cramon-Taubadel (Chapter 7, this volume) include only the contemporaneous value of the health-information index. Conversely, ten lags are introduced in the Scottish model (see Santarossa and Mainland, Chapter 13, this volume).[12] In this study, a flexible approach is adopted. A 12-period distributed lag is specified ($h = 1, \ldots, 12$) in Equation 8.4. The choice of a 12-month lifespan of health information is based on several experiments with the data. The decision to eliminate the contemporaneous effect is related to the sampling methodology of the French NFS.[13]

Finally, the sociodemographic variables selected to model the heterogeneity in household behaviour are age and education of household head, household size and number of household members participating in the labour market. Education of the head of the household is divided into four levels: primary school, before baccalaureate, baccalaureate or equivalent and after baccalaureate. The fourth level is the reference category. Seasonal dummy variables are also introduced and the fourth quarter is the reference quarter. Table 8.3 gives a description and summary statistics of the variables used in the estimation.

Estimation Results

The system (Equation 8.1) is estimated using the iterated linear least squares estimator (ILLE) for conditionally linear systems (see Blundell and Robin, 1999). First, to test and correct for endogeneity, the log of total food expenditure is instrumented by the log of income.[14] Secondly, the index $a(p)$ is replaced by the Stone index $\sum_i \bar{w}_i \ln p_i$, where $\bar{w}_i$ is the mean share of food group i, and the modified system is estimated by linear least squares. This provides an initial value for every parameter to start the iterations. Each iteration consists of re-evaluating $a(p)$ and $b(p)$ using the estimates from the previous step, and of re-estimating the system (Equation 8.1) using these evaluations. The iterations are repeated until all numerical solutions converge.

Some households are observed with zero expenditures for specific food groups. Three explanations of zero expenditures are usually given:

Table 8.3. Description of the variables.

Variable	Mean	Standard deviation
Budget share of beef	0.131	0.135
Budget share of pork	0.043	0.076
Budget share of other meats	0.044	0.087
Budget share of meat products	0.110	0.107
Budget share of poultry	0.062	0.097
Budget share of eggs	0.024	0.037
Budget share of fish	0.034	0.066
Budget share of milk	0.022	0.051
Budget share of cheese	0.108	0.096
Budget share of yoghurt	0.031	0.052
Budget share of butter and animal fats	0.049	0.063
Budget share of oils	0.014	0.035
Budget share of vegetables	0.089	0.085
Budget share of fruits	0.099	0.098
Budget share of grain products	0.140	0.131
Unit value of beef (fr. kg^{-1})	49.17	13.74
Unit value of pork (fr. kg^{-1})	32.95	8.66
Unit value of other meats (fr. kg^{-1})	45.96	16.11
Unit value of meat products (fr. kg^{-1})	40.61	13.00
Unit value of poultry (fr. kg^{-1})	27.06	9.70
Unit value of eggs (fr. $unit^{-1}$)	0.73	0.20
Unit value of fish (fr. kg^{-1})	36.02	16.25
Unit value of milk (fr. l^{-1})	2.80	0.69
Unit value of cheese (fr. kg^{-1})	32.61	10.64
Unit value of yoghurt and dairy desserts (fr. $unit^{-1}$)	1.53	0.84
Unit value of butter and animal fats (fr. kg^{-1})	24.22	4.63
Unit value of oils (fr. kg^{-1})	11.44	3.97
Unit value of vegetables (fr. kg^{-1})	7.72	3.35
Unit value of fruits (fr. kg^{-1})	8.37	3.07
Unit value of grain products (fr. kg^{-1})	8.86	2.84
Total food expenditure (fr. $week^{-1}$)	230.37	161.51
Household income (fr. $year^{-1}$)	89,406.98	67,806.92
Global health index	8.102	6.228
Age of head of household	49.076	17.173
D = 1 if education level of head of household is primary school[a]	0.507	0.500
D = 1 if education level of head of household is < bac.	0.263	0.440
D = 1 if education level of head of household is bac. or equivalent	0.113	0.316
D = 1 if education level of head of household is > bac.	0.117	0.322
Household size	2.822	1.491
Number of household's active members	1.177	0.906
D = 1 if survey done in quarter 1	0.283	0.451
D = 1 if survey done in quarter 2	0.274	0.446
D = 1 if survey done in quarter 3	0.223	0.416
D = 1 if survey done in quarter 4	0.220	0.414

[a] D for dummy variable. Reference categories printed in italic.
bac., baccalaureate.

corner solutions, personal preferences and infrequency of purchases. If the reason for zero expenditures is infrequency of purchases, an instrumental variable estimator can be used to solve the problem (Keen, 1986). However, this method only applies to linear expenditure systems, not to non-linear models (see Meghir and Robin, 1992). Heien and Wessells (1990) proposed a two-step procedure for dealing with zero expenditures, essentially a modified Heckman (1979) method. This procedure has often been used in applied studies because it is easy to implement. Recently, Shonkwiler and Yen (1999) pointed out an inconsistency in the Heien and Wessells (1990) estimator. They developed an alternative estimation procedure for a system of equations with limited dependent variables and used a Monte Carlo simulation to prove the superiority of their procedure. Alasia and Soregaroli (2001) applied Shonkwiler and Yen's technique to a LAIDS. To our knowledge, no application for a non-linear case exists. In addition, none of these methods seems to account for the problem of endogeneity of total (food) expenditure. Having said this, research in progress focuses on the feasibility of integrating the estimation of a non-linear demand system with endogenous regressors in the Shonkwiler and Yen's procedure.

Two models of the system (Equation 8.1) are estimated; first, model 1 without the use of the health-information index and, secondly, model 2 including the health-information index as specified in Equation 8.4. For both versions of the model, convergence of the ILLE is achieved at a high level of tolerance within 11 iterations.[15]

Estimation results show that total food expenditure is a significant determinant of the demand for 14 of the 15 food groups considered. The effects of quadratic expenditure terms are less important than those of linear expenditure terms. However, they are significant for 13 of 15 food groups, suggesting that the choice of the QUAIDS specification is appropriate.[16] Concerning the exogeneity of log total food expenditure, it is clearly rejected in ten of the 15 estimated equations.

Expenditure and Price Elasticities

Table 8.4 presents expenditure and uncompensated own-price elasticities from models 1 and 2. All elasticity estimates are evaluated at sample means. The models are estimated under the implicit assumption of separability of food from other consumer goods and services and conditional elasticities are calculated. Expenditure elasticities are all positive and significant. Table 8.4 shows that all meats and meat products, poultry, eggs, fish, cheese, fruits and vegetables react more strongly to increases in total food expenditure (elasticity greater than unity) than dairy products, butter, oils and grain products (elasticity lower than unity). Other meats have the highest expenditure elasticity;

Table 8.4. Expenditure, uncompensated own-price, and health information elasticities.

	Expenditure		Own price		Health
Food group	Model 1	Model 2	Model 1	Model 2	Model 2
Beef	1.242**	1.186**	−0.763**	−0.781**	−0.301**
Pork	1.022**	1.008**	−0.697**	−0.722**	−0.162**
Other meats	1.835**	1.785**	−0.519**	−0.541**	−0.354**
Meat products	1.041**	1.065**	−0.895**	−0.924**	0.125**
Poultry	1.205**	1.220**	−0.931**	−0.927**	−0.022
Eggs	1.305**	1.118**	−0.262*	−0.209*	−0.333**
Fish	1.410**	1.415**	−0.703**	−0.708**	−0.077
Milk	0.650**	0.710**	−0.737**	−0.618**	1.030**
Cheese	1.042**	1.056**	−0.647**	−0.648**	0.032
Yoghurt	0.777**	0.851**	−0.866**	−0.853**	0.718**
Butter	0.591**	0.546**	−0.189**	−0.293**	−0.220**
Oils	0.641**	0.632**	−0.935**	−0.901**	0.281**
Vegetables	1.259**	1.253**	−1.010**	−1.017**	−0.105**
Fruits	1.074**	1.068**	−0.710**	−0.798**	0.033
Grain products	0.538**	0.560**	−1.104**	−1.177**	0.178**

**, Significant at the 1% level; *, significant at the 5% level.

grain products and butter have the lowest. All uncompensated own-price elasticities are significant with the expected negative sign. Surprisingly, grain products appear to be the most sensitive to price changes. Except for vegetables and grain products, the demands for all other food groups are price-inelastic. The comparison of the results obtained from the two models shows that including the health-information index (model 2) has a limited effect on estimated expenditure and price elasticities. There are some differences in the expenditure elasticities of eggs, yoghurt and milk and in the own-price elasticities of butter, milk and fruits. Nevertheless, the similarities observed between the other coefficients make these differences acceptable. Consequently, the results reported in the following are based on model 2. Table 8.5 presents the uncompensated price elasticities. Several cross-price effects are statistically significant. However, these cross elasticities are fairly small in absolute value. Although the negative sign in many of them suggests complementary relations, it should be borne in mind that these are gross relationships. When one looks at the compensated elasticities reported in Table 8.6, it is clear that most goods are net substitutes. The highest substitution effects are observed between grain products and milk, grain products and fish and butter and oils. We also note that all the compensated own-price elasticities are negative, as the theory requires.

Table 8.5. Uncompensated price elasticities, model 2.

	Price of														
Food group	Beef	Pork	Other meats	Meat products	Poultry	Eggs	Fish	Milk	Cheese	Yoghurt	Butter	Oils	Vege-tables	Fruits	Grain products
Beef	−0.781**	−0.022	0.005	−0.051**	−0.043**	−0.008	−0.011	−0.049**	−0.040**	−0.056**	0.040**	−0.011	−0.123**	−0.079**	0.044*
Pork	−0.041	−0.722**	−0.074**	−0.074	0.013	−0.037	−0.136**	−0.037	−0.072*	−0.021	0.132**	0.059**	−0.028	0.026	0.005
Other meats	−0.057	−0.120**	−0.541**	−0.224**	−0.072*	−0.049**	0.021	−0.075**	−0.146**	−0.032	−0.107**	−0.087**	−0.035	−0.201**	−0.061
Meat products	−0.045*	−0.033*	−0.052**	−0.924**	0.002	−0.027*	−0.040**	0.010	−0.036*	0.048**	−0.027	0.003	0.051**	0.074**	−0.069**
Poultry	−0.095**	0.001	−0.025	−0.013	−0.927**	−0.034**	0.020	0.028*	−0.056*	−0.010	0.079**	−0.003	−0.132**	−0.085**	0.033
Eggs	−0.038	−0.077	−0.057	−0.132*	−0.084**	−0.209*	−0.136**	0.092*	−0.088*	0.034	−0.445**	0.011	0.033	−0.114**	0.090
Fish	−0.076	−0.215**	0.040	−0.181**	0.029	−0.108**	−0.708**	−0.059**	−0.082*	−0.016	−0.323**	0.017	−0.047	0.049	0.265**
Milk	−0.212**	−0.057	−0.083*	0.082	0.102**	0.100*	−0.057*	−0.618**	−0.062	−0.064*	−0.157**	−0.017	−0.048	−0.077	0.458**
Cheese	−0.032	−0.032*	−0.025*	−0.035*	−0.022	−0.018*	−0.013	−0.021*	−0.648**	−0.022**	−0.151**	−0.003	0.080**	−0.140**	0.027
Yoghurt	−0.181**	−0.022	−0.003	0.184**	0.002	0.030	0.002	−0.049**	−0.051	−0.853**	−0.125**	0.003	0.037	−0.020	0.195**
Butter	0.167**	0.121**	−0.025	0.004	0.122**	−0.163**	−0.145**	−0.060**	−0.226**	−0.060**	−0.293**	0.112**	−0.118**	0.159**	−0.140**
Oils	−0.014	0.172**	−0.155**	0.065	0.024	0.026	0.054*	−0.021	0.022	0.012	0.381**	−0.901**	−0.079**	−0.073	−0.144**
Vegetables	−0.198**	−0.026*	0.004	0.047**	−0.098**	0.006	−0.013	−0.026**	0.083**	0.002	−0.122**	−0.026**	−1.017**	−0.047**	0.177**
Fruits	−0.088**	0.009	−0.051**	0.082**	−0.043**	−0.025**	0.026**	−0.026**	−0.153**	−0.013	0.064**	−0.020**	−0.025*	−0.798**	−0.006
Grain products	0.109**	0.020*	0.032**	0.012	0.050**	0.024**	0.070**	0.063**	0.067**	0.044**	−0.046**	−0.012*	0.138**	0.045**	−1.177**

**, Significant at the 1% level; *, significant at the 5% level.

Table 8.6. Compensated price elasticities, model 2.

	Price of														
Food group	Beef	Pork	Other meats	Meat products	Poultry	Eggs	Fish	Milk	Cheese	Yoghurt	Butter	Oils	Vege- tables	Fruits	Grain products
Beef	−0.634**	0.029*	0.049**	0.074**	0.026	0.018*	0.024*	−0.023*	0.082**	−0.019*	0.106**	0.008	−0.027*	0.034*	0.251**
Pork	0.084*	−0.679**	−0.037	0.032	0.072**	−0.015	−0.106**	−0.014	0.032	0.011	0.189**	0.075**	0.053*	0.122**	0.181**
Other meats	0.165**	−0.043	−0.476**	−0.036	0.032	−0.010	0.074**	−0.035	0.038	0.025	−0.007	−0.058**	0.109**	−0.031	0.251**
Meat products	0.087**	0.013	−0.012	−0.812**	0.064**	−0.003	−0.009	0.034**	0.074**	0.082**	0.033*	0.020*	0.137**	0.176**	0.117**
Poultry	0.056	0.053**	0.020	0.115**	−0.855**	−0.007	0.056**	0.055**	0.070**	0.028*	0.148**	0.017	−0.033	0.031	0.246**
Eggs	0.101*	−0.029	−0.016	−0.014	−0.019	−0.184*	−0.102**	0.117**	0.028	0.069*	−0.382**	0.029	0.124**	−0.007	0.286**
Fish	0.100*	−0.154**	0.092**	−0.032	0.112**	−0.077**	−0.666**	−0.027	0.064	0.029	−0.244**	0.040**	0.067*	0.184**	0.512**
Milk	−0.124*	−0.026	−0.057	0.157**	0.143**	0.115**	−0.036	−0.602**	0.011	−0.042	−0.118**	−0.005	0.009	−0.009	0.583**
Cheese	0.099**	0.013	0.014	0.076**	0.040**	0.006	0.018	0.002	−0.539**	0.011	−0.092**	0.014*	0.166**	−0.039**	0.212**
Yoghurt	−0.075*	0.015	0.029	0.274**	0.052*	0.049**	0.027	−0.030	0.036	−0.826**	−0.077**	0.017	0.106**	0.061*	0.344**
Butter	0.235**	0.145**	−0.005	0.061*	0.154**	−0.151**	−0.129**	−0.047**	−0.170**	−0.043**	−0.263**	0.120**	−0.074**	0.211**	−0.044
Oils	0.064	0.199**	−0.132**	0.132**	0.061	0.040	0.073**	−0.007	0.087*	0.032	0.417**	−0.891**	−0.028	−0.013	−0.034
Vegetables	−0.042*	0.028*	0.050**	0.179**	−0.024	0.034**	0.024*	0.002	0.212**	0.041**	−0.052**	−0.006	−0.916**	0.072**	0.396**
Fruits	0.045*	0.055**	−0.012	0.194**	0.019	−0.002	0.057**	−0.002	−0.043**	0.020*	0.124**	−0.002	0.061**	−0.697**	0.181**
Grain products	0.179**	0.045**	0.053**	0.071**	0.082**	0.036**	0.087**	0.075**	0.125**	0.062**	−0.014	−0.003	0.183**	0.099**	−1.079**

**, Significant at the 1% level; *, significant at the 5% level.

Health-information Elasticities

The elasticities with respect to the health-information index are reported in the last column of Table 8.4. The results show that the fat and cholesterol information has a negative impact on the demand for beef, other meats, eggs, butter (values ranging from −0.22 to −0.35), and, to a lesser extent, on the demand for pork (−0.16) and vegetables (−0.10). A positive impact is observed on meat products, milk, yoghurt, oils and grain products (0.13, 1.03, 0.72, 0.28 and 0.18, respectively). The demand for poultry, fish, cheese and fruits seems to be unaffected by the dissemination of health information.

In most cases, the results appear to be plausible. The negative elasticities obtained for meats, eggs and butter are as expected for products with high contents of saturated fatty acids. Conversely, it is surprising to obtain a negative effect on vegetable demand. However, this effect is very low in absolute value. As in Chang and Kinnucan (1991), the demand for oils gains from increases of health information at the expense of the demand for butter. This result is consistent with the knowledge about the positive role of polyunsaturated fats in preventing cardiovascular diseases (see the composition of the oils group in Table 8.2). Our results concerning the positive and rather high effect of health information on the demand for milk and yoghurt confirm Chern's results (Chapter 2, this volume) with dairy and related products. As in the USA, the French consumption of whole milk has declined to the benefit of the consumption of semi-skimmed and skimmed milk.[17] Rather surprisingly, we do not find a positive effect of health information on the demand for poultry at the expense of the demand for red meat, as in Kinnucan *et al.* (1997). The positive elasticity of the demand for meat products is also somewhat unexpected. As in the case of dairy products, the positive elasticity may reflect the shifts in the demand from unhealthy meat products (sausages, salami, etc.) to low-fat products (cooked ham with fat and rind removed, etc.).[18]

Sociodemographic Effects

Table 8.7 presents the effects of the sociodemographic variables on the demand for food. Overall, the sociodemographic variables significantly influence the expenditure share of most food groups. However, the effects differ in their range of magnitude. For example, the effect of the age of the household head is weak.

The impact of household size is introduced to reflect economies of scale in food consumption. It appears that an increase in household size reduces the share of food expenditure on most protein-rich foods, fruits and vegetables. Conversely, it positively influences the budget share of

Table 8.7. Sociodemographic and seasonal effects, model 2.

	Share of														
Variable	Beef	Pork	Other meats	Meat products	Poultry	Eggs	Fish	Milk	Cheese	Yoghurt	Butter	Oils	Vege-tables	Fruits	Grain products
Age	0.0002**	−0.0001**	0.0001**	−0.0007**	0.0000	−0.0001**	0.0004**	−0.0003**	−0.0001**	−0.0005**	0.0002**	0.0001**	0.0002**	0.0003**	0.0004**
Education 1	0.0112**	0.0149**	−0.0013	0.0125**	0.0111**	−0.0026**	−0.0060**	−0.0012	−0.0215**	−0.0091**	0.0024**	0.0036**	−0.0193**	−0.0333**	0.0385**
Education 2	0.0091**	0.0108**	0.0009	0.0143**	0.0092**	−0.0021**	−0.0040**	0.0004	−0.0184**	−0.0060**	0.0008	0.0027**	−0.0119**	−0.0253**	0.0193**
Education 3	0.0074**	0.0053**	0.0017	0.0078**	0.0065**	−0.0009	−0.0015	0.0001	−0.0093**	−0.0049**	0.0013	0.0013*	−0.0067**	−0.0139**	0.0059**
Household size	−0.0073**	0.0017*	−0.0085**	−0.0032**	−0.0019*	−0.0016	−0.0036**	0.0066**	−0.0067**	0.0000	0.0082**	0.0030**	−0.0088**	−0.0075**	0.0296**
Active members	0.0039**	0.0009*	−0.0008	0.0023**	−0.0023**	−0.0031**	−0.0008*	−0.0034**	0.0023**	−0.0016**	0.0018**	−0.0002	−0.0033**	−0.0030**	0.0071**
Quarter 1	0.0025	−0.0013	−0.0009	−0.0091**	−0.0017	0.0009	−0.0002	0.0012	−0.0054**	0.0018	0.0000	0.0006	0.0149**	−0.0037*	0.0004
Quarter 2	−0.0105**	−0.0052**	0.0006	−0.0147**	−0.0013	−0.0001	−0.0027*	−0.0006	−0.0034	0.0021*	−0.0005	−0.0007	0.0349**	0.0061**	−0.0039
Quarter 3	−0.0072**	0.0013	−0.0037*	−0.0002	−0.0050**	−0.0013	−0.0053**	0.0004	−0.0053**	0.0022*	−0.0038**	−0.0007	0.0052**	0.0179**	0.0056*

**, Significant at the 1% level; *, significant at the 5% level.

rather inexpensive foods, such as grain products, pork, fats and milk. This last result may be linked to the presence of young children in large households.

The number of household members who are participating in the labour market is expected to negatively affect the expenditure share of food consumed at home to the benefit of food eaten away from home. We do not find this result. Several budget shares are positively influenced but we cannot conclude that these food groups correspond to convenience foods.

More interesting is the impact of the education of the household head. The influence of education is evaluated holding all other characteristics and the index of health information constant.[19] With reference to the highest level of education, the expenditure shares of beef, pork, meat products, poultry, oils and grain products increase as the level of education decreases. For example, in the case of grain products, less educated households have almost a 4 percentage-point higher budget share as compared with households with the highest level of education. We also find a positive effect for households with a primary school education on the budget share of butter. Households with only a primary education for the head of the household have a 3 percentage-point lower budget share for fruits than households with more than a baccalaureate. In the same way, a negative influence of the two first levels of education is observed for the expenditure share of fish and eggs. With a few exceptions (eggs, cheese), these results suggest that households with only a primary education for the head of the household tend to have diets higher in fat and cholesterol and lower in fruits and vegetables.

Finally, as expected, seasonal factors influence food consumption. For example, the expenditure share of vegetables is significantly lower in the autumn. The first quarter negatively affects the fruit budget share, while the second and third quarters favour fruit consumption.

Summary and Conclusions

This study evaluates the effects of health information on French food demand. A demand system that includes a measure of health information is estimated. The QUAIDS is used to describe the food-consumption behaviour regarding 15 food groups with different fat content. The French NFS for the period 1978–1991 in combination with a monthly health-information index is used.

Plausible price and expenditure elasticities are found. Furthermore, health information about fat and cholesterol plays an important role in explaining French food-consumption patterns. In particular, health information decreases the consumption of beef, pork, other meats, eggs and butter, but increases the consumption of oils and dairy products. The degree of household sensitivity to topics such as diet and health depends on demographic variables. Thus, it would be interesting to analyse the effect of health information on subgroups of households and the results could be used to target nutrition-related health campaigns.

Notes

[1] See Hercberg and Tallec (2000).
[2] European contract No. FAIR5–CT97–3373. Nutrition, health and the demand for food.
[3] French Quarterly National Accounts are not disaggregated enough concerning meat products.
[4] This is the average response rate over the period 1985–1991. During the period, the response rate was decreasing from 72.8 to 63.5%.
[5] The average number of households per month over the period 1978–1991 is the following:

Jan.	Feb.	Mar.	Apr.	May	Jun.	Jul.	Aug.	Sep.	Oct.	Nov.	Dec.
649.3	556.7	607.6	626.5	581.8	541.9	449.9	348.1	628.2	590.1	582.2	235.5

[6] A detailed description of the 15 food groups is available from the author on request.
[7] For example, butter and oils form two different groups, whereas usually they are aggregated because of their functional equivalence.
[8] Deaton assumes weak separability and constant within-group relative prices.
[9] They also mention 'complicated matrix multiplication' (Huang and Lin, 2000: 19).
[10] The highest coefficient was obtained for the group alcoholic beverages (not considered in the present study). At 0.22, it was significantly inferior to the 0.34 coefficient obtained by Huang and Lin (2000) for the group of dairy products. Moreover, a majority of studies found weak quality effects.
[11] The use of an index based on English-language articles in a study of French consumption is questionable. However, the information gap originating in different languages is diminishing. French media closely follow and report on international research results published in English.
[12] However, computing an information index as a weighted average of the number of articles, with the weights decreasing over time, allows the suppression of the problem of the index performing as a trend.
[13] Household purchases were collected for 7 days running in a given month, at the beginning as well as at the end of the month. We assume that there is a 1-month lag before a new article has an effect on food purchases.
[14] Blundell and Robin (1999) proposed an augmented regression approach in order to account for endogenous regressors.
[15] The homogeneity restriction is imposed on both models. Symmetry is tested and rejected by the data. However, it is imposed in the final model. Complete estimates are not presented but they are available from the author on request.

[16] Banks *et al.* (1997) consider only one aggregate food group and conclude that the standard AIDS model provides a sufficient description of Engel curve behaviour for this single food group.
[17] The trend in the share of semi-skimmed and skimmed milk of total fluid milk over the period 1980–1991 is given in the table below:

1980	1981	1982	1983	1984	1985	1986	1987	1988	1989	1990	1991
0.44	0.45	0.48	0.51	0.55	0.59	0.57	0.60	0.64	0.66	0.68	0.71

Source: Centre National Interprofessionnel de l'Economie Laitière.
[18] There exists a whole gamut of low-fat pork products that benefit from the considerable progress accomplished in animal nutrition; in fact, the health advantages of cooked ham are often mentioned. Unfortunately, no statistics are available to confirm this information.
[19] In this context, the links between education and health information are not accounted for. Several papers on the demand for nutrients have addressed this issue. For instance, Variyam *et al.* (1997) and Variyam (2000) apply statistical methods to US data that allow them to 'isolate the impact of nutrition knowledge and awareness of diet–disease relationships from individual characteristics on the intake' of fats and cholesterol (Variyam, 2000: 283).

References

Alasia, A. and Soregaroli, C. (2001) Food demand with incomplete markets: a model for Mozambique. Paper presented at the 71st EAAE seminar, Zarogoza, Spain.

ANS (1991/92) *Annuaire Statistique de la France – résultats de 1991*. Vol. 96. Institut National de la Statistique et des Etudes Economiques et les Services Statistiques des Ministères, Paris, 824 pp.

ANS (1999) *Annuaire Statistique de la France – résultats de 1997*. Vol. 102. Institut National de la Statistique et des Etudes Economiques et les Services Statistiques des Ministères, Paris, 985 pp.

ANS (2000) *Annuaire Statistique de la France – résultats de 1998*. Vol. 103. Institut National de la Statistique et des Etudes Economiques et les Services Statistiques des Ministères, Paris, 1000 pp.

Ayadi, M., Baccouche, R., Goaied, M. and Matoussi, M. (1995) *Spatial Variation of Prices and Household Demand Analysis in Tunisia*. Working Paper, Faculté des Sciences Economiques et de Gestion, Université de Tunis III, Tunis, Tunisia.

Banks, J., Blundell, R.W. and Lewbel, A. (1997) Quadratic Engel curves and consumer demand. *Review of Economics and Statistics* 79, 527–539.

Bertrand, M. (1993) *Consommation et lieux d'achat des produits alimentaires en 1991*. INSEE Résultats–Consommation–Modes de Vie no. 54–55, Service Conditions de Vie des Ménages, Institut National de la Statistique et des Etudes Economiques, Paris, 299 pp.

Blundell, R.W. and Robin, J.M. (1999) Estimation in large and dissagregated demand systems: an estimator for conditionally linear system. *Journal of Applied Econometrics* 14, 209–232.

Brown, D.J. and Schrader, L.F. (1990) Cholesterol information and shell egg consumption. *American Journal of Agricultural Economics* 72, 548–555.

Capps, O. Jr and Schmitz, J.D. (1991) A recognition of health and nutrition factors in food demand analysis. *Western Journal of Agricultural Economics* 16, 21–35.

Cases, L. (1999) *La Consommation des ménages en 1998*. INSEE Résultats Consommation–Modes de Vie no. 99–100, Service des Comptes Nationaux, Institut National de la Statistique et des Etudes Economiques, Paris, 266 pp.

Chang, H.S. and Kinnucan, H.W. (1991) Advertising, information, and product quality: the case of butter. *American Journal of Agricultural Economics* 73, 1195–1203.

Chern, W.S. and Zuo, J. (1995) Alternative measures of changing consumer information on fat and cholesterol. Paper presented at the AAEA annual meeting, Indianapolis, Indiana.

Chern, W.S., Loehman, E.T. and Yen, S.T. (1995) Information, health risk beliefs, and the demand for fats and oils. *Review of Economics and Statistics* 76, 555–564.

Combris, P. (1992) Changements structurels: la cas des consommations alimentaires en France de 1949 à 1988. *Economie et Prévision* 102–103, 221–245.

Crawford, I., Laisney, F. and Preston, I. (1996) *Estimation of a Household Demand System Using Unit Values Data*. Working Paper no. 96/16, Institute for Fiscal Studies, London, 36 pp.

Deaton, A. (1987) Estimation of own- and cross-price elasticities from household survey data. *Journal of Econometrics* 36, 7–30.

Deaton, A. (1988) Quality, quantity and spatial variation of price. *American Economic Review* 78, 418–430.

Deaton, A. (1990) Price elasticities from survey data: extensions and Indonesian results. *Journal of Econometrics* 44, 281–309.

Deaton, A. and Grimard, F. (1991) *Demand Analysis for Tax Reform in Pakistan*. Discussion paper no. 151, Research Program in Development Studies, Centre for International Studies, Woodrow Wilson School, Princeton University, Princeton, New Jersey.

Favier, J.C., Ireland-Ripert, J., Toque, C. and Feinberg, M. (1995) *Répertoire général des aliments (REGAL): tables de composition*, 2nd edn. CNEVA-CIQUAL and TEC & DOC, INRA editions, Paris, 897 pp.

Fulponi, L. (1989) The almost ideal demand system: an application to food and meat groups for France. *Journal of Agricultural Economics* 40, 82–92.

Heckman, J.J. (1979) Sample selection bias as a specification error. *Econometrica* 47, 153–161.

Heien, D. and Wessells, C.R. (1990) Demand systems estimation with microdata: a censored regression approach. *Journal of Business and Economic Statistics* 8, 365–371.

Hercberg, S. and Tallec, A. (2000) *Pour une politique nutritionnelle de santé publique-enjeux et propositions*. 1st edn. Collection Avis et Rapports, ENSP, Ministère de l'Emploi et de la Solidarité-Haut Comité de la Santé Publique, Rennes, France, 275 pp.

Huang, K.S. and Lin, B.H. (2000) *Estimation of Food Demand and Nutrient Elasticities from Household Survey Data*. Technical Bulletin no. 1887, Economic Research Service, US Department of Agriculture, Washington, DC, 30 pp.

Keen, M.J. (1986) Zero expenditure and the estimation of Engel curves. *Journal of Applied Econometrics* 1, 277–286.

Kinnucan, H.W., Xiao, H., Hsia, C.J. and Jackson, J.D. (1997) Effect of health information and generic advertising on U.S. meat demand. *American Journal of Agricultural Economics* 79, 13–23.

Meghir, C. and Robin, J.M. (1992) Frequency of purchase and the estimation of demand system. *Journal of Econometrics* 53, 53–85.

Nichèle, V. and Robin, J.M. (1999) *About Quality and Quantity in Consumer Expenditure Surveys: a Fully Structural Specification.* Working Paper, Laboratoire de Recherche sur la Consommation, Institut National de la Recherche Agronomique, Ivry-sur-Seine, France.

Shonkwiler, J.D. and Yen, S.T. (1999) Two-step estimation of a censored system of equations. *American Journal of Agricultural Economics* 81, 972–982.

Variyam, J.N. (2000) Role of demographics, knowledge, and attitudes: fats and cholesterol. In: Frazao, E. (ed.) *America's Eating Habits.* Bulletin no. 750, Economic Research Service, US Department of Agriculture, Washington, DC, pp. 281–294.

Variyam, J.N., Blaylock, J. and Smallwood, D. (1997) *Diet–Health Information and Nutrition: the Intake of Dietary Fats and Cholesterol.* Technical Bulletin no. 1855, Economic Research Service, US Department of Agriculture, Washington, DC, 53 pp.

The Impact of Nutrient Intake on Food Demand in Spain

9

ANA M. ANGULO,[1] JOSE M. GIL,[2] AZUCENA GRACIA[2] AND MONIA BEN KAABIA[1]

[1]*University of Zaragoza, Zaragoza, Spain;*
[2]*Agricultural Research Service (DGA), Zaragoza, Spain*

Introduction

Health information has become increasingly available to consumers through sources such as health professionals (visits to doctors and nutritionists) and the mass media. These information sources typically suggest that appropriate nutrient intakes combined with moderate and regular physical exercise can reduce the risk of food-related diseases, such as coronary disease, cancer, stroke and diabetes (Kantor, 1999). The relationship between food, diet and health has also raised the interest among agricultural economists dealing with food-demand analysis. There are two generally accepted approaches for incorporating nutritional factors into demand analyses.

The first approach measures the effect on food demand of consumers' knowledge regarding the nutritional content of food and regarding the relationships between the nutritional content and the probability of suffering certain diseases. Using this approach, consumers' knowledge variables are difficult to create. When dealing with time-series data, they are usually approximated by indices created either from the number of relevant articles appearing in newspapers or in medical journals or from expenditures on generic and/or brand advertising in the mass media. Brown and Schrader (1990) pioneered the inclusion of 'informed consumers' by constructing an index to investigate how consumer information affected US shell-egg consumption. Several attempts have been made to improve the index and to include various food products (e.g. Chern and Zuo, 1995; Kinnucan *et al.*, 1997; Kim and Chern, 1999). With respect to cross-sectional analyses, several studies have been carried out using two US

Department of Agriculture surveys: the Continuing Survey of Food Intakes by Individuals (CSFII) and the Diet Health Knowledge Survey (DHKS) (e.g. Jensen *et al.*, 1992; Variyam *et al.*, 1998; Nayga, 2000).

The second approach measures the impact of income, prices and sociodemographic variables on nutrient intake. A 'direct' and an 'indirect' measurement technique have been used. 'Direct' measurements are obtained through a regression of nutrient intake on relevant variables (e.g. Adrian and Daniel, 1976; Nayga, 1994; Subramanian and Deaton, 1996; Chesher, 1998). 'Indirect' measurements require estimating the effects of relevant variables on food-product demand (by estimating a demand system). Then, nutrient-intake effects are calculated by applying nutrient conversion factors to the resultant food demand (Ramezani *et al.*, 1995; Huang, 1999; Xiao and Taylor, 1995).

In spite of this research interest, to our knowledge there has not been any formal attempt to study the relationship between food demand and consumer-perceived health issues in Spain. Most food-demand analyses have only included expenditure, prices and, when using cross-sectional data, sociodemographic variables. Using cross-sectional data, Chung and López (1988) and Chung (1994) conducted regional analyses of food consumption. Furthermore, Laajimi and Albisu (1997) and Gracia and Albisu (1998) estimated Spanish demand for different types of meat and fish. Finally, Manrique and Jensen (1997, 1998) analysed the Spanish household demand for convenience meat products and for food away from home. Using time-series models, the dynamic responses of Spanish consumers have been the main focus of some studies. Molina (1994) and Gracia *et al.* (1998) estimated the demand for main food aggregates using different dynamic versions of the almost ideal demand system (AIDS) and the generalized addilog demand system (GADS). Another avenue of research is to analyse food-consumption patterns across European countries. Angulo *et al.* (1997) tested for differences in income and price elasticities for several food aggregates between Spain, Italy and France (Mediterranean countries) and Great Britain and Denmark (non-Mediterranean), while Angulo *et al.* (2001) analysed whether calorie and income elasticities among EU countries (Norway also included) converged in the long run.

This study attempts to provide a new approach to analysing food demand in Spain by incorporating nutritional factors. From the methodological point of view, we are aiming to provide a new analytical method using microdata. Microdata have several important advantages. First, microdata yield substantially greater precision in the parameter estimates than aggregate data (Orcutt *et al.*, 1968). Secondly, and perhaps more importantly, some of the relevant explanatory variables exist at the micro level in a form that cannot readily be aggregated. Moreover, microdata represent the quantities of goods purchased directly by consumers for at-home consumption, while time-series data are frequently based on supply-disappearance data. Finally, microdata yield more information about demand relationships than macrodata when individual demand functions are non-linear (Lau, 1982).

Unfortunately, micro-level surveys in Spain are not so informative as those available in the USA (such as the CSFII and DHKS). For example, there is no information that can be used to analyse the effect of consumers' health knowledge on food demand. Thus, the second approach discussed above seems to be more useful in our case. We assume that the consumer's utility is a function of nutrients instead of goods. This hypothesis is supported by the fact that it is realistic to assume that consumers are starting to think more in terms of nutrients than in terms of food products, while simultaneously taking into account prices and disposable income. However, goods and not nutrients are available in the market and we derive demand functions with food quantities as dependent variables, and income, prices and nutrients as explanatory variables. The model is applied to the Spanish demand for food using the Quarterly National Expenditure Survey for 1995.

The structure of the chapter is as follows. First, some descriptive data on food consumption in Spain are presented. Secondly, the theoretical model and the data set are described. Finally, the main results are presented and some concluding remarks are offered.

Food Consumption in Spain

Table 9.1 shows the annual per capita consumption of different food items for 1985, 1990, 1995 and 1998. In the last column, the annual variation rates (AVR) in percentages have been calculated between 1985 and 1998. The consumption of most products has remained quite stable. The main exceptions are the notable increases in processed vegetables and fruits, with an AVR of 4.54%, and pork (2.65%). The consumption of similar and competitive products has simultaneously followed negative trends (fresh fruits −1.60%, poultry −1.58% and beef −1.19%). The consumption of sugar (−3.39%), eggs (−1.89%) and fats and oils (−1.50%) has also shown a remarkable decrease during the period analysed.

We make three important general observations based on the figures in Table 9.1. First, the changes are quite smooth, suggesting that consumption has reached a maximum level. Gil *et al.* (1995) have empirically confirmed this result. Secondly, the decreases in the consumption of fresh fruits and vegetables and the simultaneous increases in the consumption of processed fruits and vegetables suggest that consumer demand is shifting towards more convenient products, mainly due to the increasing participation of women in the labour force

Table 9.1. Annual per capita consumption (kg). (From: Ministerio de Agricultura Pesca y Alimentación, several years.)

Food item	1985	1990	1995	1998	AVR (%)[a]
Meat	66.9	66.2	61.2	65.3	−0.18
Beef	11.0	9.6	10.0	9.3	−1.19
Pork	9.3	9.4	8.6	12.5	2.65
Poultry	20.5	17.8	16.2	16.3	−1.58
Other meats	26.1	29.4	26.4	27.2	0.32
Fish	20.8	19.0	18.2	18.0	−1.04
Eggs	18.0	15.2	11.3	13.6	−1.89
Cheese	6.0	5.8	6.7	6.2	0.26
Fluid milk	121.6	106.7	115.9	113.3	−0.52
Vegetables and fruits	185.9	187.2	155.2	163.4	−0.93
Fresh vegetables	66.5	69.1	55.3	60.4	−0.71
Fresh fruits	108.9	105.3	84.5	86.3	−1.60
Processed fruits and vegetables	10.5	12.8	15.5	16.7	4.54
Cereals	91.2	80.1	79.7	80.9	−0.87
Sugar	13.6	10.1	8.8	7.6	−3.39
Fats and oils	23.2	19.0	17.8	18.7	−1.50
Other foods	61.1	54.1	57.1	50.1	−1.38

[a] Annual variation rate for 1985 to 1998 is calculated as

$$\frac{Consumption_{1998} - Consumption_{1985}}{Consumption_{1985} \times 13} \times 100$$

and to changes in lifestyles. Finally, reduced consumption of sugar, eggs and fats and oils suggests that consumers are increasingly aware about health issues. These products provide the highest percentages of cholesterol and saturated fats (the most 'dangerous' nutrients with regard to the diet–health relationship).

Table 9.2 displays annual per capita expenditures for major food items, using constant 1998 US dollars.[1] While food expenditure decreases for all listed products, the greatest rates of decrease correspond to sugar (−4.80%), eggs (−4.28%), fats and oils (−3.62%) and beef (−3.36%).

Socio-economic variables are also important for food expenditures. Of the various socio-economic variables collected in the Quarterly National Expenditure Survey, the most discriminating are education level, size of the town where the household is living and age. Results from Table 9.3 indicate that the level of education of the head of the household affects food-expenditure structure considerably. A higher level of education is associated with higher expenditure shares on meat, fish and fruits and vegetables but lower shares on cereals and potatoes. As the size of the hometown increases, fish and fruits and vegetables expenditure shares increase while the shares decrease for cereals and

Table 9.2. Annual per capita constant expenditures of major food items (in 1998 US dollars). (From: Ministerio de Agricultura Pesca y Alimentación, several years.)

Food item	1985	1990	1995	1998	AVR (%)[a]
Meat	478.1	379.0	314.5	321.2	−2.52
Beef	128.6	89.1	82.0	72.5	−3.36
Pork	69.0	49.6	42.3	53.4	−1.74
Poultry	71.7	49.9	38.5	42.2	−3.16
Other meats	208.8	190.5	151.7	153.1	−2.05
Fish	131.6	108.8	91.1	95.2	−2.13
Eggs	51.2	33.3	19.0	22.7	−4.28
Cheese	57.9	45.9	45.3	40.3	−2.33
Fluid milk	109.4	79.2	70.1	63.7	−3.21
Vegetables and fruits	256.4	227.1	171.4	181.7	−2.24
Fresh vegetables	87.8	82.2	61.7	72.4	−1.35
Fresh fruits	129.6	109.4	83.3	80.9	−2.89
Processed fruits and vegetables	39.0	35.5	26.5	28.5	−2.07
Cereals	212.3	168.1	155.0	152.7	−2.16
Sugar	19.7	11.2	9.0	7.4	−4.80
Fats and oils	76.1	51.6	48.6	40.3	−3.62
Other foods	32.4	23.0	30.4	19.8	−3.00
Total food at home	1425.1	1127.3	954.4	945.0	−2.59
Food away from home	[b]	387.9	376.4	410.5	0.73[c]

[a] Annual variation rate for 1985 to 1998 is calculated as

$$\frac{Expenditure_{1998} - Expenditure_{1985}}{Expenditure_{1985} \times 13} \times 100$$

[b] Not available.

[c] Annual variation rate for 1990 to 1998.

potatoes. This last result may be explained by the greater availability of a variety of fresh foods in big cities. Finally, age is measured by the percentage of household members over 14 years of age. It is interesting to note that as this percentage increases, the relative importance of fruits and vegetables also increases, while the proportion allocated to cereals and potatoes and dairy products and eggs decreases.

The Model

We assume a representative consumer whose preference structure is defined by nutrient intakes instead of goods, implying a movement towards the frameworks used by Lancaster (1966, 1971), Rosen (1974) and Pudney (1981).

Table 9.3. Food-expenditure structure in 1995 (%). (From: own calculations based on the Quarterly National Expenditure Survey.)

	Cereals and potatoes	Meat	Fish	Fruits and vegetables	Dairy products and eggs	Oils and other foods
General	17.0	21.8	11.6	20.3	13.0	16.4
Education						
Unschooled	20.5	19.2	10.0	20.2	13.2	16.9
Primary school	16.6	22.1	12.0	20.1	13.0	16.2
Secondary school	15.7	23.5	12.0	20.0	12.4	16.3
Post-secondary degree	12.1	23.1	13.1	22.6	13.7	15.3
Size of the town						
0–10,000	18.2	21.7	10.9	19.5	13.0	16.8
10,001–50,000	19.6	21.2	10.2	18.3	13.4	17.1
50,001–500,000	16.0	22.2	12.4	20.9	13.0	15.5
> 500,000	13.4	21.7	12.6	23.4	12.2	16.7
% of members over 14						
25–50	18.9	21.5	11.0	16.9	15.5	16.2
50–75	18.1	22.3	10.7	18.5	14.5	15.9
> 75	16.7	21.7	11.8	20.9	12.5	16.5

Let $x = (x_1, x_2, \ldots, x_n)'$ represent n different food products and $z = (z_1, z_2, \ldots, z_r)'$ represent r types of nutrients. The preference structure of a representative consumer (defined for nutrient intakes) satisfies the assumptions of traditional economic theory and can be expressed by:

$$U = U(z) \tag{9.1}$$

A linear relationship is assumed between z and x through the consumption technology, B, such that:

$$z = Bx \tag{9.2}$$

where $B = [b_{ij}]$ is an $r \times n$ matrix of elements along the ith row, which shows the amount of nutrient i provided by one unit of each of the n food products, and along the jth column, which shows the amount of each of the r nutrients provided by one unit of the jth product.

Furthermore, we must consider the consumer budget constraint:

$$p'x = m \tag{9.3}$$

with $p = (p_1, p_2, \ldots, p_n)'$ denoting the vector of unit prices of the n food products and m denoting total food expenditure.

The optimization problem consists of maximizing the utility function (Equation 9.1) subject to the technology constraint (Equation 9.2) and the budget constraint (Equation 9.3). The most appropriate solution can be found as follows. First, we find the 'efficiency consumption frontier' in the nutrient space by combining the technology and budget restrictions. This frontier represents the upper boundary of the set of nutrient vectors that a consumer can achieve by combining goods. The frontier is independent of preferences and its essential properties are identical for all consumers, regardless of income. The first step results in the new optimization problem:

$$\text{Max } U = U(z) \quad \text{s.t.} \quad \pi' z = m \tag{9.4}$$

where $\pi = (\pi_1, \pi_2, \ldots, \pi_r)'$ represents the vector of shadow or implicit prices for nutrients z.

In order to solve Equation 9.4, we must bear in mind that, although the π vector is not directly observable, it can be estimated by using the hedonic price equation:

$$p_j = \sum_{i=1}^{r} \left(\frac{1}{\lambda} \frac{\partial U}{\partial z_i} \right) b_{ij} = \sum_{i=1}^{r} \pi_i \, b_{ij} \quad \text{for } j = 1, 2, \ldots, n \tag{9.5}$$

where λ is the Lagrangian multiplier, which can be interpreted as the marginal utility of money. From Equation 9.5, we note that market prices paid by the consumer equal the sum of the marginal values of the different nutrients.

Next, we introduce the consumer preferences to obtain the optimal choice of each nutrient:

$$z_i = d_i(\pi_1, \pi_2, \ldots, \pi_r, m) \quad \text{for} \quad i = 1, 2, \ldots, r \tag{9.6}$$

Given the optimal combination of nutrients, the optimal choice of food products can be estimated as the food-demand functions:

$$x_j = f_j\,(p_1, p_2, \ldots, p_n, m, b_{11}, b_{12}, \ldots, b_{1n}, b_{21}, b_{22}, \ldots, b_{2n}, \ldots, b_{r1}, b_{r2}, \ldots, b_{rn}) \tag{9.7}$$

We can either estimate the demand for nutrients using Equation 9.6 or the demand for food, taking into account consumer awareness about nutrient intake from food products, using Equation 9.7. In this chapter, we follow the second approach since consumers buy food products and not nutrients.

Functional Form

We use the GADS (Bewley, 1986) because of its desirable characteristics and the ease of estimation. Moreover, this system guarantees the non-negativity property of the estimated average budget shares, satisfies the aggregation restriction and is flexible enough to impose homogeneity, symmetry and negativity restrictions by linear restrictions on the estimated parameters.

Given Equation 9.7, the GADS model takes the form:

$$w_j = \frac{e^{\lambda_j}}{\sum_{p=1}^{n} e^{\lambda_p}} \text{ where } \lambda_j = \alpha_j + \beta_j \ln m + \sum_{k=1}^{n} \gamma_{jk} \ln p_k + \sum_{i=1}^{r}\sum_{k=1}^{n} \mu_{j,ik} \ln b_{ik} \tag{9.8}$$

and w_j represents the average budget share of the jth good.

From Equation 9.8, income, η, uncompensated price, e, compensated price, ζ, and nutrient elasticities, υ, can be calculated as follows:

$$\eta_j = 1 + \beta_j - \sum_{p=1}^{n} w_p \beta_p \qquad \text{for } j = 1, 2, \ldots, n \tag{9.9}$$

$$e_{jk} = \gamma_{jk} - \delta_{jk} - \sum_{p=1}^{n} w_p \gamma_{pk} \qquad \text{for } j, k = 1, 2, \ldots, n \tag{9.10}$$

$$\xi_{jk} = e_{jk} + w_k \eta_j \qquad \text{for } j, k = 1, 2, \ldots, n \tag{9.11}$$

$$\upsilon_{j,ik} = \mu_{j,ik} - \sum_{p=1}^{n} w_p \mu_{p,ik} \qquad \text{for } j, k = 1, 2, \ldots, n \text{ and } i = 1, 2, \ldots, r \tag{9.12}$$

where $\delta_{jk} = 1$ when $j = k$ and zero otherwise and $\upsilon_{j,ik}$ is the elasticity of the jth product with respect to the content of the ith nutrient per unit of the kth product.

The model (Equation 9.8) is difficult to estimate and some transformations are needed. First, the model is linearized by taking logs:

$$\ln w_j = \alpha_j + \beta_j \ln m + \sum_{k=1}^{n} \gamma_{jk} \ln p_k + \sum_{i=1}^{r}\sum_{k=1}^{n} \mu_{j,ik} \ln b_{ik} - \ln \sum_{p=1}^{n} e^{\lambda_p} \tag{9.13}$$

Then, after premultiplying Equation 9.13 with $\sum_i \bar{w}_j$, where $\bar{w}_j$ denotes the budget share of the jth good at a given data point, we can calculate the weighted geometric mean of the average budget shares as:

$$\sum_{j=1}^{n} \bar{w}_j \ln w_j = \sum_{j=1}^{n} \bar{w}_j \alpha_j + \sum_{j=1}^{n} \bar{w}_j \beta_j \ln m + \sum_{j=1}^{n} \bar{w}_j \sum_{k=1}^{n} \gamma_{jk} \ln p_k + \sum_{j=1}^{n} \bar{w}_j \sum_{i=1}^{r}\sum_{k=1}^{n} \mu_{j,ik} \ln b_{ik} - \sum_{j=1}^{n} \bar{w}_j \ln \sum_{p=1}^{n} e^{\lambda_p} \tag{9.14}$$

Subtracting Equation 9.14 from Equation 9.13 and rearranging terms, the final linear version of the GADS is obtained:

$$\ln\left(\frac{w_j}{w^+}\right) = a_j + b_j \ln m + \sum_{k=1}^{n} g_{jk} \ln p_k + \sum_{i=1}^{r}\sum_{k=1}^{n} \varphi_{j,ik} \ln b_{ik} \qquad (9.15)$$

where:

$$\ln w^+ = \sum_{j=1}^{n} \overline{w}_j \ln w_j; a_j = \alpha_j - \sum_{p=1}^{n} \overline{w}_p \alpha_p;$$

$$b_j = \beta_j - \sum_{p=1}^{n} \overline{w}_p \beta_p = \overline{\eta}_j - 1; g_{jk} = \gamma_{jk} - \sum_{p=1}^{n} \overline{w}_p \gamma_{pk}$$

$$= \overline{e}_{jk} + \delta_{jk}; \text{and } \varphi_{j,ik} = \mu_{j,ik} - \sum_{p=1}^{n} \overline{w}_p \mu_{p,ik} = \overline{\upsilon}_{j,ik}$$

Subtracting $\ln(p_j/m)$ from both sides of Equation 9.15 directly yields a useful version of the GADS in which parameters represent the elasticities shown in Equations 9.9–9.12:

$$\ln\left(\frac{q_j}{w^+}\right) = a_j + \overline{\eta}_j \ln m + \sum_{k=1}^{n} \overline{e}_{jk} \ln p_k + \sum_{i=1}^{r}\sum_{k=1}^{n} \overline{\upsilon}_{j,ik} \ln b_{ik} \qquad (9.16)$$

where q_j represents the per capita consumption of the jth good and $\overline{\eta}_j$, $\overline{e}_{jk}$ and $\overline{\upsilon}_{j,ik}$ are η_j, e_{jk} and $\upsilon_{j,ik}$ and evaluated at $w_j = \overline{w}_j$.

If we want to test the restrictions of demand theory (i.e. homogeneity, symmetry and negativity), further transformations are necessary. Multiplying Equation 9.16 by $\overline{w}_j$ and rearranging terms:

$$\overline{w}_j \ln\left(\frac{q_j}{w^+}\right) = a_j^* + \overline{\theta}_j \ln\left(\frac{m}{p^+}\right) + \sum_{k=1}^{n} \overline{\pi}_{jk} \ln p_k + \sum_{i=1}^{r}\sum_{k=1}^{n} \overline{\kappa}_{j,ik} \ln b_{ik} \qquad (9.17)$$

where $a_j^* = \overline{w}_j a_j$; $\overline{\theta}_j = \overline{w}_j \overline{\eta}_j$; $\overline{\pi}_{jk} = \overline{w}_j \overline{e}_{jk} + \overline{w}_j \overline{w}_k \overline{\eta}_j$; $\overline{\kappa}_{j,ik} = \overline{w}_j \upsilon_{j,ik}$; and $\ln P^+ = \sum_{j=1}^{n} \overline{w}_j \ln p_j$. The $\overline{\theta}_j$ and $\overline{\pi}_{jk}$ parameters are the marginal budget shares and the Slutsky parameters, respectively, defined at $\overline{w}_1, \overline{w}_2, \ldots, \overline{w}_n$. Finally, the translation method of Pollak and Wales (1981) is used to introduce socio-economic variables. Hence, the final demand system is:

$$\overline{w}_j \ln\left(\frac{q_j}{w^+}\right) = a_j^{**} + \sum_{d=1}^{s} \rho_{jd} S_{jd} + \overline{\theta}_j \ln\left(\frac{m}{P^+}\right) + \sum_{k=1}^{n} \overline{\pi}_{jk} \ln p_k + \sum_{i=1}^{r}\sum_{k=1}^{n} \overline{\kappa}_{j,ik} \ln b_{ik} \qquad (9.18)$$

where S_{jd} represents the socio-economic characteristics affecting the allocation of food expenditure.

Adding up is satisfied by construction and implies the following restrictions:

$$\sum_{j=1}^{n} a_j^{**} = 0; \sum_{j=1}^{n} \rho_{jd} = 0 \;\; \forall\, d; \sum_{j=1}^{n} \bar{\theta}_j = 1; \sum_{j=1}^{n} \bar{\pi}_{jk} = 0 \;\; \forall\, k; \text{ and } \sum_{j=1}^{n} \bar{\kappa}_{j,ik} = 0 \;\; \forall\, i,k.$$

Homogeneity and symmetry require $\sum_{k=1}^{n} \bar{\pi}_{jk} = 0\ \forall j$ and $\bar{\pi}_{jk} = \bar{\pi}_{kj}\ \forall\, k, j$. Finally, the negativity condition holds if, for any $\alpha \neq 0$, $\alpha' \bar{\Pi} \alpha \leq 0$, where $\bar{\Pi} = [\bar{\pi}_{jk}]$. Hence, $\bar{\Pi}$ must be a negative semi-definite matrix.[2] If all $\bar{\kappa}_{j,\, ik}$ are zero, nutrients have no impact on consumers' utility. Since the basic GADS model results from imposing such restrictions, a likelihood ratio test can be constructed to test whether nutrients have an impact on utility.

Data

The Spanish Quarterly Household National Survey collects information from a stratified random sample of 3200 households each quarter. The main information collected relates to expenditure and the quantities consumed of different food products. Additional information on a limited number of household characteristics is collected, including the level of education and the main activity of the head of the household, household income, household size, age and sex of family members, and town size. Information is collected from each household during a week within each quarter. Theoretically, one household remains in the survey for eight quarters. However, in practice, only a few households stay in the sample for the maximum period. For this reason, we have included all households that participated during 1 complete year (1995) and we have aggregated their purchases in order to obtain yearly food consumption. The main advantage of this approach is that many zero responses due to infrequency of purchase are eliminated and more realistic consumption figures are obtained from each household.

Food products have been aggregated into six broad categories: (i) cereals and potatoes; (ii) meat; (iii) fish; (iv) fruits and vegetables; (v) dairy products and eggs; and (vi) oils, sugar and other food products. The following nutrients have been considered: (i) carbohydrates; (ii) lipids; (iii) proteins; (iv) fibre; (v) vitamins; and (vi) minerals. Cereals and potatoes are rich in carbohydrates, fibre and minerals. Meat is rich in proteins, lipids, vitamins and minerals. Fish is rich in proteins, vitamins and minerals (differences from meat are found in the lipid content). Fruits and vegetables are rich in carbohydrates, fibre, vitamins and minerals. Dairy products and eggs are rich in lipids, proteins and minerals. Finally, oils, sugar and other food products are rich in carbohydrates and lipids.

The nutrient content, b_{ij}, is defined as the average nutritional content per unit of food consumed: that is, they have been calculated by dividing the total amount of nutrient by the quantity of food consumed, making use of nutrient conversion tables elaborated by Andújar *et al.* (1983). As the composition of each food group within each household is different, the nutrient content also varies from household to household.

Finally, since prices were not recorded, unit values have been taken as proxies of the market prices after adjusting them for quality effects. This adjustment comes out of the estimation of a hedonic price equation, which relates price to certain household sociodemographic variables in order to approximate the household selection of quality. Cowling and Raynor (1970) and Gao *et al.* (1997), among others, have used similar price adjustments.

Model Estimation and Calculated Elasticities

The basic GADS model without nutrient variables and the model in Equation 9.18 are estimated, using full-information maximum likelihood (FIML). Weak separability of preferences, homogeneity and symmetry are imposed. The equation for the group oils, sugar and other foods is deleted in order to avoid a singularity of the variance–covariance matrix due to the adding-up restriction. The likelihood ratio test statistics for model selection between basic GADS and Equation 9.18 take a value of 1862.4. This clearly surpasses the critical value at the 5% level of significance ($\chi^2(95) = 118.75$). Therefore, the nutrient composition of food is relevant in the decision-making process and should be incorporated in food-demand analyses. Thus, from now on, we shall concentrate only on results obtained from the model that includes nutrients. The negativity condition is analysed by calculating the eigenvalues of matrix $\bar{\Pi}$ in Equation 9.18. Results indicate that this condition is satisfied, since the calculated eigenvalues are: 0, −0.021, −0.028, −0.083, −0.115 and −0.150. Estimated parameters are shown in Table 9.4.

The only socio-economic variables that significantly affect consumption are: (i) the level of education of the household head (unschooled, primary school, secondary school and post-secondary degree); (ii) the town size (less than 10,000 inhabitants, between 10,001 and 50,000, between 50,001 and 500,000, and more than 500,000 inhabitants); and finally, (iii) the household composition, measured by the percentage of household members over 14 years of age.

Most of the estimated parameters are significant at the 5% level of significance and the goodness of fit of each equation is high, given the

Table 9.4. Parameter estimates and goodness of fit.[a]

Parameter	Cereals and potatoes	Meat	Fish	Fruits and vegetables	Dairy products and eggs
a_j^{**}	−3.208*	3.036*	2.395*	−2.471*	−1.295*
ρ_j, primary school	−0.026*	0.009	0.011*	0.002	0.003
ρ_j, secondary school	−0.035*	0.012*	0.004	0.002	−0.003
ρ_j, post-secondary degree	−0.069*	0.018*	0.004	0.011*	0.020*
ρ_j, towns with less than 10,000 inhabitants	0.043*	0.007	−0.013*	−0.040*	0.006
ρ_j, towns with between 10,001 and 50,000 inhabitants	0.048*	−0.002	−0.009*	−0.044*	0.007
ρ_j, towns with between 50,001 and 500,000 inhabitants	0.022*	0.004	0.002	−0.028*	0.002
ρ_j, percentage of members over 14 years old	0.005	−0.010	0.003	0.084*	−0.035*
θ_j	0.106*	0.197*	0.100*	0.195*	0.103*
$\bar{\pi}_{j1}$	−0.076*				
$\bar{\pi}_{j2}$	0.034*	−0.109*			
$\bar{\pi}_{j3}$	−0.006	0.003	−0.024*		
$\bar{\pi}_{j4}$	0.022*	0.039*	0.012*	−0.104*	
$\bar{\pi}_{j5}$	0.034*	0.029*	0.000	0.004	−0.058*
$\bar{\kappa}_j$, carbohydrates, cereals and potatoes	−0.029	0.034	0.987*	−0.007	0.222
$\bar{\kappa}_j$, fibre, cereals and potatoes	1.518*	0.254	0.205*	−1.296*	0.775*
$\bar{\kappa}_j$, minerals, cereals and potatoes	0.718*	−1.186*	−0.860*	0.051	−0.280*
$\bar{\kappa}_j$, lipids, meat	0.661*	−1.362*	−0.217*	0.532*	0.280*
$\bar{\kappa}_j$, proteins, meat	0.555*	1.318*	−0.980*	0.629*	−0.105
$\bar{\kappa}_j$, vitamins, meat	−1.078*	1.079*	0.190	−0.764*	−0.385*
$\bar{\kappa}_j$, minerals, meat	−0.673*	1.058*	0.647*	−1.015*	−0.472*
$\bar{\kappa}_j$, proteins, fish	0.574*	−0.674*	1.416*	0.790*	−0.216
$\bar{\kappa}_j$, vitamins, fish	−0.023	−0.026	0.264*	−0.046	−0.016
$\bar{\kappa}_j$, minerals, fish	−0.017	0.026	0.028	−0.033	0.018
$\bar{\kappa}_j$, carbohydrates, fruits and vegetables	0.000	−0.128*	−0.129*	0.176*	−0.065*
$\bar{\kappa}_j$, fibre, fruits and vegetables	0.134*	0.258*	0.133*	0.500*	0.051
$\bar{\kappa}_j$, vitamins, fruits and vegetables	0.055*	−0.095*	−0.091	0.035	−0.044
$\bar{\kappa}_j$, minerals, fruits and vegetables	−0.110*	−0.039	0.009	0.175*	0.007
$\bar{\kappa}_j$, lipids, dairy products and eggs	1.216*	0.923*	0.981	−0.537*	−2.392*
$\bar{\kappa}_j$, proteins, dairy products and eggs	−0.871*	−0.655*	−0.707	0.475*	1.054*
$\bar{\kappa}_j$, minerals, dairy products and eggs	0.210	0.134	0.135	−0.027	0.751*
$\bar{\kappa}_j$, carbohydrates, oils, sugar and others	0.005	0.001	0.003	−0.001	0.011*
$\bar{\kappa}_j$, lipids, oils, sugar and others	0.003	0.013*	0.011*	0.015*	0.014*
Goodness of fit (R^2)	0.642	0.553	0.573	0.602	0.620

[a] The reference households are those headed by an unschooled person and living in a large city (more than 500,000 inhabitants).

* Indicates that parameters are statistically significant at the 5% level, using White's (1980) heteroscedasticity-consistent covariance matrix.

cross-sectional nature of the data. As expected, the signs of all expenditure parameters are positive and the signs of the own-price parameters are negative. Finally, most cross-price parameters are positive, while nutrient-intake parameters do not have a very clear pattern.

As shown in Table 9.4, the level of education has a significant influence on the consumption of the products analysed. Households headed by an unschooled person (the reference group) have the highest consumption of cereals and potatoes but the lowest of meat, fish, and fruits and vegetables. As regards the town size, results indicate that those living in the biggest cities consume more fish and fruits and vegetables, while their consumption of cereals and potatoes is considerably lower. Finally, an increase in the share of household members over 14 years of age has a positive influence on fruit and vegetable consumption but a negative influence on the consumption of dairy products and eggs.

In food-demand analyses, the most interesting results are the demand elasticities. One of the main features of the estimated model is that it enables us to calculate the price and expenditure elasticities, while the nutritional composition of food is also included. The significance of the elasticities can be estimated using standard errors calculated by mathematical approximation.[3]

The expenditure and price elasticities calculated from the model (Equation 9.18) are shown in Table 9.5. In the second column, the average expenditure share of each product is presented. Note that the own-price elasticities are uncompensated while the cross-price elasticities are compensated elasticities. For comparison purposes, the expenditure and price elasticities calculated from parameter estimates of the basic GADS (without nutrients) are shown in Table 9.6. Note again that these are uncompensated own-price and compensated cross-price elasticities.

With respect to expenditure elasticities, we observe that, when nutrients are included, elasticities for healthy products, such as meat, fish, and fruits and vegetables, increase while those for the less healthy products, such as dairy products and eggs, decrease considerably: that is, when nutrients are included, consumers allocate a higher proportion of their total food-expenditure increases to products that provide healthier nutrients and a lower proportion to less healthy alternatives. Furthermore, results indicate that all products, except for the group of oils, sugar and other foods, can be considered necessities in relation to the total food expenditure.

Except for the group dairy products and eggs, uncompensated own-price elasticities are lower when nutrients are included. Again, these results are consistent with expectations, since the consumption of healthy products is less sensitive to own-price changes than in the case of the less healthy products. The demand for all products is inelastic with respect to price.

The two most significant differences when comparing the compensated cross-price elasticities in Tables 9.5 and 9.6 are the following. First, while fish and cereals and potatoes are complementary

Table 9.5. Expenditure shares and expenditure and price elasticities in model including nutrients.[a]

	Expenditure		Price elasticities					
Food category	Shares (%)	Elasticities	1	2	3	4	5	6
Cereals and potatoes (1)	17.0	0.626*	−0.552*	0.198*	−0.036	0.128*	0.199*	−0.042
Meat (2)	21.8	0.904*	0.155*	−0.697*	0.016	0.180*	0.132*	0.017
Fish (3)	11.6	0.867*	−0.054	0.029	−0.305*	0.108*	0.001	0.119*
Fruits and vegetables (4)	20.3	0.961*	0.107*	0.193*	0.061*	−0.706*	0.021	0.128*
Dairy products and eggs (5)	13.0	0.797*	0.261*	0.222*	0.001	0.032	−0.552*	−0.067*
Oils, sugar and other foods (6)	16.4	1.818*	−0.043	0.023	0.084*	0.159*	−0.053*	−0.468*

[a] Uncompensated own-price elasticities and compensated cross-price elasticities calculated at mean values. * Indicates that elasticities are statistically significant at the 5% level.

Table 9.6. Expenditure and price elasticities in model not including nutrients.[a]

	Expenditure	Price elasticities					
Food category	elasticities	1	2	3	4	5	6
Cereals and potatoes (1)	0.665*	−0.831*	0.111*	−0.055*	0.238*	0.383*	0.041
Meat (2)	0.881*	0.086*	−0.961*	0.039	0.078*	0.145*	0.021
Fish (3)	0.774*	−0.081*	0.074	−0.675*	0.000	0.053	0.140*
Fruits and vegetables (4)	0.951*	0.199*	0.084*	0.001	−0.882*	0.231*	0.174*
Dairy products and eggs (5)	0.923*	0.502*	0.244*	0.047	0.363*	−0.427*	0.152*
Oils, sugar and others foods (6)	1.786*	0.043	0.027	0.099*	0.216*	0.120*	−0.798*

[a] Uncompensated own-price elasticities and compensated cross-price elasticities calculated at mean values. * Indicates that elasticities are statistically significant at the 5% level.

when nutrients are not included, they turn out to be independent when nutrients are taken into account. Secondly, dairy products and eggs are substitutes with respect to fruits and vegetables if nutrients are not included, while they are independent when the model with nutrients is used. Neglecting the magnitudes, most of the elasticities are positive and significant and we can conclude that there is a general substitutability relationship among food products in Spain.

In Table 9.7, nutrient elasticities are calculated at mean values. In general terms, a distinction can be made between own-nutrient and cross-nutrient demand elasticities. The first set, printed in bold type, shows the percentage change in the demand for a food item due to a 1% change in the particular nutrient in that food item. The cross-nutrient demand elasticities show the percentage change in the demand for a food item due to a 1% change in the nutrients found in other food groups.

The own-nutrient elasticities suggest that the demand for cereals and potatoes is positively affected by the fibre and mineral content, while the effect of carbohydrate content is negative. The amount of proteins, minerals and vitamins in meat has a positive impact on its demand, while the lipid content has a negative influence. For fish, own-nutrient elasticities indicate that the demand for fish is positively influenced by its own content of proteins and vitamins. The demand for fruits and vegetables is positively affected by the content of all nutrients considered. Dairy products and eggs are positively affected by their own content of proteins and minerals, while lipids exert a negative influence. Lastly, the demands for oils, sugar and other products are negatively affected by their own content of carbohydrates and lipids.

The typical relationship in cross-nutrient elasticities reveals that increases (decreases) in a valuable nutrient provided by a product negatively (positively) affect the demand for other products providing the same nutrient. For instance, an increase (decrease) in the fibre content of cereals and potatoes produces a significant decrease (increase) in the demand for fruits and vegetables. The same type of reaction occurs in relation to the protein content of meat and the demand for fish; the vitamin content of fish and the demand for the group oils and other foods; the lipid content of milk and eggs and the demand for meat and fish; or the carbohydrate and lipid contents of oils, sugar and other foods and the demand for the remaining groups. The typical relationship is not always observed, since some food items are used in specific combinations to produce different dishes and the upper limits for nutrients are difficult to set. In other words, cross-nutrient elasticities across food groups can be positive or negative depending on how each food item is used. Therefore, the interpretation of the cross-nutrient elasticities is extremely difficult.

Table 9.7. Nutrient elasticities.

Food/nutrient	Cereals and potatoes	Meat	Fish	Fruits and vegetables	Dairy products and eggs	Oils and other foods
Cereals and potatoes						
Carbohydrates	−**0.171***	0.156	8.539*	−0.033	1.716	−7.369*
Fibre	**8.931***	1.166*	1.770*	−6.375*	5.984*	−8.890*
Minerals	**4.226***	−5.445*	−7.444*	0.249	−2.162*	9.508*
Meat						
Lipids	3.889*	−**6.255***	−1.877*	2.617*	2.163*	0.648
Proteins	3.263*	**6.051***	−8.481*	3.093*	−0.813*	−8.644*
Vitamins	−6.345*	**4.955***	1.640*	−3.761*	−2.970*	5.854*
Minerals	−3.959*	**4.857***	5.598*	−4.991*	−3.643*	2.775*
Fish						
Proteins	3.375*	−3.094*	**12.250***	3.884*	−1.666*	−11.531*
Vitamins	−0.133*	−0.121	**2.281***	−0.224*	−0.125	−0.932*
Minerals	−0.101	0.120	**0.240**	−0.164	0.135*	−0.127
Fruits and vegetables						
Carbohydrates	−0.002	−0.586*	−1.119*	**0.864***	−0.501*	0.896*
Fibre	0.790*	1.183*	1.152*	**2.460***	0.392*	−6.568*
Vitamins	0.322*	−0.437*	−0.791*	**0.171***	−0.343*	0.864*
Minerals	−0.645*	−0.180	0.078	**0.861***	0.057	−0.260
Dairy products and eggs						
Lipids	7.153*	4.237*	8.486*	−2.640*	−**18.458***	−1.166*
Proteins	−5.126*	−3.007*	−6.116*	2.338*	**8.137***	4.293*
Minerals	1.233*	0.614	1.170	−0.131	**5.797***	−7.345*
Oils, sugar and other foods						
Carbohydrates	0.030*	0.003	0.023*	−0.004	0.081*	−0.110*
Lipids	0.019	0.062*	0.099*	0.073*	0.105*	−**0.345***

* Indicates that elasticities are statistically significant at the 5% level.

Conclusions

In this chapter, a new approach for analysing the demand for food is proposed in order to take into account consumers' increasing concern about the importance of an adequate diet in terms of nutrient intake. The consumer's utility is defined as a function of nutrient intake instead of food quantities. Nutrient restrictions are also included in the consumer optimization problem, together with the traditional budget restriction. Using the GADS model, a demand system is specified and accurate expenditure, price and nutrient elasticities are derived. Only the most relevant socio-economic variables are included.

The calculated expenditure and price elasticities are consistent with a priori expectations. When considering nutrients, consumers allocate a larger proportion of their total food-expenditure increases to those products that provide the healthiest nutrients. Moreover, the consumption of healthy products is less sensitive to own-price changes. Finally, as regards nutrient elasticities, there is some very interesting new information offered by the model about whether some of the nutrients provided by a product exert a significant influence on its own demand.

Further research can be aimed in three main directions. First, given available data, more disaggregated food categories can be analysed. Secondly, comparative studies with other countries are of interest. Thirdly, given a panel data set, we can study the evolution of food demand in order to anticipate future market conditions.

Notes

[1] The exchange rate used is US$1 = 149.40 pesetas (the Spanish currency).
[2] Since the rank of the Π matrix is $(n - 1)$ in the demand system, the negative semi-definite condition will be satisfied if one eigenvalue is zero and the rest are negative. If negativity is not maintained, it can be imposed either through Bayesian techniques (Chalfant *et al.*, 1991; Hasegawa *et al.*, 1999) or through the classical framework (e.g. Barten and Geyskens, 1975; Moschini, 1998; Ryan and Wales, 1998).
[3] If we denote the parameters vector of the model as ϑ, $\hat{\vartheta}$ as the maximum likelihood estimator of ϑ with variance–covariance matrix $\hat{\Sigma}$, $\hat{E}_f = \hat{E}_f(\hat{\vartheta})$ as the fth elasticity, expressed as a function of model parameters and J_f as the Jacobian of the transformation from $\hat{\vartheta}$ to $\hat{E}_f$, then the variance of $\hat{E}_f$ can be approximated by the expression Var $(\hat{E}_f) \approx J'_f \hat{\Sigma} J_f$, where J_f can be evaluated at the maximum likelihood estimates and at the sample mean of exogenous variables.

References

Adrian, J. and Daniel, R. (1976) Impact of the socioeconomic factors on consumption of selected food nutrients in the United States. *American Journal of Agricultural Economics* 58, 31–38.

Andújar, M.M., Moreiras-Varela, O. and Gil, F. (1983) Tablas de composición de alimentos. Unpublished manuscript, Consejo Superior de Investigaciones Científicas, Madrid, Spain.

Angulo, A.M., Gil, J.M. and Gracia, A. (1997) A test of differences in food demand among European consumers: a dynamic approach. In: Wierenga, B., van Tilburg, A., Grunert, K.G., Steenkamp, E.M. and Wedel, M. (eds) *Agricultural Marketing and Consumer Behavior in a Changing World*. Kluwer Academic Publishers, Boston, Massachusetts, pp. 275–294.

Angulo, A.M., Gil, J.M. and Gracia, A. (2001) Calorie intake and income elasticities in EU countries: a convergence analysis using cointegration. *Papers in Regional Science* 80, 165–187.

Barten, A.P. and Geyskens, E. (1975) The negativity condition in consumer demand. *European Economic Review* 6, 227–260.

Bewley, R.A. (1986) *Allocation Models: Specification, Estimation and Applications*. Ballinguer, Cambridge, Massachusetts, 368 pp.

Brown, D. and Schrader, L.F. (1990) Cholesterol information and shell eggs consumption. *American Journal of Agricultural Economics* 72, 548–555.

Chalfant, J.A., Gray, R.S. and White, K.J. (1991) Evaluating prior beliefs in a demand system: the case of meat demand in Canada. *American Journal of Agricultural Economics* 73, 476–490.

Chern, W.S. and Zuo, J. (1995) Alternative measures of changing consumer information on fat and cholesterol. Paper presented at the American Agricultural Economics Association's Annual Meeting, Indianapolis, Indiana.

Chesher, A. (1998) Individual demands from household aggregates: time and age variation in the composition of diet. *Journal of Applied Econometrics* 13, 505–524.

Chung, C.F. (1994) A cross-section demand analysis of Spanish provincial food consumption. *American Journal of Agricultural Economics* 76, 513–521.

Chung, C.F. and López, E. (1988) A regional analysis of food consumption in Spain. *Economics Letters* 26, 209–213.

Cowling, R.G.D. and Raynor, A.J. (1970) Price, quality and market shares. *Journal of Political Economy* 78, 1292–1309.

Gao, X.M., Richards, T.J. and Kagan, A. (1997) A latent variable model of consumer taste determination and taste change for complex carbohydrates. *Applied Economics* 29, 1643–1654.

Gil, J.M., Gracia, A. and Pérez, L. (1995) Food consumption and economic development in the European Union. *European Review of Agricultural Economics* 22, 385–399.

Gracia, A. and Albisu, L.M. (1998) The demand for meat and fish in Spain: urban and rural areas. *Agricultural Economics* 19, 359–366.

Gracia, A., Gil, J.M. and Angulo, A.M. (1998) Spanish food demand: a dynamic approach. *Applied Economics* 30, 1399–1405.

Hasegawa, H., Kozumi, H. and Hashimoto, N. (1999) Testing for negativity in a demand system: a Bayesian approach. *Empirical Economics* 24, 211–223.

Huang, K.S. (1999) Effects of food prices and consumer income on nutrient availability. *Applied Economics* 31, 367–380.

Jensen, H.H., Kesavan, T. and Johnson, S.R. (1992) Measuring the impact of health awareness on food demand. *Review of Agricultural Economics* 14, 299–312.

Kantor, L.S. (1999) A comparison for the U.S. food supply with the food guide pyramid recommendations. In: Frazao, E. (ed.) *America's Eating Habits: Changes and Consequences*. Agricultural Information Bulletin No. 750, Economic Research Service, USDA, Washington, DC, pp. 71–95.

Kim, S. and Chern, W.S. (1999) Alternative measures of health information and demand for fats and oils in Japan. *Journal of Consumer Affairs* 33, 92–109.

Kinnucan, H.W., Xiao, H., Hsia, C.J. and Jackson, J.D. (1997) Effect of health information and generic advertising on U.S. meat demand. *American Journal of Agricultural Economics* 79, 13–23.

Laajimi, A. and Albisu, L.M. (1997) La demande de viandes et de poissons en Espagne: une analyse micro-économique. *Cahiers d'Economie et Sociologie Rurales* 42–43, 71–91.

Lancaster, K. (1966) A new approach to consumer theory. *Journal of Political Economy* 74, 132–157.

Lancaster, K. (1971) *Consumer Demand: a New Approach*. Columbia University Press, New York, 177 pp.

Lau, L. (1982) A note on the fundamental theorem of exact aggregation. *Economics Letters* 9, 119–126.

Manrique, J. and Jensen, H.H. (1997) Spanish household demand for convenience meat products. *Agribusiness* 13, 579–586.

Manrique, J. and Jensen, H.H. (1998) Working women and expenditures on food away-from-home and at-home in Spain. *Journal of Agricultural Economics* 49, 321–333.

Ministerio de Agricultura, Pesca y Alimentación (several years) *La alimentación en España*. MAPA, Madrid, Spain.

Molina, J.A. (1994) Food demand in Spain: an application of the almost ideal system. *Journal of Agricultural Economics* 45, 252–258.

Moschini, G. (1998) The semiflexible almost ideal demand system. *European Economic Review* 42, 349–364.

Nayga, R.M. (1994) Effects of socioeconomic and demographic factors on consumption of selected food nutrients. *Agricultural and Resource Economics Review*, 171–182.

Nayga, R.M. (2000) Schooling, health knowledge and obesity. *Applied Economics* 32, 815–822.

Orcutt, G.H., Watts, H.W. and Edwards, J.B. (1968) Data aggregation and information loss. *American Economic Review* 58, 773–787.

Pollak, R.A. and Wales, T.J. (1981) Demographic variables in demand analysis. *Econometrica* 49, 1533–1558.

Pudney, S. (1981) Instrumental variable estimation of a characteristic model of demand. *Review of Economic Studies* 48, 417–433.

Ramezani, C.A., Rose, D. and Murphy, S. (1995) Aggregation, flexible forms, and estimation of food consumption parameters. *American Journal of Agricultural Economics* 77, 525–532.

Rosen, S. (1974) Hedonic prices and implicit markets: product differentiation in perfect competition. *Journal of Political Economy* 82, 34–55.

Ryan, D.L. and Wales, T.J. (1998) A simple method for imposing local curvature in some flexible consumer demand systems. *Journal of Business and Economic Statistics* 16, 331–338.

Subramanian, S. and Deaton, A. (1996) The demand for food and calories. *Journal of Political Economy* 104, 133–162.

Variyam, J.N., Blaylock, J., Smallwood, D. and Basiotis, P.P. (1998) USDA's healthy eating index and nutrition information. *Technical Bulletin* 1866, 1–21.

White, H. (1980) A heteroscedasticity-consistent covariance matrix estimator and a direct test for heteroscedasticity. *Econometrica* 48, 817–838.

Xiao, Y. and Taylor, J.E. (1995) The impact of income growth on farm household nutrient intake: a case-study of a prosperous rural area in Northern China. *Economic Development and Cultural Change* 43, 805–819.

Health Information and Unstable Effects from Autocorrelation

10

KYRRE RICKERTSEN AND
DADI KRISTOFERSSON
Agricultural University of Norway, Ås, Norway

Introduction

Rickertsen and von Cramon-Taubadel (Chapter 3, this volume) reached no clear conclusions regarding the effects of health information on demand. The effects varied from one European country to another, were often insignificant and were sometimes unexpected. The results are, however, conditional on the joint hypothesis of no autocorrelation, as well as other aspects of model specification (Alston and Chalfant, 1991). In the presence of autocorrelation, the point estimates remain unbiased but the variance and the t statistics are invalid. Autocorrelation is a serious problem in the estimation of the country-specific models. Since there is no generally 'best' correction method, different methods for correcting for autocorrelation were applied in the various countries involved. The purpose of this chapter is to investigate the fragility in estimated health-information effects for choice of correction mechanism using Norwegian data.

When testing for serial correlation, it is common to test within each equation by using, for example, a Breusch–Godfrey test. However, equation-by-equation tests do not take into account interactions between the equations in the system; hence a system test is preferred (Godfrey, 1988). The use of system tests has also been advocated in, for example, McGuirk *et al.* (1995) and Edgerton *et al.* (1996). We perform both types of tests.

Trends in Consumption

The consumption of major food items in Norway is reported in Table 10.1. Disappearance data are constructed from supply and utilization

Table 10.1. Annual per capita consumption (kg of food and l of milk) in Norway. (From: Norwegian Agricultural Economics Research Institute, Statistics Norway, Food and Agriculture Organization, Agricultural Budget Commission and the Norwegian Egg Producers Association.)

	Disappearance data					Survey data	
Food item	1980	1985	1990	1995	1998	1980	1998
Meat	52.1	49.9	50.7	58.5	60.3	46.7	46.4
Beef	19.8	18.0	18.0	20.0	20.7	14.1	12.9
Pork	21.7	20.2	19.7	23.3	23.8	11.0	10.6
Lamb	5.4	5.9	5.7	6.2	5.5	4.1	2.5
Poultry	2.8	3.1	4.7	6.5	7.7	2.0	5.0
Other meats	2.5	2.6	2.5	2.5	2.6	15.5	15.4
Fish[a]	22.0	21.8	22.6	24.9	25.2	20.0	18.3
Eggs	10.8[b]	11.8[c]	11.0	10.8	10.3	9.5	7.4
Cheese	11.8	12.0	12.8	13.2	14.2	10.2	10.6
Fluid milk	165.5	157.9	151.2	133.5	128.6	147.8	101.2
Whole	142.6	106.0	44.7	32.9	32.6	124.3	28.2
Low-fat	0.0	27.2	79.7	78.4	72.0	0.0	55.1
Non-fat	22.9	24.7	26.8	22.1	24.0	23.5	17.9
Vegetables and fruits	193.4[b]	206.1[c]	196.7	211.0	212.9	155.6	160.2
Vegetables	44.7[b]	49.3[c]	55.0	60.0	60.7	33.7	40.0
Fruits	74.9[b]	78.5[c]	71.7	83.0	77.1	63.8	79.5
Potatoes	73.8[b]	78.3[c]	70.0	68.0	75.1	58.1	40.7
Grains and cereals	105.7	102.7	113.2	115.0	122.4	75.4	74.4
Sugar	42.3	42.4	41.3	44.3	45.3	17.6	9.2
Fats and oils	19.5	17.6	16.3	15.6	16.6	17.6	12.3
Butter	5.6	4.6	4.0	4.2	4.0	3.4	1.1
Oils and margarines	13.9	13.0	12.3	11.4	12.6	14.2	11.2
Other foods	[d]					11.5	19.1

[a] Extraction rate is set to 50% in the disappearance data.
[b] Data for 1979.
[c] Data for 1984.
[d] Blanks indicate data not available.

figures and cover consumption of food for use at home, away from home and as an ingredient in other products. Household-survey data include only quantities of food purchased for use at home. The consumption of most foods is higher according to disappearance than according to household-expenditure-survey data. The consumption is about a third greater for most foods, and the differences are even higher for sugar and cereals. Loss and waste in processing and distribution and consumption away from home cause most of the discrepancy. The difference for sugar is mainly caused by the use of sugar as an ingredient in soft drinks (not included in the table) and processed foods (included in the group 'other foods' in the survey data).

The two data types also show different trends in consumption of some foods. According to the disappearance data, the consumption of meat and fish increased, while it was relatively stable according to the survey data. These differences are mainly due to increased consumption away from home and increased sales of meats and fish trimmed for skin, bones and fats. Therefore, the calculated extraction rate for fish is 50% in the disappearance data. Furthermore, a substantial home production of fish is not included, making the consumption figures a low estimate. The different trend in sugar consumption may be explained by increased sales of processed foods and soft drinks containing sugar.

The fluid-milk consumption declined and the structure of consumption changed after the introduction of low-fat milk in 1984, with trends away from whole milk and butter and towards reduced-fat milk and cheese. The consumption of meat and vegetables increased, while the consumption of fats and oils decreased. Compared with other countries, for example, Germany or France, the Norwegian diet includes less meat, eggs, cheese, fruits, vegetables and fats, and more milk and fish.

Norwegian food-demand studies published in English include Edgerton *et al.* (1996) and Rickertsen (1998a). Meat demand is studied in Rickertsen (1996), Gustavsen (1999) and Rickertsen *et al.* (2003), while the demand for vegetables is analysed in Rickertsen *et al.* (1995) and Rickertsen (1998b).

The Almost Ideal Demand System Incorporating Health Information

The linear approximate almost ideal demand system (LA/AIDS) of Deaton and Muellbauer (1980) is used as in Rickertsen and von Cramon-Taubadel (Chapter 3, this volume). The expenditure share for the ith good in period t, w_{it}, is given by:

$$w_{it} = \alpha_i + \sum_{j=1}^{n} \gamma_{ij} \ln p_{jt} + \beta_i \left(\ln x_t - \ln P_t \right) \qquad (10.1)$$

where $\ln P_t$ is Stone's price index:

$$\ln P_t = \sum_{k=1}^{n} w_{kt} \ln p_{kt} \tag{10.2}$$

and p_j is the per unit price of the jth good, while x denotes per capita total expenditure. All prices are normalized to 1 in 1995 to calculate a corrected index, as discussed in Moschini (1995).

The effects of health information are included through a modification of the constant term, α_i, in Equation 10.1. There may exist a lag from the time a scientific study is published until its content reaches the general public and we use the free distributed-lag structure:

$$\alpha_i = \alpha_{i0} + \lambda_i \sum_{h=0}^{2} \omega_h GIA_{t-h} \tag{10.3}$$

where GIA is the adjusted global index described in Rickertsen and von Cramon-Taubadel (Chapter 3, this volume). The lag weights are restricted to sum to 1 ($\Sigma_h \omega_h = 1$, where h denotes the number of lags) and an article is assumed to have an effect for a maximum of 3 years. Adding up, homogeneity and symmetry restrictions are imposed and Equations 3.9–3.11 are used for calculating the elasticities.

The data and the two-stage model discussed in Rickertsen and von Cramon-Taubadel (Chapter 3, this volume) are also used. In the first stage, the demand for five groups of food (meat and fish; dairy products and eggs; oils and fats; cereals, fruits, vegetables and potatoes; and other foods and beverages) and one non-food group is estimated. The second stage includes four groups of meats (beef and veal, pork, lamb and poultry) and fish and fish products. National account data provided by Statistics Norway are used to estimate stage one. Stage two is estimated using disappearance data collected by the Norwegian Agricultural Economics Research Institute and consumer price-index data. The analysis covers the period 1966–1995.

Corrections and Tests for Autocorrelation

We apply six corrections for autocorrelation. First, Berndt and Savin (1975) suggested a general correction for autocorrelation in singular demand systems. A similar correction is used in the Scottish model (Rickertsen and von Cramon-Taubadel, Chapter 3, and Santarossa and Mainland, Chapter 13, this volume) and has previously been used in, for example, Moschini and Moro (1994) and Piggott *et al.* (1996). The vector of errors in the share equations are assumed to follow the scheme $\mathbf{u_t} = \mathbf{Ru_{t-1}} + \mathbf{v_t}$, where the $\mathbf{v_t}$ terms are independent $N(0, \Sigma)$ random vectors. The $\mathbf{R}$ matrix is restricted such that $\boldsymbol{\iota}'\mathbf{R^*} = 0$ where $\mathbf{R^*}$ is an n

by $(n - 1)$ matrix with elements $\rho^*_{ij} = \rho_{ij} - \rho_{in}$. The constant term in Equation 10.3 is modified to:

$$\alpha_{i0} = \alpha_{i00} + (\rho_{i1} - \rho_{in})u_{1,\,t-1} + \ldots + (\rho_{i,n-1} - \rho_{in})u_{n-1,\,t-1} + \upsilon_{i,t} \tag{10.4}$$

This model is denoted by ρ_{ij} in the tables.

Secondly, a more restrictive version of the Berndt–Savin correction is estimated. In this version, the **R** matrix is restricted to have identical diagonal parameters and zero off-diagonal parameters (e.g. Cashin, 1991). The correction modifies the constant term in Equation 10.3 such that:

$$\alpha_{i0} = \alpha_{i00} + \rho u_{i,\,t-1} \tag{10.5}$$

This model is referred to as ρ in the tables.

Thirdly, the Bewley transformation, advocated by Wickens and Breusch (1988), is used in the German model (Rickertsen and von Cramon-Taubadel, Chapter 3, and Wildner and von Cramon-Taubadel, Chapter 7, this volume) and has been used in, for example, Kesavan *et al.* (1993). It can be derived from an autoregressive distributed-lag model. We assume that there are no lagged effects of the regressors and only one lag is included for the differenced lagged expenditure shares, such that the constant term in Equation 10.3 is modified to:

$$\alpha_{i0} = \alpha_{i00} + \sum_{j=1}^{n} \theta_{ij} \Delta w_{j,\,t} \tag{10.6}$$

where $\Sigma_j \theta_{ij} = 0 \;\forall\, i$ is imposed for identification. The estimated parameters are directly interpreted as long-run parameters. Since the current period's expenditure shares are included as regressors, the system is estimated by three-stage least squares (3SLS), using the vector of $w_{j,\,t-1}$ and other predetermined variables as instruments for $\Delta w_{j,\,t}$. This model is denoted by Δw in the tables.

Fourthly, the partial adjustment model used in the Spanish model (Rickertsen and von Cramon-Taubadel, Chapter 3, this volume) and previously in, for example, Alessie and Kapteyn (1991) and Edgerton *et al.* (1996) modifies Equation 10.3 such that:

$$\alpha_{i0} = \alpha_{i00} + \sum_{j} \theta_{ij} w_{j,\,t-1} \tag{10.7}$$

where $\Sigma_j \theta_{ij} = 0 \;\forall\, i$ is imposed for identification. This model has behavioural implications. When the vector of lagged budget shares is substituted into the expenditure function, it can be interpreted as the cost of habits. This model is referred to as θ_{ij} in the tables.

Fifthly, a restricted version of the partial adjustment model is the diagonal partial adjustment model, where all the diagonal parameters are assumed identical and all off-diagonal parameters are set to zero (e.g. Yen and Chern, 1992). The constant term in Equation 10.3 is modified such that:

$$\alpha_{i0} = \alpha_{i00} + \theta w_{i,t-1} \tag{10.8}$$

This model is denoted by θ in the tables. No extra restrictions are needed, but the adding-up restriction for the constant term is modified such that:

$$\sum_{j=1}^{n} \alpha_{j00} + \theta = 1 \tag{10.9}$$

Sixthly, autocorrelation may be present in the partial adjustment models above and the general Berndt–Savin correction can be applied to these models. This correction was used in the Norwegian model (Rickertsen and von Cramon-Taubadel, Chapter 3, this volume) and previously in, for example, Holt and Goodwin (1997). The correction modifies the constant term in Equation 10.3 such that:

$$\begin{aligned}\alpha_{i0} = \alpha_{i00} + \sum_{j} \theta_{ij} w_{j,t-1} + (\rho_{i1} - \rho_{in}) u_{1,t-1} + \ldots \\ + (\rho_{i,n-1} - \rho_{in}) u_{n-1,t-1} + \upsilon_{i,t}\end{aligned} \tag{10.10}$$

This model is referred to as ρ_{ij} & θ_{ij} in the tables.

First-order autocorrelation is tested equation by equation by using the Breusch–Godfrey score test (see, for example, Ruud, 2000: 464). This test is applicable in the presence of lagged endogenous variables. The test is strictly valid only in a single-equation framework and the results must be interpreted with some caution.

A system-wide Breusch–Godfrey test (Godfrey, 1988) is also used. In this test, the hypothesis that $\rho^*_{ij} = \rho_{ij} - \rho_{in} = 0$ for $i, j = 1, \ldots, n-1$ is tested in the system:

$$\begin{aligned}u_{i,t} = x_t'\beta + (\rho_{i1} - \rho_{in}) u_{1,t-1} + \ldots \\ + (\rho_{i,n-1} - \rho_{in}) u_{n-1,t-1} + \upsilon_{i,t}\end{aligned} \tag{10.11}$$

The singularity of the LA/AIDS model implies that only $n - 1$ errors can be used as dependent variables. This pattern of autocorrelation is removed by the general Berndt–Savin correction (ρ_{ij}) and the test is not applicable to the models in Equations 10.4 and 10.10.

The P values of tests for first-order autocorrelation are shown in Table 10.2. Autocorrelation is a severe problem in the static model, especially in stage one. All correction mechanisms result in an improvement concerning equation-by-equation autocorrelation. While the Bewley transformation removed most of the single-equation autocorrelation, the resulting health, own-price and total expenditure elasticities were frequently insignificant. Since 3SLS is used, the insignificant elasticities may be due to insufficient quality of the instruments. The least successful corrections for autocorrelation are the most restrictive ones (ρ and θ). This is hardly surprising since it is unlikely that the autocorrelation coefficient, ρ, or the partial adjustment

Table 10.2. *P* values of tests for first-order autocorrelation.

Model	Static	ρ	ρ_{ij}	Δw	θ	θ_{ij}	ρ_{ij} & θ_{ij}
Stage one							
Meat and fish	0.01	0.43	0.62	0.20	0.51	0.13	0.82
Dairy products and eggs	0.01	0.25	0.35	0.04	0.28	0.98	0.85
Carbohydrates	0.00	0.92	0.73	0.36	0.08	0.80	0.57
Fats and oils	0.02	0.95	0.22	0.63	0.54	0.68	0.96
Other foods	0.00	0.63	0.45	0.16	0.83	0.47	0.82
Other goods	0.00	0.62	0.66	0.77	0.40	0.75	0.57
System test	0.00	0.00		0.00	0.00	0.00	
Stage two							
Beef and veal	0.25	0.75	0.88	0.38	0.79	0.26	0.38
Pork	0.40	0.04	0.31	0.24	0.83	0.17	0.26
Lamb	0.26	0.01	0.44	0.19	0.01	0.03	0.53
Poultry	0.00	0.02	0.83	0.99	0.04	0.12	0.06
Fish	0.22	0.66	0.95	0.98	0.80	0.14	0.21
System test	0.00	0.00		0.01	0.00	0.00	

term, θ, is the same for every share equation. The general Berndt–Savin correction relaxes the restrictions that the off-diagonal elements are zero and the diagonal elements are all the same. Furthermore, it does not impose any arbitrary restrictions for identification ($\Sigma_j\theta_{ij} = 0 \ \forall \ i$) such as the general partial adjustment model. Consequently, it is not unexpected that equation-by-equation first-order autocorrelation is only completely removed by the general Berndt–Savin corrections, ρ_{ij} and ρ_{ij} & θ_{ij}. Moreover, with the exception of these models, system autocorrelation is present. Therefore, the general Berndt–Savin correction, either alone or in combination with the general partial adjustment model, is preferred.

The Effects of Health Information

Health-information elasticities for 1995 and *P* values for tests for no effects of health information are shown in Table 10.3. For comparison, we include results from all the models. The R^2 values for the partial adjustment model, estimated with the Berndt–Savin correction (ρ_{ij} & θ_{ij}), are also reported. As we can see, the R^2 values are high.

Conclusions regarding the total effects of health information are sensitive to the choice of the correction method for autocorrelation, and misleading conclusions can easily be made using an inadequately corrected model. Using the Bewley transformation (Δw) or the partial adjustment model (θ_{ij}), we could have concluded that there are no effects

Table 10.3. Health-information elasticities, R^2 values and tests for no information effects.[a]

Model	Static	ρ	ρ_{ij}	Δw	θ	θ_{ij}	ρ_{ij} & θ_{ij}	R^2
Stage one								
Meat and fish	**−0.31**	**−0.13**	**−0.12**	−0.62	0.01	0.02	0.03	0.99
Dairy products and eggs	−0.17	−0.14	−0.09	0.09	−0.01	−0.03	−0.05	0.98
Carbohydrates	**−0.36**	**−0.30**	**−0.27**	0.41	−0.11	0.00	−0.02	0.99
Fats and oils	**0.53**	0.21	**0.37**	−0.36	0.19	0.18	**0.67**	
Other foods	0.02	**0.09**	**0.11**	0.15	**0.08**	**0.12**	**0.10**	0.98
Other goods	**0.04**	0.02	0.01	0.00	0.00	**−0.01**	−0.01	0.99
No information effects[b]	0.00	0.00	0.00	0.02	0.00	0.00	0.00	
Stage two								
Beef and veal	0.03	0.04	0.07	−0.09	0.02	0.03	−0.02	0.83
Pork	**0.05**	0.04	0.02	−0.01	0.04	0.05	0.04	0.88
Lamb	−0.13	−0.10	**−0.13**	0.03	−0.09	−0.10	−0.09	0.87
Poultry	**0.63**	**0.64**	**0.65**	0.59	**0.50**	−0.03	**0.51**	0.97
Fish	**−0.16**	**−0.17**	**−0.18**	0.05	**−0.13**	**−0.08**	**−0.08**	
No information effects[b]	0.00	0.00	0.00	0.44	0.00	0.21	0.00	

[a] Elasticities that are significant at the 5% level are printed in bold type.
[b] *P* values.

of health information in stage two. However, in the two models without autocorrelation (ρ_{ij} and ρ_{ij} & θ_{ij}), we find effects of health information.

The next issue concerns the commodity-specific effects. The different correction methods yield significantly different own-health information elasticities, emphasizing the sensitivity of the results. For example, the Bewley transformation (Δw) results in no commodity-specific effects, but we find seven commodity-specific effects when the Berndt–Savin (ρ_{ij}) correction is used. The results demonstrate that conclusions regarding the effects of health information depend critically on the selected correction for autocorrelation. However, the directions of the significant effects are quite stable. When several models find a significant information effect, the effect goes in the same direction, except for the group 'other goods', for which there were significant positive as well as negative effects.

Previous studies of Norwegian food demand (Edgerton *et al.*, 1996; Rickertsen, 1998a) found habit persistence to be important, suggesting that the autocorrelation-corrected partial adjustment model (ρ_{ij} & θ_{ij}) may be the appropriate model for Norwegian data. The results from this model are also reported by Rickertsen and von Cramon-Taubadel (Chapter 3, this volume) and the estimated parameters are shown in the Appendix to this chapter. In this model, we find a negative effect on fish

Table 10.4. Uncompensated own-price elasticities.[a]

Model	Static	ρ	ρ_{ij}	Δw	θ	θ_{ij}	ρ_{ij} & θ_{ij}
Stage one							
Meat and fish	**−0.24**	**−0.59**	**−0.47**	1.46	**−1.08**	**−1.03**	**−0.97**
Dairy products and eggs	0.14	0.13	0.11	−0.25	**−0.56**	0.04	0.06
Carbohydrates	**−0.83**	**−1.18**	**−1.17**	4.61	**−1.22**	**−1.43**	**−1.37**
Fats and oils	0.17	0.44	0.22	−0.02	−0.22	−0.36	**−1.29**
Other foods	−0.14	**−0.18**	**−0.22**	0.11	**−0.51**	**−0.41**	**−0.58**
Other goods	**−0.98**	**−1.08**	**−1.12**	**−1.00**	**−1.03**	**−0.93**	**−0.94**
Stage two							
Beef and veal	**−0.51**	**−0.54**	**−0.46**	**−0.54**	**−0.68**	**−0.52**	**−0.70**
Pork	**−0.58**	**−0.57**	**−0.61**	**−0.59**	**−0.60**	**−0.66**	**−0.57**
Lamb	**−0.38**	−0.31	**−0.43**	−0.37	**−0.54**	−0.32	**−0.56**
Poultry	**−0.45**	**−0.43**	**−0.71**	**−0.92**	**−0.57**	**−0.76**	**−0.45**
Fish	**−0.53**	**−0.51**	**−0.55**	**−1.15**	**−0.61**	**−0.68**	**−0.91**

[a] Elasticities that are significant at the 5% level are printed in bold type.

demand and positive effects on the demand for poultry, the group fats and oils and the group other foods. The positive effect of health information on poultry demand is in line with previous studies. The negative effect on fish demand and the positive effect on the demand for fats and oils are more surprising but may be partly explained by the commodity aggregation. The fish group consists of fresh fish and some less healthy processed fish products. Furthermore, as mentioned above, the consumption data for fish are less reliable. The group fats and oils consists of less healthy butter and animal fats, which are high in saturated and monounsaturated fats, as well as more healthy vegetable oils, which are high in polyunsaturated fats. The importance of commodity aggregation for fats and oils is also supported by Nichèle (Chapter 8, this volume), who finds that the demand for oils gains from increases of health information at the expense of the demand for butter.

Price and Total Expenditure Effects

Uncompensated own-price elasticities for 1995 are presented in Table 10.4. The significant own-price elasticities are negative, as expected. However, especially in stage one, there is considerable variation in the numerical values for several of the goods. In the preferred model (ρ_{ij} & θ_{ij}), the own-price elasticities are significantly different from zero, except for those of dairy products.

Table 10.5. Total expenditure elasticities.[a]

Model	Static	ρ	ρ_{ij}	Δw	θ	θ_{ij}	ρ_{ij} & θ_{ij}
Stage one							
Meats and fish	**0.34**	**0.14**	0.07	0.77	**0.63**	**0.74**	**0.94**
Dairy products and eggs	**−0.30**	**−0.39**	**−0.44**	**−0.55**	**0.43**	**0.74**	0.71
Carbohydrates	**−0.42**	**−0.51**	**−0.65**	−1.07	**0.39**	**1.42**	**1.48**
Fats and oils	**−2.82**	**−2.52**	**−2.41**	−2.65	−0.55	−0.11	0.72
Other foods	**0.73**	**0.64**	**0.58**	**0.69**	**0.84**	**1.33**	**1.36**
Other goods	**1.21**	**1.24**	**1.26**	**1.23**	**1.10**	**0.97**	**0.95**
Stage two							
Beef and veal	**1.40**	**1.41**	**1.45**	**1.24**	**1.37**	**1.27**	**0.98**
Pork	**1.05**	**1.03**	**1.07**	**0.98**	**0.95**	**1.10**	**1.00**
Lamb	−0.19	−0.19	−0.02	0.43	0.18	0.08	0.40
Poultry	**1.90**	**2.05**	**1.68**	−0.98	**1.65**	0.43	**1.61**
Fish	**0.60**	**0.61**	**0.50**	**1.14**	**0.71**	**0.85**	**1.16**

[a] Elasticities that are significant at the 5% level are printed in bold type.

There is also variation in the expenditure elasticities presented in Table 10.5. Negative expenditure elasticities are possible but not expected for broader aggregates of foods, and we interpret such findings as indications of misspecification rather than unexpected results. According to this interpretation, all the models without the partial adjustment factor perform badly and confirm our belief in the selected model.

Conclusions

It is frequently necessary to correct for autocorrelation in demand systems and several correction mechanisms are commonly applied. Single-equation test results indicate that the diagonal Berndt–Savin correction (ρ) and the diagonal partial adjustment model (θ) are least successful in removing autocorrelation in our data set. This is hardly surprising, given their restrictive nature, and the results suggest that these restrictive corrections should be avoided. Equation-by-equation autocorrelation is only completely removed by the general Berndt–Savin correction (ρ_{ij} and ρ_{ij} & θ_{ij}). A system test indicates that most of the corrections are unsuccessful in removing autocorrelation at a system level, although it is removed in each equation. By construction, the general Berndt–Savin correction removes the type of system autocorrelation that we tested for and this correction is selected as the most appropriate one. Habits have been found to be important in explaining Norwegian food demand, and a partial adjustment model that is estimated by using the general Berndt–Savin correction (ρ_{ij} & θ_{ij}) is chosen as the preferred model. The selected model is supported by several negative total-expenditure elasticities in the other specifications.

The estimated health-information, price and expenditure elasticities are sensitive to choice of correction mechanism. Our results suggest that one likely explanation for some of the unexpected findings discussed in Rickertsen and von Cramon-Taubadel (Chapter 3, this volume) is autocorrelation. Some health effects are, however, quite robust across the different corrections. We find positive effects on the demand for poultry and the group other foods. An unexpected negative effect on the demand for fish and an unexpected positive effect on the demand for fats and oils indicate that there are additional problems.

Acknowledgements

Financial support for this research has been received from the EU, contract FAIR5–CT97–3373 and is greatly appreciated. We gratefully acknowledge the comments of two anonymous reviewers.

References

Alessie, R. and Kapteyn, A. (1991) Habit formation, interdependent preferences and demographic effects in the almost ideal demand system. *Economic Journal* 101, 404–419.

Alston, J.M. and Chalfant, J.A. (1991) Unstable models from incorrect forms. *American Journal of Agricultural Economics* 73, 1171–1181.

Berndt, E.R. and Savin, N.E. (1975) Estimation and hypothesis testing in a singular equation system with autoregressive disturbances. *Econometrica* 43, 937–957.

Cashin, P. (1991) A model of the disaggregated demand for meat in Australia. *Australian Journal of Agricultural Economics* 35, 263–284.

Deaton, A. and Muellbauer, J. (1980) An almost ideal demand system. *American Economic Review* 70, 312–326.

Edgerton, D.L., Assarsson, B., Hummelmose, A., Laurila, I.P., Rickertsen, K. and Vale, P.H. (1996) *The Econometrics of Demand Systems: With Applications to Food Demand in the Nordic Countries.* Kluwer Academic Publishers, Dordrecht, 290 pp.

Godfrey, L.G. (1988) *Misspecification Tests in Econometrics.* Cambridge University Press, Cambridge, 252 pp.

Gustavsen, G.W. (1999) The BSE crisis and the reaction of Norwegian consumers. *Cahiers d'Économie et Sociologie Rurales* 50, 22–34.

Holt, M.T. and Goodwin, B.K. (1997) Generalized habit formation in an inverse almost ideal demand system: an application to meat expenditure in the US. *Empirical Economics* 22, 293–320.

Kesavan, T., Hassan, Z.A., Jensen, H.H. and Johnson, S.R. (1993) Dynamics and long-run structure in US meat demand. *Canadian Journal of Agricultural Economics* 41, 139–153.

McGuirk, A., Driscoll, P., Alwang, J. and Huang, H. (1995) System misspecification testing and structural change in the demand for meats. *Journal of Agricultural and Resource Economics* 20, 1–21.

Moschini, G. (1995) Units of measurement and the Stone index in demand system estimation. *American Journal of Agricultural Economics* 77, 63–68.

Moschini, G. and Moro, D. (1994) Autocorrelation specification in singular equation systems. *Economic Letters* 46, 303–309.

Piggott, N.E., Chalfant, J.A., Alston, J.M. and Griffith, G.R. (1996) Demand response to advertising in the Australian meat industry. *American Journal of Agricultural Economics* 78, 268–279.

Rickertsen, K. (1996) Structural change and the demand for meat and fish in Norway. *European Review of Agricultural Economics* 23, 316–330.

Rickertsen, K. (1998a) The demand for food and beverages in Norway. *Agricultural Economics* 18, 89–100.

Rickertsen, K. (1998b) The effects of advertising in an inverse demand system: Norwegian vegetables revisited. *European Review of Agricultural Economics* 25, 129–140.

Rickertsen, K., Chalfant, J.A. and Steen, M. (1995) The effects of advertising on the demand for vegetables. *European Review of Agricultural Economics* 22, 481–494.

Rickertsen, K., Kristofersson, D. and Lothe, S. (2003) Effects of health information on Nordic meat and fish demand. *Empirical Economics* (in press).

Ruud, P.A. (2000) *An Introduction to Classical Econometric Theory*. Oxford University Press, New York, 951 pp.

Wickens, M.R. and Breusch, T.S. (1988) Dynamic specification, the long-run and the estimation of transformed regression models. *Economic Journal* 98, 189–205.

Yen, S.T. and Chern, W.S. (1992) Flexible demand systems with serially correlated errors: fat and oil consumption in the United States. *American Journal of Agricultural Economics* 74, 689–697.

Appendix

Table A10.1. Estimated and recovered parameters from partial adjustment model estimated with Berndt and Savin's autocorrelation correction (t values in parentheses).

	$\alpha_{i\,00}$	γ_{ij} 1	2	3	4	5	6	β_i	λ_i	ω_h 1	2	3	θ_{ij} 1	2	3	4	5	6
Stage 1																		
Meat and	0.298	0.001	−0.008	0.005	0.008	−0.001	−0.006	−0.003	0.002	0.283	−0.033	0.750	0.573	−0.400	−0.319	−0.382	−0.284	0.812
fish	(5.13)	(1.08)	(−4.38)	(2.61)	(2.81)	(−0.14)	(−6.98)	(−0.44)	(1.40)	(20.80)	(−1.94)	(43.11)	(7.90)	(−3.51)	(−1.95)	(−4.02)	(−4.82)	(2.52)
Dairy products	0.302	−0.008	0.031	0.004	−0.014	−0.013	0.000	−0.008	−0.001				0.124	−0.145	−0.452	−0.260	−0.299	1.032
and eggs	(3.23)	(−4.38)	(8.84)	(1.37)	(−2.74)	(−1.60)	(0.31)	(−0.76)	(−0.75)				(1.09)	(−0.79)	(−1.69)	(−1.68)	(−3.16)	(1.99)
Carbohydrates	0.468	0.005	0.004	−0.015	−0.009	0.020	−0.004	0.020	−0.001				−0.241	−0.555	0.382	−0.513	−0.475	1.402
	(5.81)	(2.61)	(1.37)	(−4.97)	(−1.98)	(2.61)	(−3.33)	(2.05)	(−0.45)				(−2.42)	(−3.53)	(1.68)	(−3.90)	(−5.79)	(3.15)
Fats and	0.601	0.008	−0.014	−0.009	0.040	−0.029	0.004	0.032	0.009				−0.452	−0.788	−0.332	−0.263	−0.568	2.404
oils	(4.16)	(2.81)	(−2.74)	(−1.98)	(2.59)	(−1.79)	(1.35)	(1.81)	(3.04)				(−2.59)	(−2.89)	(−0.83)	(−1.13)	(−3.80)	(3.06)
Other foods	−0.780	−0.001	−0.013	0.020	−0.029	0.016	0.007	−0.039	−0.011				0.038	1.938	0.789	1.586	1.741	−6.094
	(−2.85)	(−0.14)	(−1.60)	(2.61)	(−1.79)	(0.60)	(2.07)	(−1.23)	(−1.92)				(0.12)	(3.63)	(1.02)	(3.59)	(6.23)	(−4.06)
Other goods	0.111	−0.006	0.000	−0.004	0.004	0.007	−0.001	−0.001	0.003				−0.042	−0.050	−0.068	−0.168	−0.115	0.443
	(3.52)	(−6.98)	(0.31)	(−3.33)	(1.35)	(2.07)	(−1.06)	(−0.26)	(3.84)				(−1.05)	(−0.83)	(−0.75)	(−3.18)	(−3.56)	(2.47)

	$\alpha_{i\,00}$	γ_{ij} 1	2	3	4	5	β_i	λ_i	ω_h 1	2	3	θ_{ij} 1	2	3	4	5
Stage 2																
Beef and	−0.163	0.096	−0.112	−0.022	0.008	0.030	−0.007	−0.008	0.538	0.240	0.222	0.861	0.864	−1.325	−0.649	0.249
veal	(−1.30)	(2.47)	(−4.94)	(−1.92)	(1.76)	(1.81)	(−0.13)	(−0.45)	(4.82)	(2.54)	(2.12)	(4.59)	(3.88)	(−2.67)	(−1.93)	(0.95)
Pork	0.446	−0.112	0.149	0.017	−0.007	−0.046	0.001	0.015				−0.323	−0.046	0.105	0.351	−0.087
	(5.59)	(−4.94)	(8.44)	(1.94)	(−1.71)	(−4.67)	(0.05)	(1.33)				(−2.78)	(−0.32)	(0.33)	(1.59)	(−0.52)
Lamb	0.169	−0.022	0.017	0.032	−0.017	−0.009	−0.049	−0.007				−0.115	−0.164	0.396	−0.027	−0.091
	(4.44)	(−1.92)	(1.94)	(2.80)	(−3.23)	(−1.27)	(−2.75)	(−1.19)				(−1.89)	(−2.41)	(2.65)	(−0.26)	(−1.18)
Poultry	0.142	0.008	−0.007	−0.017	0.019	−0.002	0.020	0.017				−0.104	−0.151	−0.068	0.573	−0.249
	(8.43)	(1.76)	(−1.71)	(−3.23)	(5.39)	(−0.67)	(1.85)	(5.30)				(−3.57)	(−4.98)	(−0.96)	(9.81)	(−7.46)
Fish	0.406	0.030	−0.046	−0.009	−0.002	0.027	0.035	−0.017				−0.319	−0.504	0.893	−0.247	0.177
	(7.39)	(1.81)	(−4.67)	(−1.27)	(−0.67)	(2.80)	(1.47)	(−2.15)				(−3.81)	(−5.10)	(4.10)	(−1.66)	(1.58)

Relative Impacts of Health Information and Advertising on Commodity Markets: US Meats

11

HENRY W. KINNUCAN,[1] ØYSTEIN MYRLAND,[2] AND LAXMI PAUDEL[3]

[1]*Auburn University, Auburn, Alabama, USA;*
[2]*University of Tromsø, Tromsø, Norway;*
[3]*University of Georgia, Athens, Georgia, USA*

Introduction

Health information has been shown to have opposite effects on beef and poultry demand (Kinnucan *et al.*, 1997). In particular, American consumers have responded to growing medical research that links dietary cholesterol to heart disease by increasing their consumption of poultry products, largely at the expense of beef (Table 11.1). At the same time, the US beef and pork industries have invested significant funds in generic advertising (over $60 million in 1999 alone). This raises the question of whether expensive advertising campaigns can be effective in markets that are regularly exposed to messages from health authorities that conflict with the advertising messages.

Empirical evidence on the effects of the beef and pork campaigns is mixed. For example, studies by Brester and Schroeder (1995) and Kinnucan *et al.* (1997) suggest that generic advertising has had little, if any, effect on US meat demand. Conversely, studies by Ward (1999) and Ward and Lambert (1993) suggest that beef advertising has increased demand sufficiently to be profitable for US beef producers. For similar evidence for Australia, see Piggott *et al.* (1995). Coulibaly and Brorsen (1999) attempt to reconcile these conflicting findings by addressing weaknesses in data and model specification. However, there is a more basic reason why meat-advertising effects may be weak. The reason comes from attribution and related theories (to be discussed later), which posit that consumers respond disproportionately to negative information. For example, Richey *et al.* (1967) find that it takes about five instances of positive information to neutralize one instance of negative information.

Table 11.1. Per capita consumption of meats and real prices, USA, 5-year averages, 1966–2000.[a] (From: US Department of Agriculture, Economic Research Service (USDA, 2000).)

	Beef		Pork		Poultry		
Time period	Q (kg)	P ($ kg^{-1})	Q (kg)	P ($ kg^{-1})	Q (kg)	P ($ kg^{-1})	All (kg)
1966–1970	36.6	4.79	32.3	3.91	20.7	1.98	89.7
1971–1975	38.4	5.24	30.6	4.13	22.0	1.93	91.1
1976–1980	38.9	4.44	29.1	3.96	24.7	1.67	92.8
1981–1985	35.4	4.31	30.0	3.21	29.1	1.36	94.5
1986–1990	32.9	4.10	29.3	3.17	36.3	1.36	98.4
1991–1995	30.3	4.04	30.4	2.99	43.5	1.28	104.2
1996–2000	31.1	3.83	29.9	3.30	47.9	1.60	108.9

[a] Prices are expressed in 1982–1984 dollars.
Q, quantity; P, price.

If this relationship is valid in a food-choice context, as empirical evidence suggests (Chang and Kinnucan, 1991; Richards and Patterson, 1999), one would expect advertising elasticities for products that are susceptible to negative health-based information (e.g. beef) to be small in relation to substitute products (e.g. poultry) that do not suffer the same handicap. Since advertising elasticities are tiny to begin with (typically less than 0.05) and thus difficult to measure, this may explain why estimated advertising elasticities for beef are fragile. In the context of the present study, attribution theory suggests that increases in health information will have larger effects on meat demand than similar increases in advertising. At issue are welfare implications. Specifically, how are US meat producers affected by health information and what are the possibilities for advertising to offset negative effects?

Prior to model specification, we review the relevant literature to determine what is known about consumer responses to negative information from health sources versus positive information from advertising sources. Simulations based on the economic model are used to determine welfare impacts and to compute total elasticities. The chapter concludes with a summary of our main findings.

Disproportionate Influence of Negative Information

The issue of how consumers process and integrate negative information with positive information has been studied in the impression-formation literature in psychology. A robust finding is the negativity effect: that is, people place more weight on negative than positive information in forming overall evaluations of a target (Fiske, 1980; Skowronski and Carlston, 1989; Klein, 1996). For example, Maheswaran and Meyers-Levy (1990) find that, when the processing is focused on message content, negative framing is more effective than positive framing.

One reason for the negativity effect may be that negative information is considered more diagnostic or informative than positive information (Skowronski and Carlston, 1989; Maheswaran and Meyers-Levy, 1990). In this context negative information might be used by the consumer to categorize a product as low in quality, whereas positive or neutral information is less useful in such a categorizing process. Hence, negative information may simply be considered more useful or diagnostic in making decisions and is given greater weight than positive information (Ahluwalia *et al.*, 2000).

Three additional explanations are provided by Mizerski (1982). The first explanation is surprise and frequency of use. Negative information reaches the market infrequently, so there is a 'surprise value' attached to negative information. Hence, negative information tends to be regarded as more credible than positive messages from advertising, which are common. The second explanation is related to ambiguity and uncertainty in the information. In particular, negative information is posited to be less ambiguous than positive; hence it has a greater effect on behaviour. However, these two explanations have only limited empirical support (Richards and Patterson, 1999). The third explanation is based on differences in causal attribution.

Attribution theory posits that consumers are likely to place more emphasis on negative information than positive information due to the cause of the information. The psychology of 'attribution' suggests that each type of news will have a different effect on consumers, depending upon the credibility of each source of information and that source's perceived vested interests in making news widely known. Individuals process information according to its perceived cause and consider information provided by the factual performance of the entity in question more reliable than information provided by other factors (e.g. vested interests, as might be the case with advertising).

Based on this, the negative impact on demand due to negative publicity is not surprising. Publicity through health authorities and media is considered a relatively credible source of information and might therefore be considered more influential than industry-based communications. If the provider of the signals has no self-interest or the information reflects a wider concern beyond individual interest, then they are regarded as credible. Furthermore, negative as opposed to positive information is known to be more attention-getting (Fiske, 1980).

The first study to test attribution theory in a food-demand context was by Chang and Kinnucan (1991), who found that butter consumers respond more forcefully to health information than to butter advertising. More recently, Richards and Patterson (1999) focused on negative publicity stemming from food-borne disease outbreaks in a model of strawberry demand. Here, we build on this literature by extending the analysis to consider welfare impacts.

Model

The essence of health information and advertising effects from the standpoint of welfare measurement is the induced price effects. Thus, in addition to taking into account demand interrelationships, the model needs to reflect the supply side of the market. Moreover, since we are interested in farm-level impacts, the model needs to include the marketing channel. Accordingly, we express the relevant behavioural relationships and equilibrium condition in matrix notation as follows:

$$\boldsymbol{q}^* = \boldsymbol{\eta}\, \boldsymbol{p}^* + \boldsymbol{\beta}\, \boldsymbol{a}^* + \boldsymbol{\iota}\, z^* \tag{11.1}$$

$$\boldsymbol{p}^* = \boldsymbol{\nu}\, \boldsymbol{w}^* \tag{11.2}$$

$$\boldsymbol{x}^* = \boldsymbol{\varepsilon}\, \boldsymbol{w}^* \tag{11.3}$$

$$\boldsymbol{q}^* = \boldsymbol{x}^* \tag{11.4}$$

where $\boldsymbol{q}^*$ and $\boldsymbol{x}^*$ are $n \times 1$ vectors representing relative changes in quantities at the retail and farm levels of the market, respectively (e.g. the first element of $\boldsymbol{q}^*$ is dq_1/q_1); $\boldsymbol{p}^*$ and $\boldsymbol{w}^*$ are $n \times 1$ vectors representing relative changes in prices at retail and farm; $\boldsymbol{a}^*$ is an $n \times 1$ vector representing relative changes in advertising expenditure; z^* is a scalar representing the relative change in health information; $\boldsymbol{\iota}$ is an $n \times 1$ vector of health-information elasticities; $\boldsymbol{\eta}$ and $\boldsymbol{\beta}$ are $n \times n$ matrices of demand and advertising elasticities; $\boldsymbol{\varepsilon}$ is an $n \times n$ matrix of supply elasticities; and $\boldsymbol{\nu}$ is an $n \times n$ diagonal matrix of farm–retail price-transmission elasticities.

Equations 11.1–11.4 constitute an equilibrium-displacement model similar to the one used by Kinnucan and Miao (2000) to analyse the producer welfare impacts of advertising levies. In this formulation, changes in retail quantity caused by advertising or health information are assumed to result in identical changes in farm quantity, which implies that the aggregate marketing technology exhibits fixed proportions. (For a discussion of the economic implications of this restriction, see Kinnucan (1997).) In addition, we implicitly assume a closed economy. Experimentation with a larger model that included trade relationships showed that, owing to the modest trade shares for the three meat items (less than 10%), this assumption is innocuous over a wide range of plausible trade elasticities.

To facilitate identification of the price-transmission elasticities, we assume that the aggregate marketing technology exhibits constant returns to scale (CRTS) and that the supply of marketing inputs is perfectly elastic. These assumptions are common in the equilibrium-displacement modelling literature (e.g. Wohlgenant, 1993) and have been shown to be innocuous when modelling the effects of shifts in retail demand on producer welfare (Kinnucan, 1997).

The key relationships for the purposes of this analysis are the reduced-form (or 'total') elasticities corresponding to price and quantity at the farm level. The farm-price effects are obtained by substituting Equations 11.1–11.3 into Equation 11.4 and solving for $\boldsymbol{w}^*$ to yield:

$$\boldsymbol{w}^* = \boldsymbol{\Pi}\, \boldsymbol{a}^* + \boldsymbol{\Psi}\, z^* \tag{11.5}$$

where $\boldsymbol{\Pi} = (\boldsymbol{\varepsilon} - \boldsymbol{\eta}\ \boldsymbol{\nu})^{-1}\ \boldsymbol{\beta}$ is an $n \times n$ matrix of total elasticities that indicates the net effect of isolated changes in advertising expenditure on farm prices. Similarly, $\boldsymbol{\Psi} = (\boldsymbol{\varepsilon} - \boldsymbol{\eta}\ \boldsymbol{\nu})^{-1}\ \boldsymbol{\iota}$ is an $n \times 1$ vector of total elasticities that indicates the net effect of changes in health information on farm prices. The farm-quantity effects are obtained by back-substitution of Equation 11.5 into Equation 11.3 to yield:

$$\boldsymbol{x}^* = \boldsymbol{\varepsilon}\, \boldsymbol{\Pi}\, \boldsymbol{a}^* + \boldsymbol{\varepsilon}\, \boldsymbol{\Psi}\, z^* \tag{11.6}$$

The price and quantity effects may be translated into welfare effects via the formula:

$$\Delta PS_i = w_i\, x_i\, w_i^* \left(1 + \tfrac{1}{2} x_i^*\right) \tag{11.7}$$

where ΔPS_i is the change in producer surplus (quasi-rent) in the ith market. Equation 11.7's validity rests on the assumption that supply is linear in the relevant range and demand shifts are parallel. Although this assumption may not be true in any particular application, the approximation error is negligible if equilibrium displacements are small, as is the case here. For a cogent discussion of the parallel shift assumption, see Wohlgenant (1999).

Parameterization

The structural elasticities needed to implement the model are presented in Table 11.2. The demand and advertising elasticities are taken from Brester and Schroeder's (1995) study. An advantage of Brester and Schroeder's (1995) estimates over others in the literature (e.g. Kinnucan *et al.*, 1997) is that elasticity estimates are provided for both brand and generic advertising, which permits evaluation of each type of advertising. In addition, the demand system is complete in the sense that the fourth good (after beef, pork and poultry) is defined as 'all other goods'. Thus, the price elasticities are unconditional, i.e. they do not rely on the assumption that group expenditure is fixed. Moreover, all cross-price elasticities are smaller in absolute value than own-price

Table 11.2. Parameter values for beef, pork, poultry and all other goods, USA.

Item	Definition	Value[a]			
		Beef	Pork	Poultry	Other
η_{1j}	Demand elasticities for beef	−0.56	0.10	0.05	0.41
η_{2j}	Demand elasticities for pork	0.23	−0.69	0.04	0.43
η_{3j}	Demand elasticities for poultry	0.21	0.07	−0.33	0.05
η_{4j}	Demand elasticities for other goods	0.009	0.004	0.000	−0.010
β_{i1}^{B}	Brand advertising elasticities for beef	0.006	0.006	0.001	0.000
β_{i1}^{G}	Generic advertising elasticities for beef	0.006	−0.010	−0.011	0.000
β_{i2}^{B}	Brand advertising elasticities for pork	−0.013	0.033	−0.010	0.000
β_{i2}^{G}	Generic advertising elasticities for pork	0.002	−0.010	−0.010	0.000
β_{i3}^{B}	Brand advertising elasticities for poultry	0.017	0.004	0.047	−0.001
ι_i	Health-information elasticities[b]	−0.186	0.000	0.471	0.000
ε_i	Supply elasticity[c]	0.15	0.40	0.90	2.00
ϕ_i	Farm–retail price-transmission elasticity[d]	0.53	0.34	0.43	1.00
$w_i\ x_i$	Farm value (billion dollars)[e]	30.0	10.1	6.3	–
θ_i^{B}	Brand advertising intensity (%)[e]	0.007	0.570	0.480	–
θ_i^{G}	Generic advertising intensity (%)[e]	0.083	0.110	0.000	–

[a] Unless indicated otherwise, values are taken from Brester and Schroeder (1995).
[b] 'Best guess' based on Table 11.3.
[c] Values for beef and pork are the same as those used by Wohlgenant (1993); value for poultry is taken from Tomek and Robinson (1990: 61) and value for other goods is a guesstimate.
[d] Estimates for beef, pork and poultry are based on farmer's cost share computed from Elitzak (1999) for 1990–1995. The estimate for 'other goods' is set to 1 since farm-level elasticity is not defined.
[e] Farm value and intensity (advertising expenditure divided by farm value multiplied by 100) refer to 1993, the last year in Brester and Schroeder's analysis. Farm revenue data were taken from USDC (1994).

elasticities, as needed to satisfy the multi-market equilibrium condition (Hicks, 1946: 315–319 and Chapter 5).

Since Brester and Schroeder (1995) did not include health information in their model, these elasticities were obtained from external sources. In particular, two sets of elasticity estimates were considered. One set is based on Brown and Schrader's (1990) (B&S) index, which measures the cumulative number of articles published in medical journals that indicate a link between dietary cholesterol and heart disease. The other set is based on Kim and Chern's (1999) (K&C) index, which tracks articles on fat and cholesterol as recorded by Medline. As indicated in Fig. 11.1, the two indices provide a different representation of the information-diffusion process. In particular, the B&S index shows a gradual increase over time, whereas the K&C index shows variability. Since both indices are meant to measure the same

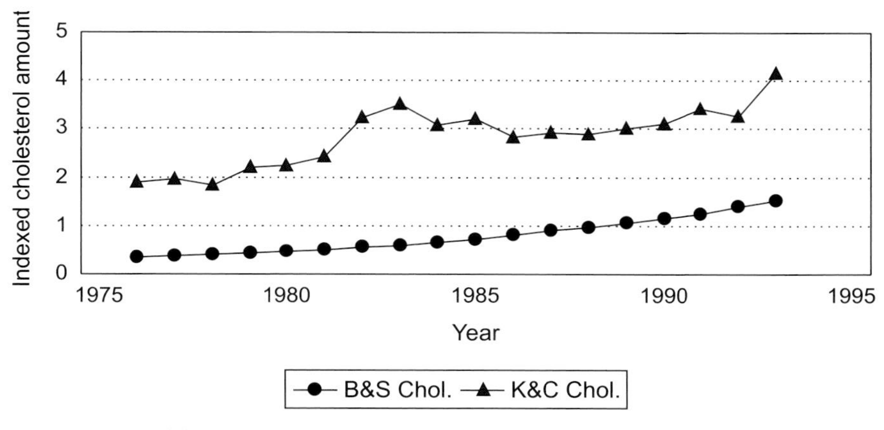

Fig. 11.1. Health indices.

Table 11.3. Health-information elasticities for Brown and Schrader's index versus Kim and Chern's index.[a]

	B&S index			K&C index	
Commodity	Rotterdam[b]	AIDS	Double log	AIDS	Double log
Beef	−0.681*	−0.1972*	−0.0846*	−0.1857*	−0.1844*
Pork	−0.195	−0.009	0.0847*	0.0406*	−0.0778*
Poultry	1.659*	0.5609*	0.1322*	0.4709*	0.4318*
Fish	1.768*	2.7619	−0.0424	0.1061	0.3043*

[a] Estimates based on 1976–1993 quarterly US data. An asterisk indicates that the coefficient's t value is greater than 1.64 in absolute value.
[b] The source is Kinnucan *et al.* (1997). Elasticities in the remaining columns were estimated by the authors using Kinnucan *et al.*'s database.

construct, it is important to know whether elasticity estimates are sensitive to the index selected.

Accordingly, we estimated a variety of meat-demand specifications using the indices as indicated in Table 11.3. The estimates are based on quarterly data for the period 1976–1993, the same data used in Kinnucan *et al.*'s (1997) analysis. In addition, the models contained the same variables as Kinnucan *et al.*'s (1997) study, with the exception of advertising, since these variables taken as a group were found by Kinnucan *et al.* (1997) to be statistically insignificant. We may note that, unlike Brester and Schroeder (1995), Kinnucan *et al.*'s (1997) study did not include brand advertising.

Estimates show a clear statistical relationship between health information and beef and poultry demand. In particular, the estimated elasticities across models and variable definitions are uniformly significant and negative for beef and positive for poultry. For pork and fish, the estimates are fragile and inconsistent. Two of the five estimates are insignificant for pork and three of the five are insignificant for fish. Moreover, pork's significant elasticities differ in sign. Accordingly, we set the health-information elasticities for fish and pork to zero. For beef, we set the elasticity to −0.186, the estimate corresponding to K&C's index as estimated by the almost ideal demand system (AIDS). This estimate is between the low (−0.085) and high (−0.681) values in Table 11.3 and thus should provide a good representation of the demand impact of this variable. For poultry, we set the elasticity to 0.471, the AIDS estimate based on K&C's index. Like beef, this elasticity is between the low (0.132) and high (1.659) estimates and thus represents an intermediate response. The health-information elasticity for 'other goods' is set to zero since this composite commodity is not directly relevant to the analysis.

The supply elasticities for beef, pork and poultry are set respectively to 0.15, 0.40 and 0.90 to be consistent with estimates in the literature (Table 11.2, footnote c). Since no estimates are available for other goods, the supply elasticity ε_4 is set to 2, our 'best guess' value. Here, we implicitly assume that meat production in the USA is specialized so that the cross-price elasticities can be taken to be zero, i.e. $\boldsymbol{\varepsilon}$ is diagonal. However, to gauge the sensitivity of results to supply response, additional simulations are performed, with the supply elasticities halved and doubled.

The farm–retail price-transmission elasticities are set equal to the farmer's share of the consumer dollar. The justification for this procedure is that the transmission elasticity converges to the farmer's share when the aggregate marketing technology exhibits CRTS and the supply curve for marketing inputs is perfectly elastic (Gardner, 1975; see also Kinnucan and Forker, 1987: 290; Table 11.2, footnote d), which is a maintained hypothesis. The price-transmission elasticity for the composite commodity is set to unity since the farm-level impact is not relevant.

Simulation

To simulate the model, we inserted the structural elasticities in Tables 11.2 and 11.3 into Equations 11.5 and 11.6 to compute the impact of a 1% increase in health information and advertising on prices and quantities. The price and quantity effects are then inserted into Equation 11.7, along with the baseline values for farm revenue in Table 11.2, to compute welfare impacts. The 1993 baseline values are used because these correspond to the last year of the econometric analyses in Brester and Shroeder's (1995) and Kinnucan *et al.*'s (1997) studies. At

Table 11.4. Partial versus total elasticities for health information and advertising, US meats.[a]

Item	Partial elasticity	Total elasticity	Ratio
Beef			
Health information	−0.186	−0.060	3.1
Brand advertising	0.0060	0.0022	2.7
Generic advertising	0.0060	0.0018	3.3
Pork			
Health information	0.000	−0.027	0.0
Brand advertising	0.0330	0.0187	1.8
Generic advertising	−0.0005	−0.0001	5.0
Poultry			
Health information	0.471	0.367	1.3
Brand advertising	0.0470	0.0449	1.1

[a] These elasticities refer to quantity effects; elasticities for price effects are presented in Table 11.5.

issue is whether the welfare impacts of health information are negative when demand interrelationships and supply response are taken into account. A related issue is whether any losses that might accrue due to health information can be offset by advertising.

Partial Versus Total Elasticities

Before addressing these issues, it is instructive to compare the total elasticities implied by the model with the partial elasticities for health information and advertising given in Tables 11.2 and 11.3. Intuitively, the partial elasticities should overstate the actual quantity effects, since they hold prices constant. For example, an increase in health information decreases the demand for beef, but also lowers its price. The lower price encourages increased beef consumption, which would tend to moderate the initial effect of the health information.

Indeed, this pattern is observed in Table 11.4, which lists the partial elasticities for own-advertising and health information along with their total counterparts extracted from $\boldsymbol{\varepsilon\Pi}$ and $\boldsymbol{\varepsilon\Psi}$ using Equation 11.6. The largest differences occur for beef, as might be expected, since beef supply is the least elastic, implying sharper price effects. In particular, the total health-information elasticity for beef is −0.060, about one-third the size of the partial (−0.186). Consequently, using the partial elasticity to predict the effect on an increase in health information on beef demand would exaggerate the actual effect threefold. A similar result is obtained for advertising. In particular, the partial advertising elasticities for brand and generic beef advertising are, respectively, 2.7 and 3.3 times larger than the total elasticities. Thus, the quantity effects of beef advertising are much smaller when prices are allowed to adjust than when prices are held constant. A similar pattern is obtained for pork and poultry, although the differences in general are not as pronounced. The one exception is for generic advertising of pork, which has a negative effect. However, this result is probably best interpreted as 'no effect', since the partial elasticity is insignificant (see Brester and Schroeder, 1995).

The total health elasticity for pork is negative (−0.027), even though the partial elasticity is zero. The reason can be traced to spillover effects: that is, an increase in health information causes a simultaneous increase in the demand for poultry and a decrease in the demand for beef. With upward-sloping supplies, the price of beef decreases relative to poultry and relative to the price of pork, which causes the pork demand curve to shift down. The downward shift is moderated by the rise in the relative price of poultry, but not sufficiently so to offset the negative effect of the lower beef price. As a consequence, owing to induced price effects, increases in health information generate a negative externality for pork producers. This highlights the dangers of ignoring the supply side of the market when modelling health-information effects.

The foregoing elasticities indicate the net effects of health information and advertising on equilibrium quantities. The complete set of total elasticities, i.e. the elasticities that indicate the net effect of increases in health information and advertising on equilibrium quantities and prices, is given in Table 11.5. Perhaps the most striking aspect of these elasticities is the large differences between the health-information and advertising elasticities. For example, a 1% increase in health information lowers the farm price of beef by 0.40%, whereas a 1% increase in brand-beef advertising expenditure raises the farm price of beef by a mere 0.014%, a 28-fold difference. A similar difference exists with respect to quantity effects. Specifically, a 1% increase in health information reduces beef quantity by 0.060%, whereas a 1% increase in brand-beef advertising increases beef quantity by a mere 0.002%.

To highlight these differences, we divided the advertising elasticities in Table 11.5 by the corresponding health-information elasticities and the results are shown in Table 11.6. For brevity, results are given for quantity effects, as ratios for price effects are identical. Focusing on own-advertising effects, the ratios indicate that health-information elasticities are 28 to 33 times larger than advertising elasticities for beef, 1.4 to 264 times larger for pork and 8.2 times larger for poultry. The ratios for beef affirm attribution theory in that health information, which is negative from beef's perspective, has a much stronger impact on demand than

Table 11.5. Complete set of reduced-form elasticities.[a]

Endogenous variable	Exogenous variable					
	z^*	ab_1^*	ag_1^*	ab_2^*	ag_2^*	ab_3^*
Farm price						
w_1^*	−0.4027	0.0144	0.0121	−0.0260	0.0040	0.0415
w_2^*	−0.0670	0.0123	−0.0121	0.0467	−0.0003	0.0155
w_3^*	0.4075	0.0028	−0.0095	−0.0094	−0.0092	0.0499
w_4^*	−0.0010	−0.0001	0.0000	−0.0000	0.0000	−0.0002
Retail price						
p_1^*	−0.2134	0.0077	0.0064	−0.0138	0.0021	0.0220
p_2^*	−0.0228	0.0042	−0.0041	0.0159	−0.0001	0.0053
p_3^*	0.1752	0.0012	−0.0041	−0.0040	−0.0039	0.0214
p_4^*	−0.0010	−0.0001	0.0000	−0.0000	0.0000	−0.0002
Quantity						
$q_1^* = x_1^*$	−0.0604	0.0022	0.0018	−0.0039	0.0006	0.0062
$q_2^* = x_2^*$	−0.0268	0.0049	−0.0048	0.0187	−0.0001	0.0062
$q_3^* = x_3^*$	0.3667	0.0025	−0.0086	−0.0085	−0.0083	0.0449
q_4^*	−0.0020	−0.0001	0.0001	−0.0001	0.0000	0.0000

[a] Subscripts are defined as follows: 1, beef; 2, pork; 3, poultry; 4, all other goods. Exogenous variables are defined as follows: *z*, health information; *ab*, brand advertising; *ag*, generic advertising.

Table 11.6. Ratio of health information to advertising elasticities.

Endogenous variable	Exogenous variable					
	z^*	ab_1^*	ag_1^*	ab_2^*	ag_2^*	ab_3^*
$q_1^* = x_1^*$	−0.0604	−27.9	−33.3	15.5	−100.0	−9.7
$q_2^* = x_2^*$	−0.0268	−5.5	5.5	−1.4	264.5	−4.3
$q_3^* = x_3^*$	0.3667	146.6	−42.7	−43.4	−44.4	8.2
q_4^*	−0.0020	17.3	−32.8	27.7	−70.4	5.2

advertising, which by definition represents only positive information. As for the poultry ratio, the fact that health information has a stronger effect than poultry advertising is consistent with a variant of attribution theory, which states that the source credibility is important: that is, individuals process information according to its perceived cause and consider information provided by the factual performance of the entity in question more reliable than information provided by other sources (e.g. vested interests). In the case of poultry, the disproportionate influence of the (positive) health information may be due to the perception that information from health authorities is more reliable than information provided by industry.

Table 11.7. Producer welfare effects of a 1% increase in K&C's health-information index versus a 1% increase in meat advertising expenditure, USA, 1993. Unit: million US dollars.

		Advertising effect				
Item	Health-information effect	Brand beef	Generic beef	Brand pork	Generic pork	Brand poultry
Beef	−120.8	4.3	3.6	−7.8	1.2	12.4
Pork	−26.8	1.2	−1.2	4.7	−0.0	1.6
Poultry	25.7	0.2	−0.6	−0.6	−0.6	3.1
All	−101.8	5.7	1.8	−3.7	0.6	17.1

Welfare Effects

As expected, based on the total elasticities, health information has a disproportionate impact on producer welfare (Table 11.7). In particular, a 1% increase in health information is associated with a loss to the beef sector of $121 million, which is 28–33 times larger than the $3.6–4.3 million gain that accrues to the beef sector from a 1% increase in beef advertising. A similar result is obtained for pork, although the differences are less pronounced, owing to the fact that health information has no direct influence on pork demand. However, as indicated earlier, health information has an indirect effect through induced price effects. These induced price effects generate a negative externality to the pork sector equal to $6.8 million per 1% increase in health information.

As for poultry, the health-information effect is positive and yet disproportionate. In particular, a 1% increase in health information is associated with a gain in the poultry sector of $25.7 million, compared with a gain of $3.1 million for a 1% increase in advertising. Thus, the health-information effect is 8.3 times larger than the advertising effect. As discussed previously, the larger influence of health information cannot be ascribed to the disproportionate influence of negative information, since health information is positive in this instance. Rather, consumers probably view information provided by health authorities as more reliable than information provided by industry. Consequently, they are more apt to respond to the health information than to the advertising, even though both are positive for poultry.

Overall, study results show a 1% increase in health information reducing welfare in the US meat sector by $102 million. By way of comparison, a 1% increase in advertising is associated with a gross welfare change of between −$3.7 million and $17.1 million, depending on product advertised and advertising type (brand or generic). This suggests that advertising is a blunt instrument for offsetting the negative effects of health information.

This is particularly true for beef, where health information's effect is both negative and direct. In this case, for industry to recoup losses associated with health information, it would need to increase advertising expenditures by at least 28 times the percentage increase in the information index. To put this in perspective, K&C's index between 1993 and 1997 (the terminal data point) increased from 5.28 to 6.20, or 17.4%. Assuming that the elasticities in this out-of-sample period are identical to the elasticities used in our simulations, to offset the effects of increased negative information during this 5-year period the beef industry would have had to increase advertising expenditures by at least 487%! Still, this does not mean that beef advertising is necessarily unprofitable. That depends on whether the demand shift is sufficient to compensate for the loss in producer surplus associated with the advertising levy. What it does mean is that advertising's ability to counteract negative publicity is limited.

Conclusions

A basic theme of this research is that consumer response to information depends on source credibility, but also on whether it is positive or negative. In particular, experimental evidence from the psychology literature suggests that it takes about five instances of positive information to neutralize one instance of negative information (Richey *et al.*, 1967). This asymmetry has important implications for food markets. In particular, it suggests that positive information conveyed in advertisements will be relatively impotent against negative information conveyed by health authorities.

Indeed, our empirical results for US meats show that consumer responses to advertising are minute, especially in relation to consumer responses to health information. For example, study results indicate that the 'total' health-information elasticity for beef, i.e. the elasticity that takes into account induced price effects, is −0.06, which is approximately 30 times larger than the total advertising elasticity for beef (0.002). A similar pattern is observed for pork, although the differences are less pronounced owing to the (apparent) fact that health information has no direct effect on pork demand. This suggests that advertising is not likely to be an effective strategy for countering health-based negative publicity. Rather, resources would probably be better spent on research to lower product costs at retail or on product redesign to make beef and pork more 'heart-healthy'. Heart-healthy products, in addition to generating benefits for beef and pork producers, might also yield a social dividend in terms of reduced cost for medical care (Gray *et al.*, 1998).

As for research implications, study results indicate that health-information elasticities that hold prices constant, i.e. partial elasticities, will overstate actual impacts. This is especially true in the case of beef, where the partial health-information elasticity (−0.186) is three times larger than the total elasticity (−0.060). In addition, a statistically significant health-information elasticity is neither a necessary nor a sufficient condition for an economically significant impact. In particular, owing to induced price effects, an increase in health information generates a negative externality for pork producers, even though the health-information elasticity for pork is zero. Thus, demand models *per se* are inadequate for drawing inferences about the economic impacts of health information. Specifically, shifts in demand caused by health information give rise to changes in relative prices, which, in turn, affect demand through second-round or feedback effects. To capture these feedback effects, the supply side of the markets needs to be specified and simulated along with the demand equations.

A caveat in interpreting our results is that the elasticities are based on rather old data (terminating in 1993). Given the dynamic nature of markets, especially with respect to information effects, it is possible that the elasticities pertaining to advertising or health information have changed, in which case the reduced-form elasticities might be misleading. Clearly, additional research is needed to confirm the elasticity estimates and to consider a wider range of impacts, including the effects on consumers. In the meantime, it seems safe to conclude that advertising's effects on meat markets are modest, both in absolute terms and especially in relation to health-information effects.

References

Ahluwalia, R., Burnkrant, R.E. and Unnava, H.R. (2000) Consumer response to negative publicity: the moderating role of commitment. *Journal of Marketing Research* 37, 203–214.

Brester, G.W. and Schroeder, T.C. (1995) The impacts of brand and generic advertising on meat demand. *American Journal of Agricultural Economics* 77, 969–979.

Brown, D.J. and Schrader, L.F. (1990) Cholesterol information and shell egg consumption. *American Journal of Agricultural Economics* 73, 548–555.

Chang, H.S. and Kinnucan, H.W. (1991) Advertising, information, and product quality: the case of butter. *American Journal of Agricultural Economics* 73, 1195–1203.

Coulibaly, N. and Brorsen, B.W. (1999) Explaining the differences between two previous meat generic advertising studies. *Agribusiness* 15, 501–515.

Elitzak, H. (1999) *Food Cost Review, 1950–1997*. Agricultural Economics Report. No 780, ERS, USDA, Washington, DC.

Fiske, S.T. (1980) Attention and weight in person perception: the impact of negative and extreme behaviour. *Journal of Personality and Social Psychology* 38, 889–906.

Gardner, B.L. (1975) The farm-retail price spread in a competitive food industry. *American Journal of Agricultural Economics* 57, 399–409.

Gray, R., Malla, S. and Stephen, A. (1998) Canadian dietary fat substitutions, 1955–93, and coronary heart disease costs. *Canadian Journal of Agricultural Economics* 46, 233–246.

Hicks, J.R. (1946) *Value and Capital*, 2nd edn. Oxford University Press, Oxford, 340 pp.

Kim, S.R. and Chern, W.S. (1999) Alternative measures of health information and demand for fats and oils in Japan. *Journal of Consumer Affairs* 33, 92–109.

Kinnucan, H.W. (1997) Middlemen behaviour and generic advertising rents in competitive interrelated industries. *Australian Journal of Agricultural Research Economics* 41, 191–208.

Kinnucan, H.W. and Forker, O.D. (1987) Asymmetry in farm-retail price transmission for major dairy products. *American Journal of Agricultural Economics* 69, 285–292.

Kinnucan, H.W. and Miao, Y. (2000) Distributional impacts of generic advertising on related commodity markets. *American Journal of Agricultural Economics* 82, 672–678.

Kinnucan, H.W., Xiao, H., Hsia, C.-J. and Jackson, J.D. (1997) Effects of health information and generic advertising on US meat demand. *American Journal of Agricultural Economics* 79, 13–23.

Klein, J.G. (1996) Negativity in impressions of presidential candidates revisited: the 1992 election. *Personality and Social Psychology Bulletin* 22, 289–296.

Maheswaran, D. and Meyers-Levy, J. (1990) The influence of message framing and issue involvement. *Journal of Marketing Research* 27, 361–367.

Mizerski, R. (1982) An attribution explanation of the disproportionate influence of unfavorable information. *Journal of Consumer Research* 9, 301–310.

Piggott, R.R., Piggott, N.E. and Wright, V.E. (1995) Approximating farm-level returns to incremental advertising expenditure: methods and application to the Australian meat industry. *American Journal of Agricultural Economics* 77, 497–511.

Richards, T.J. and Patterson, P.M. (1999) The economic value of public relations expenditures: food safety and the strawberry case. *Journal of Agricultural Resource Economics* 24, 440–462.

Richey, M.H., McClelland, L. and Shimkunas, A. (1967) Relative influence of positive and negative information in impression formation and persistence. *Journal of Personality and Social Psychology* 6, 322–327.

Skowronski, J.J. and Carlston, D.E. (1989) Negativity and extremity biases in impression formation: a review of explanations. *Psychological Bulletin* 105, 131–142.

Tomek, W.G. and Robinson, K.L. (1990) *Agricultural Product Prices*, 3rd edn. Cornell University Press, Ithaca, New York, 360 pp.

US Department of Agriculture (USDA) (2000) *Price Spreads for Beef and Pork, Per Capita Consumption*. Economic Research Service – stock No 90006, Washington, DC.

US Department of Commerce (USDC) (1994) *Statistical Abstract of the United States 1994*. US Government Printing Office, Washington, DC.

Ward, R.W. (1999) Evaluating the beef promotion checkoff: the robustness of the conclusions. *Agribusiness* 15, 517–524.

Ward, R.W. and Lambert, C. (1993) Generic promotion of beef: measuring the impact of the US beef checkoff. *Journal of Agricultural Economics* 44, 456–466.

Wohlgenant, M.K. (1993) Distribution of gains from research and promotion in multi-stage production systems: the case of US beef and pork industries. *American Journal of Agricultural Economics* 75, 642–651.

Wohlgenant, M.K. (1999) Distribution of gains from research and promotion in multistage production systems: reply. *American Journal of Agricultural Economics* 81, 598–600.

The Impact of Dietary Cholesterol Concerns on Consumer Demand for Eggs in the USA

12

TODD M. SCHMIT AND HARRY M. KAISER

Cornell University, Ithaca, New York, USA

Introduction

Beginning in the 1980s, the Incredible Edible Egg® suffered a barrage of negative publicity surrounding the potential link between dietary cholesterol and heart disease. The following are just a few examples of some of the negative headlines that appeared in popular periodicals:

'Fear of eggs' (Anon. *Consumer Reports*, October 1989)
'Hold the eggs and butter: cholesterol is proved deadly and our diets may never be the same' (C. Wallis, *Time*, March 1984)
'The cholesterol question: eggs and cholesterol' (M. Mikolas, *Blair and Ketchum's Country Journal*, July 1983)

It has been empirically documented in the literature that this well-publicized link has had a significant negative impact on egg consumption in the USA. For instance, separate studies by Brown and Schrader (1990) and Schmit and Kaiser (1998) found consumer perceptions about dietary cholesterol to have a negative and statistically significant impact on egg demand. Using data from 1955 to 1987, Brown and Schrader (1990) estimate that egg consumption was 16–25% lower in 1987 due to negative perceptions about cholesterol. Using similar but more recent data from 1986 to 1995, Schmit and Kaiser (1998) estimate a decrease in egg consumption of 19%, *ceteris paribus*, due to increasing negative cholesterol concerns. Thus, it appears that much of the observed change in egg demand can be explained by dietary-cholesterol concerns. These studies were conducted using time-series data prior to 1996.

In the mid- to late 1990s, the relative number of positive articles concerning eggs and cholesterol began to increase – however, not without with some scepticism. Take, for example, the following:

'It's OK to eat eggs' (F. Murray, *Let's Live*, April 1996)
'Eggs' shadowy reputation turning sunny side up' (Anon., *Environmental Nutrition*, February 1996)

Then, in 1997, there was a clear change in perceptions about cholesterol and egg consumption following a landmark study by the University of Arizona. Based on dietary studies on over 8000 subjects carried out over a 25-year period, this study concluded that the main influence on blood cholesterol levels was saturated fat in the diet rather than dietary cholesterol (Howell *et al.*, 1997).[1] Following this study, there was a substantial increase in publicity indicating that eggs could be part of a heart-healthy and nutritious diet. To illustrate, consider how different the tone is in the following headlines appearing after the study was released:

'Is cholesterol really bad for you? Dietary vs. blood cholesterol' (N. Simon, *New Choices*, May 2000)
'Sunny-side up: eggs and cholesterol' (C. Gorman, *Time*, May 1999)
'Comforting words for egg lovers' (Anon., *Good Housekeeping*, April 1998)
'Egg-ception to cholesterol?' (C. Marandino, *Vegetarian Times*, October 1997)
'Scientists ease up on fear of eggs' (G. Kolata, *New York Times*, September 1997)

To our knowledge, there has been no study done since the University of Arizona report to measure the impact of this change in scientific evidence concerning cholesterol and egg consumption.

Accordingly, the purpose of this chapter is to re-examine the impact that consumer perceptions have had on egg demand in the USA. We estimate a quarterly econometric model using US time-series data from 1987 to 2000. Per capita egg demand is hypothesized to be a function of price, income, price of substitutes, generic advertising and a cholesterol-awareness index constructed using popular-press periodicals as a proxy for consumer exposure to information about dietary cholesterol. The demand model incorporates slope and intercept dummy variables to measure whether the cholesterol-awareness elasticity differed between the years 1987–1996 and 1997–2000 to coincide with changing consumer cholesterol perceptions (hereafter called the changing-message model). This model is compared with a restricted model assuming that the demand response to cholesterol information was constant over the entire time period (hereafter called the constant-message model). Our hypothesis is that cholesterol awareness had a more negative impact on egg demand in the earlier period prior to the release of the comprehensive report. The impact of a 10% shock in the cholesterol-awareness index on producer prices and egg consumption over alternative time periods is simulated using both models.

Methodology

Simulation to evaluate changes in demand and prices requires information on producer response and price behaviour, in addition to modelling demand response. The conceptual framework of the econometric model consists of four equations: a farm supply equation, a wholesale derived demand equation, a price-linkage equation between the farm and wholesale markets and an identity relation to close the model and account for other exogenous determinants in the egg market. Descriptions of the variables used in the analysis are listed in Table 12.1.

Farm supply

On the supply side, a farm-level supply equation was estimated with national quarterly data over the time period 1987–2000. Farm production levels per capita have demonstrated a decreasing trend from 1987 to 1989, but have been modestly trending upward since (Fig. 12.1). Assuming a Cobb–Douglas production relationship and profit-maximizing firm behaviour, aggregate supply is modelled as:

$$QSF_t = \beta_0 \left(FP_t^* / FdP_t^*\right)^{\beta_1} \exp\left(\sum_{i=1}^{3} \gamma_i DUM_{i,t}\right) QSF_{t-1}^{\alpha_1} \exp\left(TREND_t\right)^{\alpha_2} \qquad (12.1)$$

where QSF_t is the farm supply of shell eggs, FP_t^* is the expected farm egg price (net of the American Egg Board (AEB) assessment, $assess_t$), and FdP_t^* is the expected feed price. Binary variables ($DUM_{i,t}$) are included to account for quarterly seasonality in production, and lagged production is included to account for rigidities in production associated with the sequential process of layer transfers from primary breeders to the laying flock as producers react to changes in relative prices. A linear trend term ($TREND_t$) is included to account for population increases and as a proxy for technological progress in production over time. As feed costs represent roughly 60% of total farm production costs, profit-maximizing farm production decisions can adequately be modelled by the egg/feed price ratio. Price expectations are assumed to follow a two-quarter lag moving-average process.[2]

Table 12.1. Variable source and description.

Variable	Abbreviation	Source	Definition
Consumer price index	CPI	E	Consumer price index, all items (2000 = 1)
Population	RESPOP	F	US resident population
Media cost index	MCI	I	Media cost index (2000 = 1)
Egg production	QSF	A, D	Farm egg production, million dozen
Wholesale egg demand	D	A, D	Wholesale egg demand (shell eggs and egg products) divided by RESPOP, eggs per capita
Egg stocks	CHSTOCKS	A, B	Change in egg storage stocks, refrigerated shell eggs and frozen egg products, million dozen
Net exports	NETEXPRT	A, G	Net exports, shell eggs and egg products, million dozen
Hatching eggs	HATUSED	A, H	Eggs used for hatching, million dozen
Farm price	FP	A, C	Producer price of market eggs divided by CPI, cents per dozen
Feed cost	FdP	A, C	Average production-weighted feed cost divided by CPI, cents per dozen
Wholesale price	WP	A, C	12-Metro area average wholesale egg price (Grade A large) divided by CPI, cents per dozen
Income	Y		US disposable personal income per capita divided by CPI, $'000 per capita
Cereal price	CBP	E	Real cereal and bakery products price index (2000 = 1)
Cholesterol index	CHOL	J	Cholesterol-awareness index (1987.1 = 1)
Seasonality	DUM1-DUM3	n/a	Quarterly binary variables
Message change	D97	n/a	Binary variable = 1 for year ≥ 1997, otherwise = 0
Trend	TREND	n/a	Cumulative trend variable (1987.1 = 1)
Wage rate	WAGE	E	Average earnings of production workers in poultry slaughter and processing divided by CPI, $ $hour^{-1}$
AEB advertising	ADV	I	AEB advertising expenditure divided by MCI

Sources:
A: *Poultry Yearbook*, 1996, Economic Research Service (ERS), USDA, No. 89007B, http://usda.mannlib.cornell.edu/data-sets/livestock/89007
B: *Cold Storage*, various issues, 1996–2001, National Agricultural Statistical Service (NASS), USDA, http://usda.mannlib.cornell.edu/reports/nassr/other/pcs-bban
C: *Agricultural Prices*, various issues, 1996–2001, NASS, USDA, http://usda.mannlib.cornell.edu/reports/nassr/price/zap-bb
D: *Chickens and Eggs*, various issues, 1996–2001, NASS, USDA, http://usda.mannlib.cornell.edu/reports/nassr/pec-bb1
E: Bureau of Labor Statistics, http://stats.bls.gov/blshome.html
F: Economagic.com, Economic Time Series Page, Economagic, LLC, http://www.economagic.com
G: Personal communication, Dave Harvey, ERS, USDA.
H: Personal communication, Milt Madison, Farm Service Agency, USDA.
I: Grey Advertising, New York.
J: Author-derived index based on cholesterol article count in *Readers Guide to Periodical Literature*, 1987–2001.
n/a, not applicable; USDA, US Department of Agriculture.

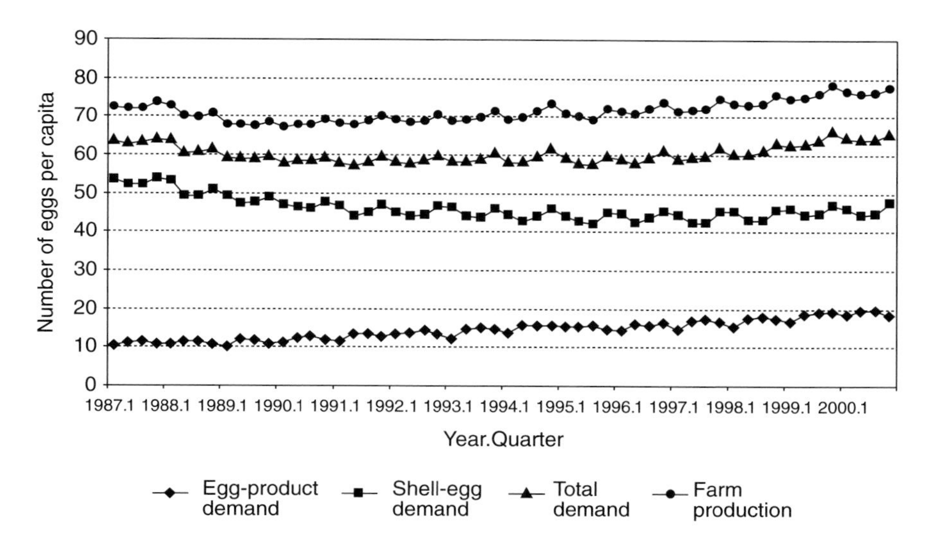

Fig. 12.1. US per capita egg production and consumption. (From: *Poultry Yearbook* (1996), No. 89007B, Economic Research Service, USDA; *Chickens and Eggs* (1996–2001), National Agricultural Statistical Service, USDA.)

After substituting lagged prices for the unobserved expected prices, taking the natural logarithm of Equation 12.1 and adding an error term, ε_t, the final supply response model can be expressed as:

$$\ln\left(QSF_t\right)=\ln\left(\beta_0\right)+\beta_1\ln\left(\overline{P_t}\right)+\sum_{i=1}^{3}\gamma_i DUM_{i,t} +\alpha_1\ln\left(QSF_{t-1}\right)+\alpha_2 TREND_t+\varepsilon_t \qquad (12.2)$$

where:

$$\overline{P_t}=\left(\frac{FP_{t-1}-assess_{t-1}}{FdP_{t-1}}+\frac{FP_{t-2}-assess_{t-2}}{FdP_{t-2}}\right)\Big/2$$

Wholesale demand

On the demand side, a wholesale per capita egg demand function was estimated based on national quarterly time-series data for the period of 1987–2000. Per capita demand was measured using commercial disappearance of eggs and egg products, which was divided by the resident population of the USA. While shell-egg demand has been relatively

Fig. 12.2. Real wholesale and farm shell-egg prices. (From: *Poultry Yearbook* (1996), No. 89007B, Economic Research Service, USDA; *Agricultural Prices* (1996–2001), National Agricultural Statistical Service, USDA.)

stable since the late 1980s, increases in egg-product consumption have been an important determinant in the upward trend in overall egg consumption (Fig. 12.1). Egg-product consumption averaged nearly one-quarter of total consumption over the study period, with the proportion of total consumption increasing from about 17% in 1987 to 30% in 2000. In general, we express wholesale demand per capita as:

$$D_t = f\left(WP_t, Y_t, CBP_t, CHOL_t, ADV_{t-s}, DUM_{i,t}, D97_t\right) \tag{12.3}$$

where D_t is per capita wholesale demand, WP_t is the wholesale price for shell eggs, Y_t is consumer per capita disposable income, CBP_t is the price of cereal and bakery products (assumed substitutes for eggs), $CHOL_t$ is an index of consumer dietary-cholesterol awareness (explained in detail below), ADV_{t-s} are lagged generic advertising expenditures for eggs, $DUM_{i,\ t}$ are quarterly binary variables to capture seasonal shifts in demand, and $D97_t$ is a binary variable segmenting the time-series data before and after 1997.

Real wholesale and farm egg prices over the study period are displayed in Fig. 12.2. Prices appear to be stabilizing after a decrease since their most recent peak in 1997. Egg demand is also influenced by changes in prices for competing products. In our model, we include a consumer price index for cereal and bakery products since these two types of breakfast products are the main competitors of eggs.

The impact of advertising on egg demand was captured through a variable reflecting AEB generic advertising expenditures. There is a large amount of literature suggesting that both current and past generic advertising efforts have an impact on current purchase behaviour (Forker and Ward, 1993; Ferrero *et al.*, 1996). To mitigate the impact of multicollinearity among the lagged advertising variables, the lag weights were approximated using a second-degree polynomial distributed lag (PDL) structure, with both endpoints restricted to zero. This structure requires an estimate of the slope of the advertising response function.[3] Generic AEB advertising expenditures are included in the wholesale demand model with a four-quarter PDL structure. Alternative lag lengths were evaluated based on previous generic advertising studies for egg products and the final lag selection was based on overall goodness of fit.

In order to account for the exposure of consumers to information about dietary cholesterol, an index similar to those developed by Brown and Schrader (1990) and McGuirk *et al.* (1995) was constructed. Brown and Schrader's quarterly index (1966 to mid-1987) was constructed by counting articles in medical journals supporting the link between dietary cholesterol and heart disease minus those refuting the link. McGuirk *et al.* (1995) argued that popular press periodicals, rather than medical journals, more accurately reflect the level of consumer awareness, and accordingly constructed an index from the 1960s to the 1980s by cumulatively counting articles (weighted by readership) cited in the *Reader's Guide to Periodical Literature* (*RGPL*) addressing the relationship between dietary cholesterol and heart disease.

There are arguably alternative approaches to estimating the level of consumer awareness to dietary cholesterol. We developed a quarterly index similar to that of McGuirk *et al.* (1995) for the period from 1975 to 2000 by counting articles (weighted by readership) in the *RGPL* associated with dietary cholesterol as an indicator of consumer awareness regarding dietary cholesterol.[4] As with McGuirk *et al.* (1995) the cholesterol awareness index ($CHOL_t$) is assumed to have a cumulative effect and can therefore be expressed as:

$$CHOL_t = \sum_{s=1}^{t} WCOUNT_s \qquad (12.4)$$

where $WCOUNT_s$ is the quarterly article count, weighted by periodical subscription levels. The index for the time period 1987–2000 (normalized to unity at the beginning of 1987) was selected to coincide with the time frame for which the remaining data were available. While the cumulative index increased relatively quickly in the mid- to late 1980s, recent years display a more gradual increase over time (Fig. 12.3). However, as mentioned earlier, the message contained in the articles in the late 1990s had largely changed.

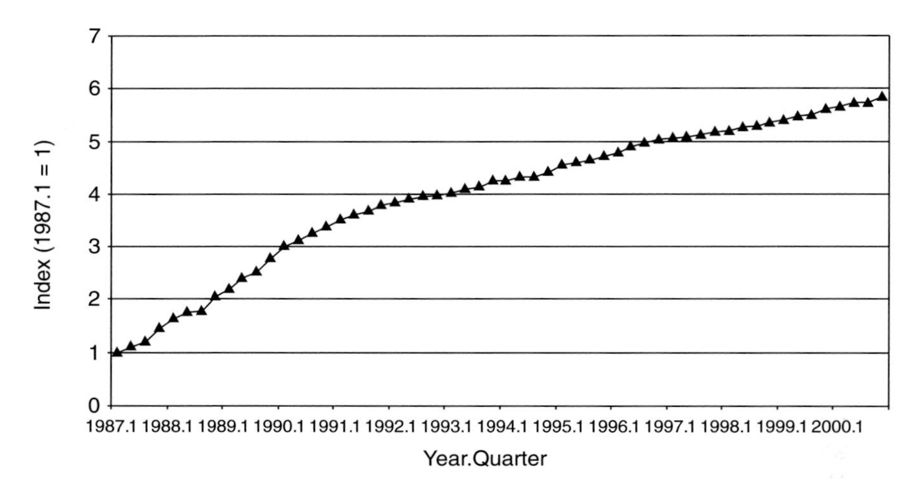

Fig. 12.3. Cholesterol-awareness index. (From: author-derived index.)

As compared with the Brown and Schrader (1990) approach, this index construction does not identify positive versus negative articles surrounding dietary cholesterol and heart disease. A careful review of the hundreds of articles used in the construction of the index would be necessary in order to make such a distinction, if in fact the distinction is exhibited in the article at all. It may be the case, particularly with popular press articles, that such a clear indication cannot be made. Some articles may be predominantly influenced in one direction or another, but some may be mixed. In any event, such a detailed distinction is beyond the scope of this study. The index is constructed, *per se*, as a measure of consumer awareness of the issue to identify the level of attention raised in society as consumers make diet decisions. The awareness level can then be increased as a result of positive, negative or mixed-message articles. However, this issue is a clear motivation for the direction of future research on this topic.[5]

In order to allow demand response to change with the release of more positive information about eggs and cholesterol, we add intercept and slope dummies with respect to the cholesterol message. This will allow us to statistically compare changes in response from the estimated elasticities before and after 1997. If in fact there has been a change of information from negative to positive, we expect the sign associated with the slope dummy to be positive.

After substitution and taking the natural logarithm of Equation 12.3, the resulting wholesale demand equation can be expressed as:

$$\ln(D_t) = \delta_0 + \delta_1 \ln(WP_t) + \delta_2 \ln(Y_t) + \delta_3 \ln(CBP_t) + \sum_{i=1}^{3} \omega_i DUM_{i,t} + \sum_{j=0}^{4} \gamma_i \ln(ADV_{t-j}) + \delta_4 \ln(CHOL_t) + \delta_5 D97_t + \delta_6 (D97_t)(\ln(CHOL_t)) + \upsilon_t \quad (12.5)$$

subject to:

$$\gamma_j = \lambda_0 + \lambda_1 j + \lambda_2 j^2 \text{ and } \gamma_{-1} = \gamma_{L+1} = 0$$

Price mark-up

In order to relate changes in demand and price at the wholesale level to the resulting price and production effects at the farm level, a farm-to-wholesale price mark-up equation was estimated. Farm egg prices are a major input cost to wholesalers and the highly correlated movement of these prices certainly reflects this (Fig. 12.2). It is interesting to note, however, that the real price margin has been relatively constant since 1987, with an average real-price differential of just under 28 cents per dozen. The price-mark-up relationship between farm and wholesale levels can be specified generally as:

$$WP_t = g(FP_t, WAGE_t, DUM_{i,t}) \quad (12.6)$$

where WP_t, FP_t and $DUM_{i,t}$ are as indicated above and $WAGE_t$ is the average hourly earnings of production workers in poultry slaughter and processing to account for a predominant variable cost in egg processing. Since eggs in storage (refrigerated shell eggs and frozen egg products) are a minor proportion of quarterly sales (i.e. less than 1% of sales on average), inventory-pricing behaviour was not assumed to be a contributing factor to wholesaler decisions. However, quarterly binary variables are included to account for any seasonal variation in wholesaler behaviour. Adding an error term, ω_t, the estimated equation can be expressed as:

$$WP_t = \phi_0 + \phi_1 FP_t + \phi_2 WAGE_t + \sum_{i=1}^{3} \kappa_i DUM_{i,t} + \omega_t \quad (12.7)$$

Identity relation

Finally, to close the model and to account for other assumed exogenous product movements, the following identity relation was applied:

$$QSF_t = QD_t + CHSTOCKS_t + NETEXPRT_t + HATUSED_t \qquad (12.8)$$

where QSF_t is defined as above, QD_t is total wholesale demand, $CHSTOCKS_t$ is the change in egg storage stocks, $NETEXPRT_t$ is the level of net exports (product and shell) and $HATUSED_t$ is the utilization of eggs for hatching. The additional components have relatively low proportions to total demand (i.e. on average *STOCKS* at 0.9%, *NETEXPRTS* at −3.0% and *HATUSED* at 14.9%) and eggs used for hatching are designated for a specialized purpose. The specification of this identity relation is necessary to simulate the econometric model and estimate the endogenous price and quantity components.

Econometric Results

For the purposes of estimation, farm supply is treated as predetermined, since producer price expectations are a function of lagged prices, and the remaining variables are also predetermined. Given that this equation has predetermined right-hand-side variables, ordinary least squares (OLS) will yield consistent estimates of the parameters. This is not the case, however, for the demand and price mark-up relations. Endogenous price variables and simultaneity associated with these two equations require an estimation approach that accounts for both the endogeneity and correlation of the equation residuals. Accordingly, we use a three stage least squares (3SLS) estimation procedure for the demand and mark-up equations, using the predetermined/exogenous variables in the model as a set of appropriate instruments.[6]

Regression results for the farm supply and mark-up equations are displayed in Table 12.2, while the demand results for the constant-message and changing-message models are displayed in Table 12.3. The initial demand equation exhibited a serial correlation problem. To correct for this, the final model incorporated a one-period autoregressive error structure, (AR(1)).

Since a primary interest involves the estimated demand response, the Box–Cox power-transformation technique was used to test the appropriateness of the double-log form for the demand equation. The results supported the use of this functional form.[7] Because of the logarithmic specification, each of the estimated parameters on the continuous variables has the interpretation of an elasticity. All equations fit the data well, with all adjusted R^2 above 0.90. Finally, the expected signs of the estimated coefficients are theoretically consistent and, in most cases, statistically significant at conventional confidence levels.

All of the variables in the farm supply equation are statistically significant at the 5% confidence level and both the short-run (0.02) and the long-run (0.20) price elasticities are highly inelastic – i.e. supply is

Table 12.2. Farm supply and price mark-up estimated equations.[a]

Variable	Parameter estimate	Standard error	P value
Farm supply[b]			
Intercept	0.7937	0.3537	0.0296
Farm price ratio	0.0216	0.0103	0.0405
DUM1	−0.0468	0.0037	< 0.0001
DUM2	−0.0315	0.0036	< 0.0001
DUM3	−0.0217	0.0033	< 0.0001
Trend	0.0006	0.0002	0.0057
Farm supply$_{-1}$	0.8914	0.0489	< 0.0001
R^2	0.9880		
DW	2.0945		
Price mark-up equation (estimated with constant-message demand model)[c]			
Intercept	22.2220	21.7897	0.2793
Farm price	1.0711	0.0287	< 0.0001
Wage rate	−0.0297	2.3213	0.9898
DUM1	−1.9603	0.9725	0.0497
DUM2	−1.5422	1.0374	0.1439
DUM3	2.2896	0.9938	0.0258
R^2	0.9762		
DW	1.8609		
Price mark-up equation (estimated with changing-message demand model)[c]			
Intercept	23.8004	21.7396	0.3131
Farm price	1.0742	0.0287	< 0.0001
Wage rate	0.1225	2.3283	0.9583
DUM1	−1.9548	0.9725	0.0503
DUM2	−1.5015	1.0372	0.1545
DUM3	2.3114	0.9939	0.0245
R^2	0.9763		
DW	1.8655		

[a] All monetary variables are converted to 2000 dollars.
[b] Farm supply is converted to its natural logarithm; the price ratio is the natural logarithm of the ratio of farm price to feed price as specified in the text and estimated in SAS using OLS.
[c] The dependent variable is the wholesale egg price. Parameter estimates are estimated simultaneously with the demand model with 3SLS using PROC MODEL in SAS.

not very sensitive to changes in price (Table 12.2). The short-run estimate is similar to that of Reberte *et al.* (1996) and Schmit *et al.* (1997), who estimated short-run price elasticities of supply at 0.02 and 0.05 for national and California egg production, respectively. Chavas and Johnson (1981) estimated a short-run price elasticity of 0.03 and a long-run price elasticity of nearly 1 over the time period of 1965–1976. While

Table 12.3. Wholesale demand per capita econometric results, restricted and unrestricted models.[a]

	Constant message			Changing message[c]		
Variable[b]	Parameter estimate	Standard error	P value	Parameter estimate	Standard error	P value
Intercept	−0.6854	3.4879	0.8452	−1.7721	2.8991	0.5445
Wholesale price	−0.0488	0.0302	0.1132	−0.0415	0.0326	0.2105
Income	0.3533	0.3000	0.2455	0.4449	0.2681	0.1048
Cereal price	0.4102	0.5013	0.4178	0.4395	0.4644	0.3497
DUM1	−0.0295	0.0033	< 0.0001	−0.0285	0.0034	< 0.0001
DUM2	−0.0420	0.0060	< 0.0001	−0.0406	0.0063	< 0.0001
DUM3	−0.0317	0.0044	< 0.0001	−0.0301	0.0044	< 0.0001
Cholesterol index	−0.1872	0.0631	0.0049	−0.1538	0.0458	0.0017
Cholesterol index × D97				0.5854	0.2293	0.0146
D97				−0.9482	0.3752	0.0156
AR(1)	0.9528	0.0452	< 0.0001	0.8335	0.2038	0.0002
Adv_L0	0.0004	0.0002	0.1169	0.0004	0.0002	0.0500
Adv_L1	0.0006	0.0004	0.1169	0.0007	0.0003	0.0500
Adv_L2	0.0007	0.0004	0.1169	0.0008	0.0004	0.0500
Adv_L3	0.0006	0.0004	0.1169	0.0007	0.0003	0.0500
Adv_L4	0.0004	0.0002	0.1169	0.0004	0.0002	0.0500
Adv_LR	0.0027	0.0017	0.1169	0.0031	0.0015	0.0500
R^2	0.9195			0.9318		
Adjusted R^2	0.9022			0.9130		
DW	2.4479			2.5051		

[a] All monetary variables are converted to 2000 dollars. The demand models are estimated simultaneously with the price mark-up equation using 3SLS with PROC MODEL in SAS.
[b] Wholesale demand per capita, prices, income and the cholesterol index are converted to their respective natural logarithms. Adv_LR represents the computed long-run advertising elasticity from the individual advertising lag components.
[c] Wald tests for $\delta_5 = 0$, $\delta_6 = 0$ and $\delta_5 = \delta_6 = 0$ were rejected at the 5% significance level with chi-square test statistics of $W_5 = 6.52$, $W_6 = 6.49$ and $W_{56} = 6.85$, respectively.

the poultry sector has become more vertically integrated, implying higher price sensitivity, egg production under contract to market firms or carried out as only one phase within vertically integrated operations may contribute less elastic responses (Salathe *et al.*, 1983). Furthermore, supply elasticities are generally lower for products with few alternative uses (Tomek and Robinson, 1990: 61); therefore, high capital investment suitable only for egg production may prohibit large production adjustments. The relatively large coefficient on the lagged-supply term is indicative of significant capacity constraints in egg production.

Both price mark-up models are displayed in Table 12.2 from the simultaneous estimation with the constant-message and changing-message demand models, respectively. Both estimation results are provided for completeness; however, given the little difference between the results, our discussion is limited to that estimated with the changing-message model. As expected, farm price is the most significant component in predicting wholesale price in the price mark-up equation. The elasticity of the wholesale price with respect to the farm price (evaluated at sample means), also known as the farm-to-wholesale price transmission elasticity, is 0.75. Given potential rigidities in processed quantities and/or other input prices remaining fixed while producer prices change, we would expect the price transmission elasticity to be below 1. This estimate appears quite reasonable compared to previous egg price transmission elasticity estimates, e.g. 0.71 by George and King (1971) and 0.72 by Chavas and Johnson (1981). Pricing behaviour appears to have some seasonal influences as well, with significant responses in the first and third quarters. The insignificant response on processor wages may reflect the limited processing that eggs undergo between the farm gate and wholesale shipments (Salathe *et al.*, 1983).

The constant-message and changing-message demand models are shown in Table 12.3. While both equations appear to fit the data well, Wald tests reject the null hypothesis that the coefficients associated with the intercept and slope dummy variables for cholesterol awareness are zero.[8] As such, we shall focus the predominance of our discussion on the changing-message model estimates.

The wholesale price is negative (−0.04), as expected, but not significant. This response was also not significant in Reberte *et al.* (1996); however, the magnitude is within the range estimated by Brown and Schrader (1990) (−0.02 to −0.17) and similar to the −0.08 estimated by Wohlgenant (1989). This is not too surprising given the increasing proportion of egg-product consumption in the total disappearance amount that is only indirectly reflected in the wholesale price of shell eggs. Per capita income and the composite price index of substitutes (cereal and bakery products) are of the right sign, but neither is significant.

Generic egg advertising expenditures by the AEB were found to have a positive and statistically significant impact on per capita egg demand. The estimated long-run advertising elasticity is 0.003, which is similar in magnitude to that of Schmit and Kaiser (1998). While the response may seem small, it is important to remember that this result is the net of the response from other factors, namely cholesterol awareness, which underwent a significant transformation during the time period evaluated. In addition, it was certainly the case that recent generic advertising efforts highlighted the results of the 1997 landmark study.

As hypothesized, the constant-message model indicates that the cholesterol-awareness index had a negative and statistically significant impact on per capita egg demand (δ_4). For each 1% increase in this index, per capita egg demand declines by 0.19%. However, when we factor in the change of consumer periodical information that occurred after 1996, we see a very different story. Prior to 1997, the cholesterol-awareness response indicated a similar negative response to that in the constant-message model (−0.15); however, after 1996, the net response was positive ($\delta_4 + \delta_6 = 0.43$). This is quite intuitive, as the tone of the periodical messages changed from negative to positive.[9]

If we concentrate on the time period from 1997 to 2000, per capita demand increased approximately 7%, none of which could be attributed to changes in cholesterol response from the constant-message model, where the effect is strictly negative. However, from the changing-message model results, we can estimate how much of this effect can be associated from changes in dietary-cholesterol response. From 1997 to 2000, the cholesterol-awareness index increased by approximately 16%. Multiplying this increase by the associated post-1996 cholesterol-index elasticity results in a predicted increase, *ceteris paribus*, in egg demand of 6.9%. Thus, it appears that nearly all of the observed change in egg demand over this time period can be explained by changes in dietary-cholesterol concerns. Not considering such significant changes in the model specification clearly leads to erroneous conclusions.

Simulation Analysis of Changing Cholesterol Perceptions

The changing-message demand specification effectively captures changing consumer perceptions about the relationship between dietary cholesterol and heart disease. To further illustrate the importance of this structural change, we simulate both the constant-message and the changing-message econometric models under alternative shocks to cholesterol-awareness index levels. Specifically, we apply three different shocks to the system of equations: one that increases the cholesterol index by 10% over the entire time period, one that increases the cholesterol index for the pre-1997 time period and one that increases the cholesterol index for the post-1996 time period. The mean simulated demand and farm price levels are then computed for each of these time periods, respectively.

The simulation procedure simultaneously incorporates supply response with the change in demand, price transmission relationships and market-clearing relations. The long-run price elasticity of supply estimated above was used in the simulation model to allow for full producer adjustments to price changes. Using a procedure similar to that in Alston *et al.* (1996), the long-run supply estimate was defined in constant-elasticity form and then equated with the predicted demand quantities. Changes in demand due to the alternative cholesterol-index levels then affect the level of production and the resulting egg prices. Specifically, wholesale demand, price mark-up and market-clearing equations were simulated with the following long-run elasticity of supply form as:

$$QSF_t = A_t\, R_t^{\varepsilon} \tag{12.9}$$

where $A_t = (QD_t^* + CHSTOCKS_t + NETEXPRT_t + HATUSED_t)/R_t^{\varepsilon}$, $R_t = FP_t - assess_t$, R_t is the real net farmer return per dozen eggs in quarter t, ε is the long-run supply elasticity and QD_t^* is the simulated demand quantity. The defined value, A_t, varies by year and ensures that, given the actual values of prices and other variables, markets clear and the supply equation passes through the quantity defined by QD^*_t.

Average per capita demand and farm price levels for the alternative cholesterol-index shock simulations are shown in Table 12.4. As expected, comparing the mean statistics for the base-model simulations (i.e. no cholesterol shock) with the data averages for the respective time periods indicate that the unrestricted demand model has better within-sample prediction ability, particularly for the post-1996 period. Apart from that, what is important here is the difference in the implications from the cholesterol-index shocks relative to the base-period simulations.

Using the constant-message demand model, a 10% increase in the cholesterol index over the entire sample period would result in a predicted 1.6% decrease in average per capita demand and a 6.3% decrease in average farm price for the period 1988–2000. Since the changing-message model accounts for changes in dietary-cholesterol perceptions after 1996, these reductions are tempered considerably. In fact, the negative effect from the first part of the time period is effectively cancelled out by the positive effect since 1997, resulting in little change in average demand and farm price over the entire time period. The smaller cholesterol elasticity for the pre-1997 time period on the changing-message model (−0.15) versus the cholesterol elasticity on the constant-message model (−0.19) also results in slightly tempered results for the 10% pre-1997 cholesterol.

Table 12.4. Average simulated egg demand and farm price under alternative models and cholesterol-index shocks.

		Model					
		Restricted			Unrestricted		
Cholesterol-index shock	Data average	Base	Shock	Percentage change	Base	Shock	Percentage change
10% increase, 1988–2000							
Quarterly demand per capita	60.27	57.77	56.87	−1.56%	60.31	60.49	0.30%
Farm price	62.88	54.26	50.82	−6.34%	63.26	63.32	0.09%
10% increase, 1988–1996							
Quarterly demand per capita	59.29	57.98	57.09	−1.54%	59.48	58.72	−1.28%
Farm price	67.60	61.76	57.83	−6.36%	68.56	64.89	−5.35%
10% increase, 1997–2000							
Quarterly demand per capita	62.46	57.31	56.40	−1.59%	62.17	64.49	3.73%
Farm price	52.27	37.38	35.03	−6.29%	51.34	59.81	16.50%

As expected, the simulation results are inversely related for the post-1996 cholesterol shock. For the time period of 1997–2000, the constant-message model demonstrated similar demand and price reductions to those in the previous simulations. However, with the changing consumer perceptions, a 10% increase in the cholesterol index post-1996 would actually increase average demand by 3.7% and average farm price by 16.5%. While the direction of effects from the simulations was expected given the econometric estimates, the simulation procedure reveals the magnitude of these differences. Certainly, imposing constant consumer perceptions about dietary cholesterol over this relatively long time period would bias downward average demand and price effects from cholesterol awareness. If the message changes, so must the model that is attempting to describe it.

Conclusions

Bob Dylan had it right when he sang 'The Times They Are A-Changing' back in 1964. However, those words apply equally well over the past two decades when considering the transformation of information surrounding the link between dietary cholesterol and heart disease. The Incredible Edible Egg® has not been immune to such fluctuations, as it suffered from a barrage of negative publicity in the late 1980s and early 1990s. However, the picture changed decisively following a landmark study in 1997 that concluded that the main influence of blood cholesterol levels was saturated fat in the diet rather than dietary cholesterol (Howell *et al.*, 1997). Following this study, there was a substantial increase in publicity, indicating that eggs could be part of a heart-healthy and nutritious diet.

An econometric model was developed to test this change of cholesterol perceptions between the pre-1997 and post-1996 time periods. The model demonstrated a significant change in response to cholesterol information from one that was negative to one that was positive. In fact, it was estimated that nearly all of the recent increases in per capita egg demand were the result of this change in information. Simulation analysis demonstrated that, if such a structural change is not properly accounted for, large downward biases result in predicted changes in egg demand and producer prices.

Notes

[1] Along with the highly publicized landmark study in the USA, it should be noted that, earlier in 1997, an article was also published in the *British Medical Journal* using a meta-analysis approach to study the effects of dietary fat and

dietary cholesterol in British diets (Clarke *et al.*, 1997). The British study used data from 82 metabolic ward studies with 395 dietary experiments among 129 groups of individuals. The results were very comparable to the larger study by Howell *et al.* (1997). ENC (1997) provides a comparison of the two studies.

[2] Alternative price-expectation assumptions were also evaluated, including a second-order Almon lag of past prices. While theoretically less restrictive but more difficult to estimate, the resulting price effects were empirically very similar to the moving-average price structure modelled here. As such, the moving-average price assumption was assumed to adequately model producer response to market price changes.

[3] The PDL structure with end-point restrictions can be written as:

$$D_t = \tau + \sum_{j=0}^{L} \gamma_j ADV_{t-j} + v_t$$

subject to $\gamma_j = \lambda_0 + \lambda_1 j + \lambda_2 j^2$ and $\gamma_{-1} = \gamma_{L+1} = 0$; L is the total lag length and all other variables are suppressed into τ for simplicity. After substituting, this reduces to:

$$D_t = \tau + \lambda_2 ADV_t^* + v_t$$

where:

$$ADV_t^* = \sum_{j=0}^{L} (j^2 - Lj - (L+1)) ADV_{t-j}$$

The individual lag-advertising parameters can then be recovered from the estimated value of λ_2, i.e. $\beta_j = \lambda_2(j^2 - Lj - (L + 1))$. An F test was conducted on the PDL end-point restrictions; a computed F statistic of 0.046 clearly cannot reject the zero end-point restriction at any reasonable significance level.

[4] Keyword descriptors included cholesterol, blood cholesterol, hypercholesterolaemia and cholesterol transport.

[5] In addition, the index as constructed in Equation 12.4 does, in some sense, assume a 100% memory retention rate, a noted restrictive hypothesis. However, as information is released, the level of awareness could reasonably be assumed to increase monotonically. Alternative specifications of the index were considered, e.g. Almon-lag structures, with little change in the empirical results. As such, the current index construction was retained.

[6] The model equations were estimated with the SAS statistical package, using the PROC MODEL procedure.

[7] The Box–Cox power transformation is expressed as:

$$y^{(\lambda)} = \begin{cases} \left(y^{\lambda} - 1\right)/\lambda & \lambda \neq 0 \\ \ln y & \lambda = 0 \end{cases}$$

Likelihood ratio tests (Judge *et al.*, 1988: 99) reject the hypothesis that the transformation is linear ($\lambda = 1$) but fail to reject the hypothesis that the transformation is logarithmic ($\lambda = 0$) at the 5% significance level. Test statistics are 5.97 and 2.77, respectively, with a chi-square critical value of 3.84.

[8] Wald tests for $\delta_5 = 0$, $\delta_6 = 0$ and $\delta_5 = \delta_6 = 0$ were rejected at the 5% significance level with chi-square test statistics of $W_5 = 6.52$, $W_6 = 6.49$ and $W_{56} = 6.85$, respectively.
[9] As noted by a reviewer, if the change in information changes people's preferences, then it may be the case that price or income responsiveness will also change. This was originally hypothesized not to be the case, given the expected insensitivities of demand to these variables, and subsequent estimation revealed that none of these interaction effects were statistically significant.

References

Alston, J.M., Chalfant, J.A., Christian, J.E., Meng, E. and Piggott, N.E. (1996) The California table grape commission's promotion program: an evaluation. In: Ferrero, J.L. and Clary, C. (eds) *Economic Evaluation of Commodity Promotion Programs in the Current Legal and Political Environment.* National Institute for Commodity Promotion Research and Evaluation, Cornell University, Ithaca, New York, pp. 33–62.

Anon. (1989) Fear of eggs. *Consumer Reports* 54, 650–652.

Anon. (1996) Eggs' shadowy reputation turning sunny side up. *Environmental Nutrition* 19, 7.

Anon. (1998) Comforting words for egg lovers. *Good Housekeeping* 226, 148.

Brown, D.J. and Schrader, L.F. (1990) Cholesterol information and shell egg consumption. *American Journal of Agricultural Economics* 72, 548–555.

Chavas, J.-P. and Johnson, S.R. (1981) An econometric model of the US egg industry. *Applied Economics* 13, 321–335.

Clarke, R., Frost, C., Collins, R., Appleby, P. and Peto, R. (1997) Dietary lipids and blood cholesterol: quantitative meta-analysis of metabolic ward studies. *British Medical Journal* 314, 112–117.

Egg Nutrition Center (ENC) (1997) Meta-analyses of plasma lipoprotein responses to changes in dietary fat and cholesterol. *Nutrition Close Up,* Special Report.

Ferrero, J., Boon, L., Kaiser, H.M. and Forker, O.D. (1996) *Annotated Bibliography of Generic Commodity Promotion Research (Revised).* NICPRE Research Bulletin 96–3, National Institute for Commodity Promotion Research and Evaluation, Department of Agricultural, Resource, and Managerial Economics, Cornell University, Ithaca, New York.

Forker, O.D. and Ward, R.W. (1993) *Commodity Advertising: the Economics and Measurement of Generic Programs.* Lexington Books, New York, 294 pp.

George, P.S. and King, G.A. (1971) *Consumer Demand for Food Commodities in the United States with Projections for 1980.* Giannini Foundation Monograph No. 26, University of California, Berkeley, California, 161 pp.

Gorman, C. (1999) Sunny-side up: eggs and cholesterol. *Time* 153, 81.

Howell, W.H., McNamara, D.J., Tosca, M.A., Smith, B.T. and Gaines, J.A. (1997) Plasma lipid and lipoprotein responses to dietary fat and cholesterol: a meta-analysis. *American Journal of Clinical Nutrition* 65, 1747–1764.

Judge, G.G., Hill, R.C., Griffiths, W.E., Lutkepohl, H. and Lee, T.C. (1988) *Introduction to the Theory and Practice of Econometrics*, 2nd edn. John Wiley and Sons, New York, 1024 pp.

Kolata, G. (1997) Scientists ease up on fear of eggs. *New York Times*, 24 September.

McGuirk, A., Driscoll, P., Alwang, J. and Huang, H. (1995) System misspecification testing and structural change in the demand for meats. *Journal of Agricultural and Resource Economics* 20, 1–21.

Marandino, C. (1997) Egg-ception to cholesterol? *Vegetarian Times* 242, 22–23.

Mikolas, M. (1983) The cholesterol question: eggs and cholesterol. *Blair and Ketchum's Country Journal* 10, 10.

Murray, F. (1996) It's OK to eat eggs. *Let's Live* 64, 14–18.

Reberte, C.J., Schmit, T.M. and Kaiser, H.M. (1996) An *ex post* evaluation of generic egg advertising in the United States. In: Ferrero, J.L. and Clary, C. (eds) *Economic Evaluation of Commodity Promotion Programs in the Current Legal and Political Environment*. National Institute for Commodity Promotion Research and Evaluation, Cornell University, Ithaca, New York, pp. 83–100.

Salathe, L.E., Price, J.M. and Gadson, K.E. (1983) The food and agricultural policy simulator: the poultry- and egg-sector submodel. *Agricultural Economics Research* 35, 23–34.

Schmit, T.M. and Kaiser, H.M. (1998) Egg advertising, dietary cholesterol concerns, and US consumer demand. *Agricultural and Resource Economics Review* 27, 43–52.

Schmit, T.M., Reberte, C.J. and Kaiser, H.M. (1997) An economic analysis of generic egg advertising in California. *Agribusiness* 13, 365–373.

Simon, N. (2000) Is cholesterol really bad for you? Dietary vs. blood cholesterol. *New Choices* 40, 76–78.

Tomek, W.G. and Robinson, K.L. (1990) *Agricultural Product Prices*, 3rd edn. Cornell University Press, Ithaca, New York, 360 pp.

Wallis, C. (1984) Hold the eggs and butter: cholesterol is proved deadly and our diets may never be the same. *Time* 123, 56–58.

Wohlgenant, M.K. (1989) Demand for farm output in complete system of demand functions. *American Journal of Agricultural Economics* 71, 241–252.

Employing an Environmental Taxation Mechanism to Reduce Fat Intake

13

JOHN M. SANTAROSSA AND DAVID D. MAINLAND

Scottish Agricultural College, Ayr, UK

Introduction

Numerous studies have established the now accepted links between dietary intake and health. Experimental, clinical and epidemiological studies and preventive trials all confirm the importance of diet in the development of coronary heart disease and other diseases. According to the British Medical Association, as many as a third of all cancer cases are diet related. The concerns about health risks associated with dietary intakes of saturated fats and cholesterol should therefore, with increasing consumer awareness, have a significant impact on the demand for foods with high contents of fats, such as red meats and, more importantly, dairy products, which are the main source of saturated fat in the UK diet (Marshall, 2000). Although there is evidence that well-designed and targeted initiatives, particularly those concerned with diet, can have some effect on reducing risk factors (Public Health Policy Unit, 1995), only modest dietary changes have occurred, with the full effect being achieved over a period of 5 years and resulting in a 10% reduction in serum cholesterol (Law *et al.*, 1994).

Attempts at altering the diet of the Scottish population by way of advertising campaigns have not proved very successful and it is difficult to find any effects of health information, as demonstrated by the inclusion of a health-information index in a Scottish food-demand model (see Rickertsen and von Cramon-Taubadel, Chapter 3, this volume). The occurrence of coronary disease results in both private and social costs in terms of productivity losses and health-care costs. This study therefore proposes a different and more radical approach to addressing the link between diet and resultant diseases by treating the latter as an externality and employing methods more commonly associated with environmental economics. By means of establishing the relationship between food expenditure, quantity consumed and nutrient intake, we investigate the effects of a fiscal-policy approach to reducing dietary fat intake.

In this study, we shall initially demonstrate the relative impact of information relating to a healthy dietary intake on food demand. Subsequently, we shall present a method for internalizing external costs. Finally, results and discussion of a simulation are presented, along with a comparison of our results with those obtained by the Center for Nutrition Policy and Promotion (CNPP, 1999) of the US Department of Agriculture (USDA). Results indicate that, in spite of a lower level of saturated fat intake at 10.65% of energy intake, this new level still does not achieve the recommended 10% of energy intake. The overall reduction in fat intake, however, is more encouraging in that both the recommended 30% of energy intake from dietary fat and the 66 g day^{-1} per person are attained.

Trends in Food Consumption

Until the early 1980s, food consumption in the average Scottish household altered little. Traditionally consisting of homemade plain meals, the Scottish diet was such that it met the dietary requirement of the majority of the population, which was involved in mostly manual occupations. These occupations were, in large part, found in sectors such as agriculture (often practised in harsh climatic conditions) or heavy industry (mining, shipbuilding, etc.). Throughout the 1980s, heavy industry saw a dramatic decline, being replaced by less physically intensive types of activities, such as the electronics sector. While in some sections of the population dietary intake remained pretty much unchanged, some others began to alter their food consumption towards either a healthier diet or more convenient food products. With regard to the healthier diet, Table 13.1 shows that consumption of whole milk declined sharply, while that of low- and non-fat milk recorded sustained increases in consumption. It is further observed that the consumption of ‘others foods’, which consist mostly of premade frozen meals, increased significantly over the period studied, marking a departure from home cooking to adopting more time-saving methods of food consumption in the home. While part of the decrease in consumption of some food items, such as beef, is explained by exogenous factors (notably the bovine spongiform encephalopathy (BSE) crisis, which resulted in consumers switching to poultry), the consumption of items such as fish and cheese remained pretty much unchanged. With respect to meat consumption, however, we observe a gradual decrease in the ‘other

Table 13.1. Consumption of major food items in Scotland. (From: National Food Survey, Department for Environment, Food and Rural Affairs.)

Food item	1980	1985	1990	1995	1999
Meat					
Beef	12.0	9.6	7.7	6.3	5.7
Pork	6.1	5.1	4.4	3.7	3.6
Poultry	9.8	9.7	11.0	11.0	12.0
Other meats	31.0	30.0	27.0	28.0	26.0
Fish	7.1	7.2	7.5	7.5	7.4
Eggs (number of)	3.7	3.2	2.2	1.8	1.7
Cheese	5.7	5.8	5.9	5.6	5.4
Fluid milk					
Whole	121.0	96.0	63.0	41.0	31.0
Low-fat	12.0	13.0	37.0	57.0	59.0
Non-fat	5.9	7.3	6.7	6.9	7.4
Vegetables and fruits					
Fresh vegetables	102.0	107.0	100.0	88.0	80.0
Fresh fruits	31.0	27.0	32.0	35.0	37.0
Processed vegetables	24.0	19.0	18.0	19.0	23.0
Processed fruits	11.0	13.0	15.0	17.0	18.0
Grains and cereals	82.0	79.0	76.0	72.0	76.0
Sugar	20.0	15.0	11.0	9.2	7.3
Fats and oils	17.0	15.0	13.0	11.0	9.7
Other foods	6.2	7.4	11.0	38.0	39.0

Figures are expressed in annual kg per capita except for eggs, which are expressed in number of eggs per capita per week. Fluid items are converted to their equivalent weight.

meats' group (which mostly consisted of products where meat content was at least 80% and of non-defined origin). In spite of a marked decrease in the consumption of fats and oils, this is not thought to be due to consumers choosing the healthy option; rather, it is the increasing reliance on ready-made meals and fast food, resulting in households abandoning home cooking. The increase in ready-made meals was a further significant factor in the declining consumption of fresh vegetables, as revealed by a correlation coefficient of −0.95 between fresh vegetables and the group 'other foods'. Overall, while the total amount of food per capita decreased from 507 kg in 1980 to 449 kg in 1999, the contribution of ready-made meals increased from 1.2% to 8.7%.

What Table 13.1 does not show, however, is that, since the mid-1980s, a persistent gap, consistently widening, has opened between low- and higher-income households as a result of shifts in industrial base and the ensuing economic boom of the 1980s, which resulted in growing inequalities in income. Another important factor that contributed to diverging trends in food consumption has been one of consumer accessibility to those food products that can provide an adequate nutrient intake. In 1998, fresh fruits and vegetables were 27% more expensive at local shops than at supermarkets, while the latter embarked on a geographical positioning strategy that aimed uniquely at maximizing profit by selectively targeting the more affluent areas. Given the limited resources available to low-income households (notably a lack of transport), they simply did not have access to affordable fresh food. Evidence from past studies has revealed the presence of large inter- and intraregional, as well as inter- and intra-urban, disparities in dietary intake, as identified by Birkin *et al.* (1996).

With respect to previous food-demand studies for Scotland, a literature search did not identify any previous work, although a number of government reports focusing on specific aspects of food consumption in Scotland have been published.

Data

Variables were constructed from data collected by a survey of households as part of the UK National Food Survey. The survey is a continuous sampling enquiry into households' food consumption and expenditure and is conducted by the Department for Environment, Food and Rural Affairs.[1] The sampling procedure is a one-stage stratified or systematic random sample and the method of data collection is a combination of face-to-face interviews and a logbook in which a nominated record keeper in each household surveyed enters the description, quantity and cost of food. The period covered for this study extends from January 1992 to December 1996 and represents monthly household average consumption and expenditure. In view of the large number of products represented in the survey (311 food items), aggregation was necessary and the following food groups were defined: meat (meats and meat products), fish (fish and fish products), dairy products (liquid milk, cheese, yoghurts and other dairy products), fruits (fresh, tinned and frozen), vegetables (fresh, tinned and frozen), cereals (bread and cereal products), fats and oils (butter, margarine, vegetable and animal oils, cooking oils and other fats), alcohol, eggs, and other foods (convenience meals, confectionery, soups, sugar, jams, etc.).

A second data set was constructed by aggregating households, with food items converted to their nutrient equivalent. This data set (comprising 3516 households) therefore gives the quantities of food consumed by each household. It also provides the data for ten nutrients, namely: carbohydrates, saturated fats, monounsaturated fats, polyunsaturated fats, fibres, vegetable proteins, animal proteins, cholesterol, minerals and vitamins. All quantity values are expressed in grams, while prices are in pence. Product prices are derived by dividing household purchase values by the quantities of purchases made for each product.

Health Information and the Almost Ideal Demand System

Public concerns about health risks associated with dietary intake of fats and cholesterol have been to date notoriously difficult to gauge, specifically when considering the potential impact of such concerns on the demand for food products with high contents of fats and cholesterol (see Rickertsen and von Cramon-Taubadel, Chapter 3, this volume). For identifying the impact of information on food consumption, the almost ideal demand system (AIDS) of Deaton and Muellbauer (1980) is used. The expenditure share for the ith good is given by:

$$w_i = \alpha_i + \sum_{j=1}^{n} \gamma_{ij} \ln p_j + \beta_i (\ln x - \ln P) \tag{13.1}$$

where $\ln P$ is a price index defined by:

$$\ln P = \alpha_0 + \sum_{j=1}^{n} \alpha_j \ln p_j + \frac{1}{2} \sum_{j=1}^{n} \sum_{k=1}^{n} \gamma_{jk} \ln p_j \ln p_k \tag{13.2}$$

p_j is the per unit price of good j and x is the per capita total expenditure. The price index $\ln P$ is frequently approximated by Stone's index:

$$\ln P = \sum_{k=1}^{n} w_k \ln p_k \tag{13.3}$$

Equations 13.1 and 13.2 define the AIDS, while Equation 13.3 defines the linear approximate almost ideal demand system (LA/AIDS). We use the linear version and follow Burton and Young (1992), who used the average expenditure share $\bar{w}_j$ in Stone's index or $\ln P^* = \sum_k \bar{w}_k \ln p_k$. The β_i parameters in Equation 13.1 determine whether goods are luxuries or necessities; $\beta_i > 0$ implies a luxury and similarly $\beta_i < 0$ implies a necessity. We denote this linear approximate version as LA/AIDS.

We imposed adding up, homogeneity and symmetry, which imply the following parametric restrictions:

$$\sum_i \alpha_i = 1; \quad \sum_i \gamma_{ij} = \sum_i \beta_i = 0 \qquad \text{(adding up)} \tag{13.4}$$

$$\sum_j \gamma_{ij} = 0 \qquad \text{(homogeneity)} \tag{13.5}$$

$$\gamma_{ij} = \gamma_{ji} \qquad \text{(symmetry)} \qquad (13.6)$$

The uncompensated own-price (e_{ii}), cross-price (e_{ij}) and expenditure elasticities (η_i) are computed as follows:

$$e_{ii} = -1 + \left(\frac{\gamma_{ii}}{w_i}\right) - \beta_i \qquad (13.7)$$

$$e_{ij} = \left(\frac{\gamma_{ij}}{w_i}\right) - \beta_i\left(\frac{w_j}{w_i}\right) \text{ for } i \neq j \qquad (13.8)$$

$$\eta_i = 1 + \frac{\beta_i}{w_i} \qquad (13.9)$$

Initial estimation of the model identified the presence of severe serial correlation and the following first-order autocorrelation structure was specified:

$$w_{it} = \rho_i w_{it-1} + \alpha_i + \sum_{j=1}^{n} \gamma_{ij}(\ln p_{jt} - \rho_{ij} \ln p_{jt-1}) + \beta_i(\ln x_t - \ln P_t^*) + \phi_i GIA_t \qquad (13.10)$$

where ρ_i and ρ_{ij} are unknown parameters. The adjusted global index (GIA) described in Rickertsen and von Cramon-Taubadel (Chapter 3, this volume) was included in the model. The complete model (Equation 13.10) was estimated and insignificant ρ_i and ρ_{ij} parameters were set to zero. This procedure was repeated until all traces of serial correlation had disappeared. In the final model, all the ρ_{ij} coefficients were zero, while the ρ_i coefficients for the lags of the dependent variables were estimated simultaneously with the other unknown parameters.

A polynomial lag structure was specified on the *GIA* index because information may influence household behaviour in two distinct stages. First, some households may react instantaneously to information. Secondly, some households may only react later, following public reports and discussions. Therefore, the preferred method is to employ a polynomial distributed lag format for the health-index variable. Testing in a way suggested by Almond (1965), the best fit was obtained from a quadratic functional form with ten lags and no end restrictions. Once the coefficients for each of the lags were recovered, health-information elasticities (h_i) were calculated as:

$$h_i = \sum_{t=1}^{T} \phi_i^t \Big/ w_i \qquad (13.11)$$

where ϕ_i^t is the tth lagged health-index coefficient for food group i.

Table 13.2. Expenditure shares and estimated health-information elasticities.

Food group	Expenditure share (%)	Elasticities	*t* values	R^2
Meat	26.28	−0.33	−0.45	0.36
Fish	4.35	0.09	0.18	0.59
Dairy products	12.65	−0.05	−0.06	0.31
Vegetables	9.95	−0.02	−0.04	0.68
Cereals	15.52	0.20	0.28	0.48
Alcohol	7.40	0.23	0.41	0.54
Eggs	1.17	−0.21	−0.34	0.52
Fats and oils	2.67	−0.33	−0.44	0.49
Fruits	6.18	0.32	1.39	–
Other foods	13.83	0.24	0.39	0.52

The health-information elasticities given in Table 13.2 suggest that, overall, the type of information considered, as well as the way in which it is being disseminated, has no significant effects on food consumption at the household level. Although price and expenditure changes will explain as much as 68% of variations in budget share (specifically in the case of vegetables), the effects of such targeted initiatives as health-information campaigns are only minimal, which tends to confirm Law *et al.*'s (1994) observation that only modest dietary changes occur. Encouragingly, however, we find that the signs of the elasticities are broadly as expected; that is to say, the ones for those food products (i.e. meat, dairy products, eggs and fats) most associated with the dietary intake of harmful nutrients are negative, while those for fish, cereals and fruits are positive. The exception is the apparent anomaly that the elasticity for vegetables turns out to be negative; however, the numerical value is very close to zero. Furthermore, there is a strong association between meat and vegetables, as the two products are most likely to be consumed together. Thus, any factors adversely affecting meat consumption will have an indirect, albeit less pronounced, negative impact on vegetable consumption.

Given the above findings that there is clearly insufficient response from consumers to health information, if one is to make significant improvements in the health of the nation, it is fair to say that more effective methods of persuasion will be required.

Methodological Approach to a Fiscal Food Policy

More commonly associated with welfare and environmental economics, the taxation approach to combating externalities is a well-trodden path. Essentially, it is a method of obtaining optimum reductions in levels of externality, where the level of the tax (also known as the Pigovian tax) should be compatible with the socially acceptable levels of the externality, as well as being proportional to the level of damage. Under such a scheme, it is expected that the normal response of those producing the externality would react in such a way as to reduce their tax liability. As a result, costs inflicted on society would also be reduced, where the utility maximization of the individual would now include an element of internalization since, if indeed consumption or use of an input becomes too costly compared with the utility derived, then a decrease in the use of that input should be encouraged.

Technically, in the context of this study, it consists of estimating a household production and a 'damage' function, given the observed quantities of the various inputs. Thus, the functions are defined by $Q = f(\mathbf{x})$ and $D = g(\mathbf{x})$ for production and damage, respectively, where Q is total output, D is total damage and $\mathbf{x}$ is a vector of all inputs i. Focusing on any one input x_i, it is possible to determine the optimum social level of utilization by equating the value of the marginal product with the value of the marginal damage of x_i. Thus, Pareto improvements will be achieved as the equality tends to a limit of zero. Algebraically, we have $P(\delta Q/\delta x_i) = \lambda(\delta D/\delta x_i)$ where P is the market price of a product, $\delta Q/\delta x_i$ is the marginal product of input x_i, $\delta D/\delta x_i$ is the marginal damage from using input x_i and λ is a monetary value that will be set at such an amount as to maintain the equality and therefore will act as the equilibrium adjustment factor *ceteris paribus*. Hence, the value of the marginal product, when set equal to the value of the marginal damage by establishing $P(\delta Q/\delta x_i) - \lambda(\delta D/\delta x_i) = 0$ (which is also the necessary condition for a maximum), will result in a socially optimal resource allocation where all externalities will be internalized. A concise discussion of the mechanism involved can be found in Perman *et al.* (1996: 198–215).

A secondary and, in our case, desirable effect of such an approach is the potential impacts on suppliers of those products of which the use causes externalities. Given the tightness of the price transmission process along the vertical distribution chain (Santarossa, 1998), consumer prices altered through taxation would inevitably be transmitted upstream, thereby causing suppliers to alter product composition.

Kristiansen *et al.* (1991) advocated a policy of dissemination of information on health issues to the scientific, agricultural, food and health sectors of the economy, along with targeted use of mass media and the taxation or subsidy of appropriate foods to improve health. However, unlike the approach proposed by Marshall (2000), where a taxation level equal to the current value-added tax (VAT) rate of 17.5% would be imposed on the whole product, we propose to tax only those inputs that are the source of externalities and only those amounts that are in excess of safe, recommended levels. Generally, it is common practice to derive a pollution/damage-generating function to gauge the

Table 13.3. Dietary goals for European population. (From: *Food Magazine* (July/September, 2000: 21).)

Item	Target
Dietary fat as % of energy	< 30
Saturated fatty acids as % of energy	< 10
Trans fatty acids as % of energy	< 2
n-6 polyunsaturates as % of energy	4–8
n-3 polyunsaturates per day	2 g linolenic and 200 mg very long chain
Carbohydrates as % of energy	> 55
Sugary intake, occasions per day	4 or fewer
Fruits and vegetables per day	> 400 g
Folates from food per day	> 400 μg
Dietary fibre per day	> 25 g (or 12.5 g per 1000 calories)
Salt (sodium chloride) per day	< 6 g (2.4 g sodium)
Iodine	150 μg (50 μg infants, 200 μg pregnant women)
Exclusive breast-feeding	About 6 months
Adult body weight	Body mass index 21–22

amount of undesirable effects an input might produce. For the purpose of this study, however, not being in possession of the necessary data to produce a damage function, we opted to rely on official guidelines, as given in Table 13.3. Of specific interest are the values for saturated fatty acids, *trans* fatty acids and dietary fat, all taken as a percentage of energy intake, which we took to be 10, 2 and 30, respectively.

Estimations for an Appropriate Level of Taxation

In order to obtain the values that would eventually be employed in setting an appropriate level of taxation, a number of estimations had to be carried out. In line with the theoretical principles outlined in the methodological approach, a household production function was estimated within a constant elasticity-of-substitution framework of the following form:

$$\ln(Q) = \gamma\left(\sum_{i=1}^{9}\beta_i v_i^{-r} + (1-\sum_{i=1}^{9}\beta_i)\nu_{10}^{-r}\right)^{-\frac{1}{r}} \quad (13.12)$$

where $\ln(Q)$ is the log of total quantity of food consumed by each household expressed in grams, γ, β_i and r are unknown parameters to be estimated and the v_i terms are vectors of nutrients corresponding to the nutrients given in the section on data (e.g. v_1 = carbohydrates, v_2 = saturated fats, etc.). The first-order derivative of Equation 13.12 is then taken in order to obtain the marginal product of each nutrient (see

Table 13.4. Parameter estimates for the CES household production function.

Variables	Coefficients	*t* values
Constant	3.23	20.94**
Carbohydrates	0.04	6.00**
Saturates	–0.22	–8.25**
Monounsaturates	0.46	8.51**
Polyunsaturates	–0.29	–7.16**
Fibres	0.29	9.29**
Vegetable proteins	0.11	5.41**
Animal proteins	–0.04	–4.58**
Cholesterol	0.30	4.44**
Minerals	0.05	3.08**
Vitamins	0.30	4.93**
ρ	–0.54	–13.99**
R^2	0.97	
DW	1.92	

** Indicates significance at the 5% level.
CES, constant elasticity of substitution; *DW*, Durbin–Watson statistic.

Hoy *et al.*, 1996: 393). Parameter estimates and *t* values for Equation 13.12 are given in Table 13.4.

A second estimation was required in order to obtain some form of price/value for each of the nutrients for the purpose of this exercise; the value of the marginal product of a nutrient must be juxtaposed to the price/cost of that nutrient. Given that we are employing data aggregated by household rather than by products, we expressed these values in terms of total household expenditure rather than some average price. The use of expenditure instead of price in Equation 13.13 below can be justified by the fact that, if we are concerned about consumer utility, and therefore consumers revealing their preferences for products through 'rational' use of household budgets, then expenditure seems a more flexible value to employ than prices. In comparison, prices are rather more inflexible in that they present the consumer with a limited choice of action, i.e. to buy or not to buy. Expenditure, on the other hand, gives a gradual scale that can be increased/decreased in parallel with consumer utility and can therefore be seen as being a more 'revealing' measure of consumer utility. Hence, in line with a hedonic price-index method where the objective is to derive implicit values, the following expenditure function was estimated:

$$\ln(PV) = \gamma\left(\sum_{i=1}^{9} \nu_i^{\beta_i} + \nu_{10}^{\left(1-\sum_i \beta_i\right)}\right) \tag{13.13}$$

Table 13.5. Parameter estimates of expenditure function.

Variables	Coefficients	*t* values
Constant	0.42	49.84**
Carbohydrates	0.01	0.11
Saturates	0.09	2.45**
Monounsaturates	0.10	1.79*
Polyunsaturates	0.08	2.21**
Fibres	0.09	2.67**
Vegetable proteins	0.10	3.48**
Animal proteins	0.18	20.70**
Cholesterol	0.01	0.41
Minerals	0.06	1.69*
Vitamins	0.20	34.73**
R^2	0.83	
DW	1.89	

** Indicates significance at the 5% level, * at the 10% level.
DW, Durbin–Watson statistic.

Table 13.6. Derivation of tax level.

	Marginal product	Value of marginal product[a]	Implicit expenditure	Tax level
Carbohydrates	−0.0000005	−0.0018	0.005	0.007
Saturated fat	0.00015	0.58	0.089	−0.49
Monounsaturated fat	−0.0004	−1.57	0.10	1.67
Polyunsaturated fat	0.0012	4.66	0.08	−4.58
Fibres	−0.0005	−1.86	0.09	1.95
Vegetable proteins	−0.0001	−0.47	0.10	0.56
Animal proteins	0.00002	0.07	0.18	0.11
Cholesterol	−0.36	−1417.80	0.015	1417.8
Minerals	−0.0003	−1.18	0.05	1.23
Vitamins	−1.68	−6516.30	0.20	6516.5

[a] Calculated at the mean household expenditure of 3885.28 pence.

where ln(PV) is the log of total household purchase value (expenditure) and, as in Equation 13.12, γ and β_i terms are unknown parameters to be estimated, while the v_i terms are the vectors of nutrients expressed in grams. The parameter estimates of Equation 13.13 are reported in Table 13.5.

We can now estimate taxation levels and resulting price increases. The marginal products from Equation 13.12 are multiplied by the average household expenditure (3885 pence) to calculate the values of the marginal product. Given that the coefficients from Equation 13.13 can be interpreted as implicit expenditure levels on each of the nutrients, the tax level is obtained by subtracting the value of the marginal product from the implicit expenditure. The numerical values involved in the calculations are given in Table 13.6.

Theoretically, an input (or nutrient in our case) is employed up to the point where the value of its marginal product is equal to its price. This is fine as far as a firm's production function is concerned but may not be the case where individual consumers are concerned. The liking for a specific product on the part of consumers may indeed reveal that some product characteristics exhibit a value of the marginal product well in excess of its 'implicit market value'. This is, in effect, what we observe in the case of saturated fat in Table 13.6, and the task is therefore one of reducing that value of the marginal product down to the level of the implicit expenditure; the resulting value (−0.49) is thus the tax to be imposed, expressed in pence per gram to be applied to any amounts of saturated fat in excess of recommended intake. With respect to the second nutrient of interest (cholesterol), a different line of reasoning is required since the marginal product of that input is negative. Logically, if consumers had perfect choice in the bundles of characteristics that are in food products, they would opt to leave cholesterol out. However, that choice does not exist since product contents are principally determined by agents located upstream in the transmission process. Hence, as an indirect way of sending a message to the food industry, we nevertheless propose to impose a tax on that amount of cholesterol which is in excess of the recommended 2% of energy intake.

Computation of the new prices involved treating each product that each household had consumed separately, following a process of: (i) converting all quantities of products consumed by households to their nutrient equivalent in grams per person per day, employing conversion factors that were obtained from McCance and Widdowson (1991); (ii) converting the amounts of fats and cholesterol consumed as a percentage of total energy obtained from each food item; (iii) calculating the excess percentage energy derived from fats and cholesterol; (iv) multiplying (iii) by the corresponding tax level, as given in Table 13.6; and (v) adding the amount obtained in (iv) to household expenditure on the food item in question. The final step enabled us to derive new budget shares, employing price elasticities obtained from an LA/AIDS model identical to that already presented, albeit this time excluding the variables for health indices. The resulting price increases and changes in expenditure shares are given in Table 13.7.

Simulation and Outcome

The estimated parameters and their *t* values in the LA/AIDS model excluding health information are presented in Table 13.8 and the

Table 13.7. Effects of price increase on expenditure shares (ES).

Food group	Price increase (%)	Original ES	Change in ES (%)
Meat	1.18	0.26	0.09
Fish	0.01	0.04	−4.65
Dairy products	4.01	0.13	−4.72
Vegetables	0.00	0.10	0.12
Cereals	0.58	0.15	−2.94
Alcohol	0.00	0.07	−0.64
Eggs	10.58	0.01	−12.56
Fats and oils	23.86	0.03	−11.29
Fruits	0.09	0.06	0.06
Other foods	0.14	0.14	0.14

associated total expenditure and Marshallian price elasticities are shown in Table 13.9. Applying the percentage price increases reported in Table 13.7 to price elasticities from Table 13.9 produced the new expenditure shares shown in Table 13.7. As expected, the expenditure share for meat shows a slight increase in spite of a 1.18% increase in the price of meat and a negative own-price elasticity. Other significant changes are a 12.6% decrease in the expenditure share of eggs (from the taxation of cholesterol) and an 11.3% decrease in that of fats and oils as a result of taxation on saturated fat. Marginal increases are recorded for the expenditure shares of vegetables (0.12%) and fruits (0.06%), as is also the case for the 'other foods'. Given the new expenditure shares, we can compute the new level of mean total household expenditure:

$$\sum_{i=1}^{10} NES_i \bar{E} \tag{13.14}$$

where NES_i is the new expenditure share for food group i and $\bar{E}$ is the mean original total household expenditure. We use the expenditure–nutrients relationship given by Equation 13.13 in an attempt to answer the question 'for which values of each nutrient would the mean household total expenditure result in that specific new value of 3819.28 pence?' We use the value of 3819.28 pence since this is the new mean household expenditure level on food and, given the expenditure-to-nutrient relationship, we can explore new levels of nutrient consumption.

Outputs of the above simulation, presented in Table 13.10, indicate that, with respect to fatty nutrients, decreases are recorded in all but one instance (monounsaturated fat from dairy products). The figures in parentheses show the percentage changes in the consumption of various nutrients. Notably, one observes a 19% decrease in saturated fat from

Table 13.8. Parameter estimates, from LA/AIDS without health information.[a]

Equation	Constant	AR term	Meat	Fish	Dairy products	Vegetables	Cereals	Alcohol	Eggs	Other foods	Fats and oils	Fruits	Exp.	R^2
Meat	0.004	0.011	0.167	−0.016	−0.024	−0.023	−0.037	−0.003	−0.002	−0.058	0.013	−0.017	0.020	0.095
	(0.045)	(0.188)	(5.235)	(−1.771)	(−1.883)	(−1.948)	(−2.184)	(−0.221)	(−0.577)	(−3.835)	(2.299)		(2.050)	
Fish	0.043	0.080		0.045	0.013	−0.016	−0.001	−0.010	0.000	−0.009	−0.010	0.004	−0.002	0.557
	(1.980)	(1.250)		(8.295)	(2.060)	(−3.036)	(−0.132)	(−2.700)	(−0.287)	(−1.275)	(−3.116)		(−0.689)	
Dairy products	0.103	0.188			0.058	−0.007	−0.001	−0.009	−0.010	−0.015	−0.008	0.001	0.002	0.239
	(2.882)	(2.581)			(3.890)	(−0.745)	(−0.034)	(−1.473)	(−3.006)	(−1.213)	(−1.292)		(0.553)	
Vegetables	0.179	0.022				0.070	−0.010	−0.020	−0.002	−0.007	0.004	0.009	0.001	0.564
	(5.248)	(0.361)				(6.933)	(−0.789)	(−3.630)	(−0.785)	(−0.680)	(0.881)		(0.295)	
Cereals	0.200	0.155					0.100	−0.006	−0.003	−0.010	−0.015	−0.018	−0.003	0.304
	(4.618)	(1.921)					(3.603)	(−0.997)	(−0.616)	(−1.341)	(−1.312)		(−0.733)	
Alcohol	0.074	0.084						0.068	0.001	−0.010	−0.001	−0.009	−0.017	0.430
	(1.140)	(1.750)						(5.365)	(−0.240)	(−1.542)	(−0.526)		(−2.033)	
Eggs	0.001	0.118							0.006	0.005	−0.002	0.008	0.002	0.424
	(0.135)	(1.388)							(3.195)	(1.337)	(−0.824)		(1.858)	
Other foods	0.198	−0.053								0.129	0.004	−0.029	−0.002	0.443
	(4.943)	(−0.797)								(6.318)	(0.685)		(−0.383)	
Fats and oils	0.038	0.003									0.016	−0.002	−0.002	0.384
	(2.768)	(0.037)									(2.956)		(−1.207)	
Fruits	0.160	–	−0.017	0.004	0.001	0.009	−0.018	−0.009	0.008	−0.029	−0.002	0.053	0.000	–

[a] Figures in parentheses are *t* values. Critical 5% *t* value with 60 d.f. is 2.00.
d.f., degrees of freedom; Exp., expenditure.

Table 13.9. Total expenditure and Marshallian price elasticities.[a]

	Meat	Fish	Dairy products	Vegetables	Cereals	Alcohol	Eggs	Other foods	Fats and oils	Fruits
Expenditure	1.08	0.96	1.02	1.01	0.98	0.96	−0.38	1.01	0.94	0.97
	(29.36)	(17.15)	(33.22)	(27.37)	(34.78)	(15.52)	(−0.56)	(157.66)	(5.64)	(43.01)
Meat price	−0.38	−0.06	−0.10	−0.10	−0.15	−0.02	−0.01	−0.23	0.05	−0.07
	(−3.18)	(−1.86)	(−2.08)	(−2.08)	(−2.33)	(−0.32)	(−0.65)	(−3.94)	(2.20)	(−30.30)
Fish price	−0.35	0.03	0.31	−0.36	−0.02	−0.22	−0.01	−0.21	−0.22	0.08
	(−1.77)	(0.25)	(2.11)	(−2.98)	(−0.10)	(−2.67)	(−0.28)	(−1.23)	(−3.10)	(22.76)
Dairy-products price	−0.19	0.10	−0.55	−0.05	−0.01	−0.07	−0.08	−0.12	−0.06	0.01
	(−1.97)	(2.04)	(−4.72)	(−0.77)	(−0.06)	(−1.50)	(−3.01)	(−1.23)	(−1.30)	(3.63)
Vegetable price	−0.23	−0.16	−0.07	−0.30	−0.10	−0.20	−0.02	−0.07	0.04	0.09
	(−2.01)	(−3.04)	(−0.76)	(−3.01)	(−0.80)	(−3.65)	(−0.79)	(−0.69)	(0.87)	(38.80)
Cereal price	−0.24	−0.01	−0.00	−0.06	−0.35	−0.04	−0.02	−0.06	−0.10	−0.12
	(−2.19)	(−0.11)	(−0.01)	(−0.76)	(−1.92)	(−0.96)	(−0.61)	(−1.27)	(−1.30)	(−63.98)
Alcohol price	0.02	−0.13	−0.09	−0.25	−0.05	−0.03	−0.00	−0.11	−0.01	−0.11
	(0.10)	(−2.47)	(−1.09)	(−3.25)	(−0.58)	(−0.15)	(−0.08)	(−1.16)	(−0.32)	(−15.58)
Egg price	−0.19	−0.05	−0.85	−0.18	−0.26	−0.03	−0.48	0.37	−0.16	0.70
	(−0.72)	(−0.33)	(−3.09)	(−0.85)	(−0.67)	(−0.33)	(−3.01)	(1.27)	(−0.84)	(147.27)
Other foods price	−0.42	−0.07	−0.11	−0.05	−0.07	−0.07	0.03	−0.07	0.03	−0.21
	(−3.90)	(−1.26)	(−1.20)	(−0.66)	(−1.29)	(−1.53)	(1.34)	(−0.43)	(0.69)	(−100.70)
Fats and oils price	0.52	−0.37	−0.28	0.16	−0.54	−0.04	−0.07	0.18	−0.40	−1.07
	(2.44)	(−3.08)	(−1.26)	(0.92)	(−1.29)	(−0.47)	(−0.81)	(0.72)	(−2.01)	(−304.70)
Fruit price	−0.27	0.05	0.02	0.15	−0.28	−0.14	0.13	−0.44	−0.03	−0.17

[a] Figures in parentheses are *t* values. Critical 5% *t* value with 60 d.f. is 2.00.
d.f., degrees of freedom.

Table 13.10. Tax effects on nutrient sourcing (figures in parentheses are the percentage changes).

	Carbo-hydrates	Saturated fats	Mono-unsaturates	Poly-unsaturates	Fibre	Vegetable proteins	Animal proteins	Cholesterol	Minerals	Vitamins
Before price increase										
Meat	14.38	9.00	10.23	2.17	0.82	1.14	21.16	0.09	1.33	0.02
Fish	1.31	0.27	0.49	0.40	0.04	0.12	3.28	0.01	0.19	0.00
Dairy products	41.45	9.16	4.26	0.51	0.09	0.10	16.01	0.05	1.87	0.01
Vegetables	62.42	0.53	0.71	1.07	5.97	5.04	0.03	0.00	1.18	0.03
Cereals	255.59	5.74	4.39	2.29	9.77	17.33	0.70	0.02	2.11	0.01
Alcohol	0.03	0.01	0.00	0.00	0.00	0.00	0.00	0.00	0.00	0.01
Eggs	0.00	0.54	0.77	0.21	0.00	0.00	2.06	0.06	0.09	0.00
Other foods	104.56	1.24	1.38	0.81	1.26	1.30	0.26	0.00	1.55	0.01
Fats and oils	1.27	8.61	7.10	3.27	0.00	0.01	0.41	0.03	0.19	0.00
Fruits	25.87	0.33	0.48	0.31	1.92	0.61	0.00	0.00	0.25	0.02
After price increase										
Meat	14.40	8.33	8.32	1.98	0.82	1.14	21.21	0.01	1.33	0.02
	(0.18)	(−7.40)	(−18.69)	(−8.46)	(0.21)	(0.16)	(0.21)	(−89.90)	(0.18)	(0.11)
Fish	1.28	0.23	0.39	0.36	0.03	0.12	3.21	0.00	0.18	0.00
	(−2.14)	(−14.51)	(−20.80)	(−10.75)	(−2.14)	(−2.14)	(−2.12)	(−90.64)	(−2.14)	(−2.11)
Dairy products	28.55	7.43	5.01	0.33	0.06	0.07	6.95	0.00	1.28	0.01
	(−31.13)	(−18.90)	(17.67)	(−34.32)	(−32.43)	(−30.61)	(−56.55)	(−92.85)	(−31.38)	(−14.73)
Vegetables	62.55	0.47	0.58	0.97	5.98	5.05	0.03	0.00	1.18	0.03
	(0.22)	(−10.86)	(−18.76)	(−8.67)	(0.21)	(0.20)	(0.21)	(−93.28)	(0.21)	(0.12)
Cereals	185.73	2.44	2.76	0.52	10.18	16.91	0.42	0.00	1.23	0.01
	(−27.33)	(−57.42)	(−37.07)	(−77.19)	(4.21)	(−2.42)	(−40.50)	(−92.59)	(−41.70)	(−11.37)

Alcohol	0.03	0.00	0.00	0.00	0.00	0.002	0.00	0.00	0.00	0.00
	(−0.19)	(−99.07)	(−97.78)	(−9.04)	(0.00)	(−0.19)	(−0.19)	(−100.00)	(−0.19)	(−0.19)
Eggs	0.00	0.36	0.48	0.14	0.00	0.00	1.45	0.01	0.07	0.00
	(0.00)	(−32.92)	(−37.39)	(−29.89)	(0.00)	(0.00)	(−29.19)	(−90.70)	(−23.03)	(−23.24)
Other foods	104.60	1.12	1.12	0.74	1.26	1.30	0.26	0.00	1.55	0.01
	(0.10)	(−9.96)	(−18.77)	(−8.76)	(0.11)	(0.10)	(0.10)	(−92.84)	(0.10)	(0.10)
Fats and oils	1.15	7.31	4.62	2.79	0.001	0.004	0.37	0.00	0.17	0.00
	(−9.27)	(−15.07)	(−34.82)	(−14.56)	(−9.27)	(−9.27)	(−9.39)	(−90.93)	(−9.28)	(−8.95)
Fruits	25.88	0.29	0.38	0.28	1.92	0.61	0.00	0.00	0.24	0.02
	(0.05)	(−12.06)	(−18.99)	(−8.81)	(0.05)	(0.06)	(0.00)	(−99.28)	(0.05)	(0.05)

dairy products and 15% from fats and oils, while the effects of price increase on cholesterol intake are rather dramatic, with an average decrease in cholesterol intake of 93%. The rather sizeable decrease in saturated fat from cereal products (57%) is explained by the fact that biscuits and cakes, which do generally have a high fat content, were included in the cereals food group. One further observes some increases, specifically carbohydrates and fibre from the consumption of meat, vegetables, other foods and fruits. Although these increases are all only modest, the simulation is moving in the right direction – that is, an apparent substitution of harmful nutrients for healthy ones.

Turning our attention to those nutrients that convert to energy, values in Table 13.11 are the corresponding values in Table 13.10 calculated as a percentage of the total energy intake per day per person prior to and after the price increases (2868 kcal and 2366 kcal, respectively). From the 'Total energy' column of Table 13.11, one notes that, before the price increase, 12% of energy intake comes from meat products, while 12% is obtained from dairy products. The largest contribution is from the cereals food group, which is not really surprising considering that the expenditure share for that food type is the second highest (15% of total expenditure) after meat (see Table 13.7, third column). The bottom half of Table 13.11 shows the new levels of energy intake after the price increase. Overall, the procedure employed predicts that energy intake from dairy products would decrease by 12%, that from cereals by 13% and that from eggs by 20%. In line with observations made in Table 13.9, increases in energy intake are recorded, namely, 12% from meat, 20% from vegetables and 19% from fruits. These increases are mostly observed to be from carbohydrates and vegetable proteins, as indicated by the percentage-change row at the bottom of Table 13.11. One further result from this analysis is that, from an original level of total energy intake of 2868 kcal per person per day, a 17.5% decrease occurs, bringing the level down to 2366 kcal. Although it does seem to be a rather dramatic decrease, one should be reminded that the average recommended daily energy intake is 2245 kcal (2550 for adult males and 1940 for adult females).

Regarding whether these results are in any way outside or indeed within any range of prediction, two issues can be addressed. The first is that, given the nutrient levels that we were trying to achieve, specifically with respect to fatty acids, if one compares our results with the benchmark values given in Table 13.3, it is fair to say that, in theory, the mechanism proposed in this study does work. Summary results, shown in Table 13.12, indicate that, overall, a 20% decrease in the consumption of fat would occur. Specifically, we observe reductions of 21% for saturated fat and monounsaturated fat, 26% for polyunsaturated fat and 91% for cholesterol. The rather sizeable percentage value for cholesterol can be explained by the fact that, from the onset, the average intake of

Table 13.11. Distribution of metabolizable energy intake as percentage of total energy intake.

	Carbo-hydrates	Saturated fats	Mono-unsaturates	Poly-unsaturates	Vegetable proteins	Animal proteins	Cholesterol	Total energy	% change in energy	% change in quantity
Before price increase										
Meat	1.88	2.82	3.21	0.68	0.16	2.95	0.030	11.73		
Fish	0.17	0.09	0.15	0.13	0.02	0.46	0.004	1.01		
Dairy products	5.42	2.87	1.33	0.16	0.01	2.23	0.020	12.05		
Vegetables	8.16	0.16	0.22	0.33	0.70	0.01	0.001	9.59		
Cereals	33.42	1.80	1.38	0.72	2.42	0.10	0.006	39.84		
Alcohol	0.00	0.00	0.00	0.00	0.00	0.00	0.000	0.01		
Eggs	0.00	0.17	0.24	0.06	0.00	0.29	0.020	0.78		
Other foods	13.67	0.39	0.43	0.25	0.18	0.04	0.001	14.97		
Fats and oils	0.16	2.70	2.23	1.03	0.00	0.06	0.010	6.19		
Fruits	3.38	0.10	0.15	0.10	0.08	0.00	0.000	3.81		
Energy intake	66.27	11.12	9.35	3.46	3.58	6.12	0.090			
After price increase										
Meat	2.28	3.17	3.16	0.75	0.19	3.58	0.004	13.15	12.09	−4.19
Fish	0.20	0.09	0.15	0.14	0.02	0.54	0.000	1.14	12.24	−3.97
Dairy products	4.52	2.83	1.90	0.13	0.01	1.17	0.001	10.57	−12.29	−32.16
Vegetables	9.91	0.18	0.22	0.37	0.85	0.01	0.000	11.54	20.34	0.21
Cereals	29.43	0.93	1.05	0.20	2.86	0.07	0.001	34.55	−13.28	−27.14
Alcohol	0.01	0.00	0.00	0.00	0.00	0.00	0.000	0.01	−33.11	−1.81
Eggs	0.00	0.14	0.18	0.06	0.00	0.25	0.002	0.62	−20.19	−57.03
Other foods	16.59	0.43	0.43	0.28	0.22	0.04	0.000	17.99	20.16	0.08
Fats and oils	0.18	2.78	1.76	1.06	0.00	0.06	0.001	5.85	−5.45	−22.59
Fruits	4.10	0.11	0.15	0.11	0.10	0.00	0.000	4.57	19.71	0.04
Energy intake	67.24	10.65	9.00	3.09	4.26	5.73	0.009			
% change in energy intake	1.46	−4.24	−3.73	−10.59	19.13	−6.39	−89.18			

Table 13.12. Changes in fat intake per person and day.

	Before tax (g)	After tax (g)	Change (%)	Dietary kcal % from[a]	
				Before tax	After tax
Total fat	82.70	65.97	−20.22	38.10	30.40
Saturated	35.43	27.99	−21.00	11.12	10.65
Monounsaturated	29.80	23.67	−20.59		
Polyunsaturated	11.02	8.13	−26.25		
Other fatty acids	6.16	6.16	0.00		
Cholesterol	0.28	0.03	−91.00	0.09	0.01

[a] Calculated as percentage of energy per day and person.

cholesterol as a percentage of energy was below recommended guidelines. However, due to the presence of certain individual products with cholesterol content well above guidelines, the effects of taxation result in an outcome somewhat above expectations when applied right across the board.

Regarding reductions in dietary calories, it seems that, as far as saturated fat is concerned, we are moving in the right direction. With a target value of 10%, our simulation achieved 10.65% of energy intake (down from 11.12%). Overall, we pretty much achieve the recommended value for total fat intake as a percentage of energy of 30% (down from 38%), which is in effect below current UK guidelines of 33–35% intake. In absolute terms, the new Eurodiet proposal advocates that fat intake should be limited to around 66 g per person per day; as shown in Table 13.12, our simulation achieves a figure of 65.97 g per person per day. The second issue is that of comparing our results with those obtained by the USDA's CNPP, as given in the 1999 Thrifty Food Plan (TFP). Described by its authors as 'a national standard for a nutritious diet at a minimal cost', the TFP attempts to set up a regime of food consumption that meets the required daily allowance of all nutrients while minimizing the cost of food purchase and constrained in such a way that no overconsumption of any one specific nutrient occurs. Results from the 1999 revised TFP indicate that the new food basket contains more fruits, vegetables, grains and dairy products but less other foods (fats, oils and sweets) and meat/meat alternatives. Our own results, in terms of quantities consumed (Table 13.11, last column), indicate that, as in those from CNPP, we also have increases in fruits and vegetables and decreases in fats and oils and meat. Conversely, however, we have decreases in dairy products, other foods and cereals. As already mentioned, our cereals food group included products with high fat content and the decrease is no reflection on reduced consumption of healthy grain products, such as wholemeal bread. Unlike the results from the TFP, our increases are rather more modest and can be explained by the difference in direction that the two studies adopted. Indeed, while the TFP concentrated on an overall dietary intake, our aim was to reduce the intake of fatty acids. As a result, purely through a substitution mechanism, we recorded those modest increases, while our decreases are more comparable to those from the TFP.

In spite of the encouraging results outlined so far, one nevertheless has to be reconciled with the fact that the policy instrument on which the exercise is based is one that is never really popular among consumers or, in some cases, politicians. Taxes have two functions – to raise revenue and to influence demand – and, although this study should ideally be in the second category in order for it to be effective, some consumers may just view it as falling into the first category. Addressing the question of equity, the fact that taxes such as the one employed in this study are regressive is not questionable. However, if the price mechanism operates as efficiently as our LA/AIDS model demonstrates and price elasticities reflect how consumers are likely to respond to price increases, then household expenditure on food would actually decrease, rather than increase. Moreover, it would do so with beneficial results, as already discussed. However, where this study may need improving is with respect to the actual nutrients that are selected for taxation. Foods are complex mixtures of nutrients (Stanley, 2000) and, while some saturated fatty acids have no effects on blood cholesterol levels, others raise blood cholesterol levels. Therefore, as an extension to this study, one might wish to consider the nature of nutrients in order to identify more clearly to which elements of food products taxation should be applied. It nevertheless remains the case that this study could make an important contribution to policy makers in assisting households to adopt a more nutritious and healthy diet.

Concluding Remarks

Having applied an analytical method not generally associated with food consumption, the ‘polluter pays’ principle was, in our humble view, rather successful at addressing a complicated public-health problem. The concept of treating coronary disease and other diet-related ailments as a straightforward externality proved to be a tool providing clear benefits. This was achieved by our simulation, especially in the light of the fact that new recommended dietary target figures are intended to be used in an EU-wide nutrition guidance. Note, however, that no consideration was given to the possibility of subsidizing products beneficial to a healthy diet, and clearly this is something that would have to be investigated, along with taxation of detrimental dietary inputs, as we have just done. As to the adoption of such a scheme ever becoming a reality, this is a sensitive issue in the UK, given that VAT or any other form of taxation is not at present extended to foodstuffs and might not be so in the foreseeable future. With respect to other forms of diet-improving interventions, such as functional or fabricated foods (Kennedy and Offutt, 2000), there is no evidence that this would work.

Over the last 20 years of the 20th century, a significant variety of so-called healthy-option food products have been marketed without resulting in a visible reduction in the rate of occurrence of diet-related diseases. What is therefore required is not the innovation of yet more such products or mass media dissemination of information on a healthy diet, but rather a radical approach to altering consumers' choice of input into their daily diet. Given that the UK food market is currently worth £96 billion a year, a temporary decrease of 1.7% in household expenditure on food would result in a loss of £32 million a week to the retail sector. Clearly, if market forces operate efficiently, the transformation of products by substituting harmful nutrients for more healthy ones by food producers is, in our view, very likely.

Acknowledgements

The Scottish Agricultural College gratefully acknowledges the financial support of the EU, research contract number FAIR5–CT97–3373: 'Nutrition, Health and the Demand for Food'.

Note

[1] Data are only available on request from the Data Archive, University of Essex, UK.

References

Almond, S. (1965) The lag between appropriation and expenditures. *Econometrica* 30, 178–196.

Birkin, M., Clarke, G., Clarke, M. and Wilson, A. (1996) *Intelligent GIS: Location Decision and Strategic Planning.* GeoInformation International, Cambridge, 54 pp.

Burton, M. and Young, T. (1992) The structure of changing tastes for meat and fish in Great Britain. *European Review of Agricultural Economics* 19, 165–180.

Center for Nutrition Policy and Promotion (CNPP) (1999) *The Thrifty Food Plan, Executive Summary*. US Department of Agriculture, Washington, DC, p. ES-2.

Deaton, A. and Muellbauer, J. (1980) An almost ideal demand system. *American Economic Review* 70, 312–326.

Food Magazine (July/September 2000) Experts review diet guidelines. 50, 21, Food Commission, London.

Hoy, M., Livernois, J., McKenna, C., Rees, R. and Stengos, T. (1996) *Mathematics for Economics*, 1st edn. Addison-Wesley, Don Mills, Ontario, 947 pp.
Kennedy, E. and Offutt, S. (2000) Commentary: alternative nutrition outcomes using a fiscal food policy. *British Medical Journal* 320, 304–305.
Kristiansen, I.S., Eggen, A.E. and Thelle, D.S. (1991) Cost effectiveness of incremental programmes for lowering serum cholesterol concentration: is individual intervention worthwhile? *British Medical Journal* 300, 1119–1122.
Law, M.R., Wald, N.J. and Thompson, S.G. (1994) By how much and how quickly does reduction in serum cholesterol concentration lower risk of ischaemic heart disease? *British Medical Journal* 308, 367–373.
Marshall, T. (2000) Exploring a fiscal food policy: the case of diet and ischaemic heart disease. *British Medical Journal* 320, 301–304.
McCance, R.A. and Widdowson, E.M. (1991) *The Composition of Foods*, 5th edn. Royal Society of Chemistry, Cambridge, 462 pp.
Perman, R., Ma, Y. and McGilvray, J. (1996) *Natural Resource and Environmental Economics*, 1st edn. Addison Wesley Longman, Harlow, 396 pp.
Public Health Policy Unit (1995) *Coronary Heart Disease in Scotland*. Scottish Office, Edinburgh, 106 pp.
Santarossa, J.M. (1998) *Strategic Behaviour and Market Deregulation in the UK Dairy Sector*. Econometrics Working Paper Series, Social Science Electronic Publishing, Washington, DC. Available at: http://papers.ssrn.com/paper.taf?ABSTRACT_ID=86598
Stanley, J.C. (2000) Taxing single nutrients is dangerous. *British Medical Journal* 320, 1469.

How do Markets Respond to Food Scares?

14

TIM LLOYD,[1] STEVEN McCORRISTON[2] AND WYN MORGAN[1]

[1] *University of Nottingham, Nottingham, UK;*
[2] *University of Exeter, Exeter, UK*

Introduction

In recent years, issues relating to food safety have risen to the top of the political agenda in many countries, particularly in Europe and North America. Much of this agenda has related to increased awareness over health, changing dietary patterns and the differences in standards between countries that lie at the heart of trade disputes. However, perhaps the most obvious feature of the increased public reaction to food safety is the issue of 'food scares'. This is particularly the case in Europe, and is exemplified by the development of the bovine spongiform encephalopathy (BSE) crisis in the UK since 1985, which has had a marked effect on the consumption of meat products in the UK and other European countries. What is particularly distinctive about the food-scare crisis in the UK is not only that it increased consumers' awareness of the health aspects of eating red meat (which was already on the increase) but also the fact that BSE has been related to variant Creutzfeld–Jakob disease (CJD), which to date has claimed more than 90 lives in the UK. The announcement in March 1996 that there was a suspected link between variant CJD in humans and BSE led to an immediate collapse in the consumption of beef products. According to estimates from a study sponsored by the UK Ministry of Agriculture, Fisheries and Food (MAFF) and the UK Treasury, the announcement of the link between variant CJD and BSE saw beef consumption fall temporarily by 40%, not only in the UK itself, but also in countries such as Germany and Italy, which had no reported cases of BSE at the time (DTZ Pieda Consulting, 1998). The domestic market was also affected by the immediate ban imposed on the export of beef products from the

UK to its traditional export markets in the EU. BSE is still a hot political topic in the UK and Europe more generally. Recently, an independent government report into the development of BSE and variant CJD in the UK, the so-called Phillips report (Phillips, 2000), surveyed the development of BSE in the UK and highlighted inadequacies by scientists, government ministers and civil servants, who failed to act quickly and effectively to protect the interests of consumers. In Europe more generally, the number of BSE cases has continued to increase and, at the time of writing, the French and German governments have moved swiftly to introduce strict measures to deal with the emerging BSE crisis in those countries. Market reports indicate an immediate fall of 50% in French beef consumption, and calls for import bans throughout the EU deepen a sense of foreboding.

To some extent, the academic literature has responded by paying greater attention to food-safety concerns. The literature has essentially focused on two themes. The first has been to revisit food-demand studies, taking into account health-driven changes in dietary patterns. Examples include Brown and Schrader (1990), Burton and Young (1996), Kinnucan *et al.* (1997) and Rickertsen and von Cramon-Taubadel (Chapter 3, this volume), among others. For example, the Burton and Young study on the impact of BSE in the UK takes a standard methodology for assessing the determinants of meat demand and includes a variable (relating to media stories) to account for BSE. Burton and Young's results showed a significant, but not very strong, impact of BSE, but it should be noted that their data precede the announcement in 1996 linking new variant CJD in humans to BSE. The other main theme in the food-safety literature has been to focus on the appropriate regulatory structure for dealing with food-safety concerns; see, for example, the symposium published in the *American Journal of Agricultural Economics* in December 1997. There has been little analysis of these issues in the UK, although the regulatory structure has now changed in response to the BSE crisis, with the establishment of an independent Food Standards Agency that is separate from the Department of the Environment, Food and Rural Affairs (DEFRA, formerly MAFF), which previously had responsibility for protecting consumer interests as well as protecting the interests of farmers.

However, to the best of our knowledge, the literature has not addressed the impact of food scares on prices at different stages of the marketing chain. This is the focus of this chapter, which studies the nature of price adjustment in the UK beef market in response to the outbreak, increased awareness and likely effects of BSE. This feature of the current study also has an important public-policy dimension, since the observation that prices at the retail level did not fall as much as prices at the wholesale and farm level was one of the key reasons for the recent inquiry by the UK Competition Commission into market power in the UK food sector, which emphasized the nature of vertical links between successive stages in the food-marketing chain.

An important feature of this study, therefore, is that we consider price adjustment at three stages (retail, wholesale and producer) of the beef-marketing chain. By doing so, we highlight not only the impact of price adjustment but also the effect of food scares on price spreads. Our data show that prices fell at all marketing stages in the 1990s but that spreads between prices have not remained constant. Specifically, whereas retail prices have fallen by 18%, wholesale and producer prices have fallen by around 40% each. During this period, all spreads have been observed to grow, but the retail–wholesale spread has grown five times more than the wholesale–producer spread. While this price decline is unsurprising in the face of heightened consumer awareness, the change in spreads is perhaps less obvious. However, as McCorriston *et al.* (1999) show, this result is consistent with the outcome of a model of vertical markets characterized by market power, in which a shock to retail demand passes through the chain where the price transmission elasticity is greater than unity.

The chapter is organized as follows. First, we briefly survey recent changes in the patterns of meat consumption in the UK and the development of the BSE crisis since 1985. Secondly, we refer to both the theoretical and the empirical literature relating to price adjustment in vertically related markets to give us a benchmark of how prices may change in markets that vary by stage of marketing/production. The literature suggests that, even though prices change in the same direction, the extent of the price changes varies according to the retail stage and hence marketing margins will change. In the case studied here, marketing margins are expected to increase even if marketing costs (due to, say, increased regulation) and the retail demand curve are shifted to the same extent. Notwithstanding this, the nature of the data relating to food scares would suggest that demand shifts are likely to dominate. Thirdly, we present the data and discuss the framework for the empirical analysis. Acknowledging the co-movement of beef prices at all stages of the chain, we adopt a co-integration framework for the empirical analysis. The results show that beef prices do not co-integrate without the inclusion of a measure of food-safety publicity – the food publicity index. The negative impact of the index differs according to the marketing level and may account for growing price spreads observed empirically.

Consumption Trends, Health Concerns and BSE

General trends

As in many developed countries, consumption of meat products has undergone considerable change in recent years. The most common

Table 14.1. Annual per capita consumption of meat in the UK (kg). (From: MAFF, 2001.)

Year	Beef	Lamb	Pork	Bacon	Red meat	Poultry	Total
1975	23.60	8.48	10.47	8.91	51.45	11.66	63.11
1980	21.30	7.69	12.55	9.00	50.54	12.71	63.25
1985	19.97	7.25	12.51	8.15	47.88	15.61	63.49
1986	20.03	6.70	12.72	8.19	47.65	17.29	64.94
1987	20.25	6.50	13.51	8.06	48.31	18.58	66.90
1988	19.34	6.71	13.97	8.00	48.02	20.44	68.46
1989	18.55	7.16	13.26	7.99	46.96	20.23	67.18
1990	17.08	7.45	13.37	7.74	45.64	21.14	66.78
1991	17.23	7.34	13.40	7.64	45.60	21.73	67.33
1992	17.13	6.51	13.28	7.27	44.19	23.03	67.22
1993	15.63	5.80	13.66	7.53	42.61	23.60	66.21
1994	16.57	6.05	13.44	7.80	43.86	24.97	68.83
1995	15.38	5.99	12.77	7.97	42.10	25.26	67.37
1996	12.58	6.22	13.35	8.01	40.17	26.92	67.09
1997	14.43	5.95	14.02	7.82	42.22	26.43	68.66
1998	15.08	6.49	14.38	8.10	44.06	28.04	72.10
1999	15.41	6.38	13.76	7.91	44.15	28.47	71.94
2000	15.98	6.54	13.26	7.80	44.25	28.30	71.88

feature of this change has been due to the switch away from red meat consumption towards the consumption of white meats, such as pork and particularly poultry, which has appealed to consumers as being less fatty, lower in cholesterol and generally ‘healthier’. Other changes that have affected meat consumption include lifestyle changes, such as the increased consumption of prepared and frozen meals and food away from home. These changes in meat consumption have been well documented in the agricultural economics literature. With respect to the UK, Burton and Young (1992) identify the role of preference changes in the consumption of meat in the UK.

Table 14.1 details changes in meat consumption (in volume terms) in the UK in the period 1975 to 2000. Focusing first on the final column, over this period total consumption of all meat products has risen by 12%. There have been significant preference shifts across meat products. Note, first of all, that the consumption of red meats (column 6) has shown a decline of 14% over this 25-year period. This represents a decrease in consumption of 32% in beef, 22% in lamb and 12% in bacon but a 27% increase in pork. Overall, consumption of red meat as a proportion of the total fell from 81 to 61%. The other most notable change shown in the table has been the increased consumption of poultry products, which has increased by 143% over the same period. Consequently, the share of poultry meat consumption in total meat consumption has increased from 19 to 39%. Although growing awareness of health concerns, and BSE in particular, is likely to have contributed to these trends, the fact that they were observed in many other countries reflects a more general shift in consumer preferences.

The development of BSE in the UK

BSE first came to the public's attention in the mid-1980s when a cow that died following symptoms of head tumours and lack of coordination was confirmed as having a new cattle disease, BSE. Nevertheless, in the early chronology of this disease, there was little concern that BSE represented a threat to human health, with the public being reassured by a government-sponsored working party (the Southwood working party) that BSE was unlikely to cause any harm to consumers. The next most significant step in the BSE crisis occurred in May 1990, following the death of a Siamese cat from a BSE-like disease, the significance of which was that BSE appeared capable of jumping between species. Although public awareness increased dramatically, the public was again reassured by government ministers and the UK Chief Medical Officer that meat was safe to eat. Although public awareness had been increasing and no doubt contributed to the decline in the consumption of red meat over the early 1990s (Burton and Young, 1992), it was not until the mid-1990s that BSE was fully recognized as a crisis, following the first human death from variant CJD in 1995, followed by the confirmation in 1996 of a link between variant CJD and BSE. Consumption of red meat fell immediately by 40% while bans on imports of beef from the UK by European countries were imposed. Since then, the link between variant CJD in humans and BSE has been confirmed and, to date, 90 deaths have been recorded in the UK. Although regulation in the meat-processing sector has intensified as a response to the BSE crisis (see below), awareness of the crisis among the public is still high, given that variant CJD is known to have a potentially long incubation period. Recent estimates even suggest that the total number of deaths in the UK may be in the thousands (*New Scientist*, 2000).

Literature on BSE in the UK

The literature on food demand in the UK has its roots in the work undertaken by researchers at the University of Manchester (Thomas, 1972). This work formed the basis of regular analyses of food demand reported by MAFF in the annual publication, *Household Food Consumption and Expenditure* (renamed *National Food Survey* since 1992) and produced annually since 1950. These publications contain a wealth of information on food expenditure, consumption and nutrient intake and are based on a survey of 6000 households in the UK evaluated by region and household characteristics. In 1994 the survey was extended to cover eating out, a topic that has received growing attention, not least because it accounts for over 25% of total food expenditure (Lund and Derry, 1985; Lund, 1998; Warde and Martens, 1998). These publications also contain occasional special analyses, such as those on demand elasticities (MAFF, 1990, 2000) and meat consumption, the methodology applied being the almost ideal demand system. Using similar methods, Burton and Young have produced a series of papers on food demand in the UK (Burton and Young, 1992; Burton *et al.*, 1995), which decompose the variation in meat and fish consumption into economic factors and shifts attributable to consumer preferences. Recent (interdisciplinary) British research in this area has been undertaken as part of the Economic and Social Research Council-funded programme, *The Nation's Diet* (Murcott, 1998), which offers an in-depth and systematic study of sociological and psychological dimensions to food choice, and underscores the complex and contrary attitudes of consumers to food in the UK. The analysis in MAFF (2000) also estimates the effect of the food scares index used in the current chapter on the expenditure shares of food groups, using an almost ideal demand system. The effect of BSE has been evaluated at the product level in Burton and Young (1996), at sector level by DTZ Pieda Consulting (1998) and in terms of the UK economy in MacDonald and Roberts (1998).

Price Adjustment: Theoretical and Empirical Issues

The focus of this chapter centres upon how markets responded to the public's growing awareness of the BSE crisis and how it may be linked with human health. More specifically, we focus on prices in the UK beef sector during the 1990s, highlighting the impact of prices at each stage of the marketing chain. This issue also became part of the political environment of the BSE crisis. As consumption fell, so too did prices (as expected). However, prices in the retail sector appeared to fall less than prices in the processing and farm sectors. This sparked claims that the market power of the retail sector was able to cushion the impact of BSE on retailers, with most of the impact being carried by farmers, and was one of the contributory factors that led to the Competition Commission inquiry into the food retailing sector that concluded in the autumn of 2000 (Competition Commission, 2000).

Theoretical issues

In the agricultural economics literature, the most popular framework for considering the impact of exogenous shocks to the food sector is in the context of the equilibrium displacement models originating with Muth (1964) and Gardner (1975). The basis of this framework is to consider the impact of price adjustment (and corresponding welfare effects) at different stages of the marketing chain and how the farm–retail spread may change as a consequence. An important feature of the Gardner framework relates to the source of the exogenous shocks. In particular, a demand-side shock directly affecting, say, the retail sector, may have a different effect on retail and farm prices than a supply-side shock originating at the farm sector.

However, one important criticism of the standard equilibrium displacement framework is that it assumes that the downstream food market is perfectly competitive. There is increasing evidence that this is not an appropriate characterization of the food sector in many developed countries. For example, Bhuyan and Lopez (1997) measured market power across 40 subsectors of the US food industry and found that, in all but three cases, the null hypothesis of competitive behaviour could be rejected. Although studies of the food sector in other developed countries are generally lacking, high levels of concentration are at least suggestive that similar phenomena would characterize the food sector in most European countries. Moreover, the current investigation by the UK Competition Commission clearly reinforces the suspicion that market power is a feature of the food sector in the UK.

What difference does this make to our understanding of price adjustment? A number of studies from the industrial organization literature have shown that imperfect competition has a considerable impact on price adjustment. Specifically, under 'reasonable' conditions, the transmission of supply shocks is reduced if the downstream sector is characterized by market power. Recently, McCorriston *et al.* (1998) have introduced market power into the Gardner framework; they have shown that, for an increase in costs at the farm level, market power in the downstream food sector will reduce the level of price adjustment compared with the competitive benchmark. Intuitively, as prices rise, the downstream firms' price–cost margins change, which cushions the impact of a cost increase on retail prices. The implication is that, in the presence of a supply shock at the farm level, the model predicts a narrowing of price spreads. The same holds true for a shift in the marketing-services supply function, which captures the increase in costs due to tighter regulatory standards. Under imperfect competition, an increase in these input costs causes retail prices to rise but this increase is proportionately less than the increase in input costs. Consequently, price spreads should narrow where processing costs increase, again due to the role played by the change in the price–cost mark-up at the downstream stage.

However, this still leaves open the issue relating to the source of the exogenous shock. As highlighted above, distinguishing between a demand and cost shock matters. McCorriston *et al.* (1999) have also considered this issue and label price adjustment due to a demand shock affecting the retail sector as a 'pass-back' effect. They show formally that market power in the downstream food sector will increase the magnitude of the price changes occurring upstream compared with the competitive benchmark. In other words, price changes at the farm level will be greater than those occurring at the retail level for a given shift in the retail demand function. As in the case of the cost changes, the role of the price–cost mark-up plays an important role in determining the extent of price adjustment, though, in the case of a demand shock, it serves to reduce farm-level prices relative to the change in retail prices. In relation to the impact on price spreads, the implication is that they should widen following a demand shock.

Over the time period of the BSE chronology, the beef sector is likely to have been affected by both demand and cost shifts affecting the evolution of prices. Most obviously, the BSE crisis would have caused shifts in the retail demand function for beef, which will have had an impact on upstream prices. However, cost-side effects may also be pertinent, given the new meat sector regulations. These have been introduced to reassure consumers of the safety of British beef in the aftermath of the crisis and are likely to have increased costs. For example, higher charges for the Meat Hygiene Service and specified risk material (SRM) disposal, reduced subsidies for rendering and increased transactions costs are widely held to have arisen from the BSE crisis (Meat and Livestock Commission, 1999). Costs have also increased due to the greater use of controlled atmosphere packaging (CAP). However, in the absence of reliable estimates, the impact of these changes is difficult to verify, not least because partial compensation was received by many of those affected.[1] Clearly, in considering the impact of food scares on the UK beef sector, there are likely to be both demand and supply factors determining prices at each vertical stage. However, as discussed in the following section, it should not necessarily be expected that these exogenous factors should have a symmetrical impact on prices (and hence price spreads) throughout the marketing chain. More specifically, while rising costs due to tighter regulatory standards will lead to a decrease in the price spreads, shifts in the retail demand function will lead to an increase in the price spreads. Moreover, even if the shift factors changed by the same amount, their effects will not cancel out, as the theoretical literature has shown that cost shifts are not necessarily equivalent to the (inverse) of demand shifts when markets are characterized by imperfect competition.

Given that the meat sector was likely to be subject to both sources of exogenous change over the 1990s, the impact on prices at each vertical stage (and, by extension, the development of price spreads) will be the net outcome from both sources of exogenous shocks to the beef sector. Moreover, even if marketing costs and the demand curve shifted by equal amounts, the shift in the retail demand function should dominate the impact of the rise in costs associated with tighter regulatory standards. Thus, unless the extent of marketing cost increases was significantly greater than the direct effect of the food scares (which is unlikely to be the case, given the nature of the food scare we are dealing with and the widespread public attention it has received since 1985), a priori, we should expect to observe a widening of the price spreads over this period in the presence of market power.

Empirical issues

There is a growing number of studies that focus on the empirical aspects of price adjustment in agricultural and related markets. Owing to the apparent co-movement in prices, many of these recent studies apply a co-integration methodology, since it: (i) accounts appropriately for the time-series properties of the data; and (ii) measures the short- and long-run nature of price adjustment. Examples of such studies include Chang and Griffiths (1998), von Cramon-Taubadel (1998) and Goodwin and Holt (1999), among others. Chang and Griffiths' and Goodwin and Holt's papers are distinguished from the earlier literature in that they focus on multiple stages of the marketing chain by separately identifying price adjustments at the farm, wholesale and retail levels, a practice that is also followed in this chapter. However, in most cases, these empirical studies confine the sources of price adjustment to be 'within' the food chain. Exogenous influences, apart from a time trend or a given structural break, are not accounted for.

In investigating the impact of food scares, it is important to recognize that this exogenous shock represents principally a demand-side shift to the food sector (even though regulation has also increased). As a result, it is reasonable to expect price adjustment in the upstream stages of the marketing chain but that the extent of these price changes would vary between stages if the downstream sectors were characterized by imperfect competition. Although the analysis of price adjustment is not a formal test of market power, higher levels of price adjustment in the upstream stages are certainly consistent with it. If the extent of price changes between each marketing stage varies, this implies that the price spread (or marketing margin) will also change. Where prices are non-stationary and potentially exhibit co-movement, a co-integration framework offers an appropriate methodology for considering the extent of price adjustment. The data and the effect of food scares on the beef market are outlined in the next section.

Data

In the empirical analysis, we employ real beef prices at the producer, wholesaler and retailer level (P_t, W_t and R_t, respectively) in England and Wales. The data have been calculated by the Meat and Livestock Commission and represent carcass weight equivalents (CWE) to facilitate direct comparison of prices at all three stages. The data are monthly and cover the period January 1990 to December 1998 (see Figs 14.1 and 14.2).[2] They are deflated by the retail price index (January 1990 base) and clearly show a trend decline over the period: retail and wholesale prices fall at a rate of 1.7% per year and producer prices by 5.4% over the sample period. Casual inspection of Fig. 14.1 also reveals that prices have a tendency to co-move, in that movements at prices in one level of the chain seem to be reflected in prices elsewhere in the chain. Other features of interest are the upturn in prices in the first half of 1993 and the rapid decline in March–April 1996. Whereas the former is accounted for by the UK Chief Medical Officer confirming that beef was safe to eat, the latter is a response to the now notorious ministerial announcement on 20 March 1996 in which the link between BSE in cattle and CJD in humans was officially recognized.

These price series are supplemented by a 'food publicity index'. This is a count of the number of articles printed per month in national broadsheet newspapers that relate to the safety of meat.[3] In general, these reports are negative in nature and reflect the concerns regarding the safety of meat, in terms of its production and processing. Articles relating to BSE dominate the index, although other similar topics are also covered, such as the health standards in abattoirs. The index reflects consumer concerns regarding the safety aspects of meat consumption and the impact of regulation on suppliers of meat. Consequently, the index will be correlated with developments that affect both the demand for and supply of beef, although, as discussed above, it is likely to be the food scares issue that dominates, given the public furore over BSE.

While by no means the first 'food scare' in the UK – for example, in 1988 *Salmonella* was linked to the consumption of eggs – the publicity surrounding the BSE crisis was unprecedented, overwhelming all concerned, including government ministers and their officials in MAFF and the Department of Health. As noted in the introduction, the recent government-sponsored BSE inquiry, known as the Phillips report, has been highly critical of the way in which government ministers, civil servants and scientists responded to this crisis. Despite estimates that the probability of contracting the disease was smaller than winning the lottery on successive occasions,[4] the public's response to the many uncertainties that surrounded BSE significantly affected consumption behaviour. The

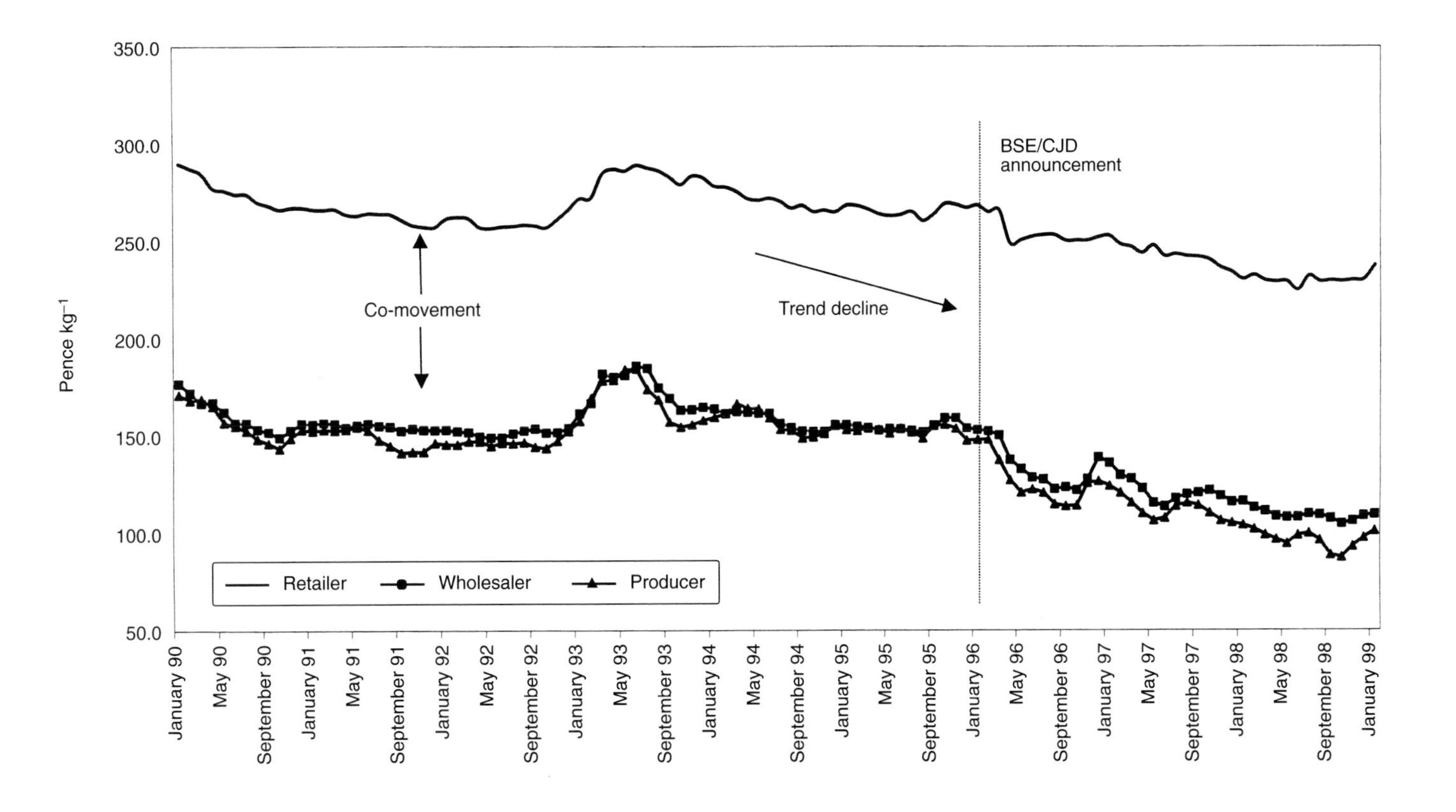

Fig. 14.1. Real UK beef prices.

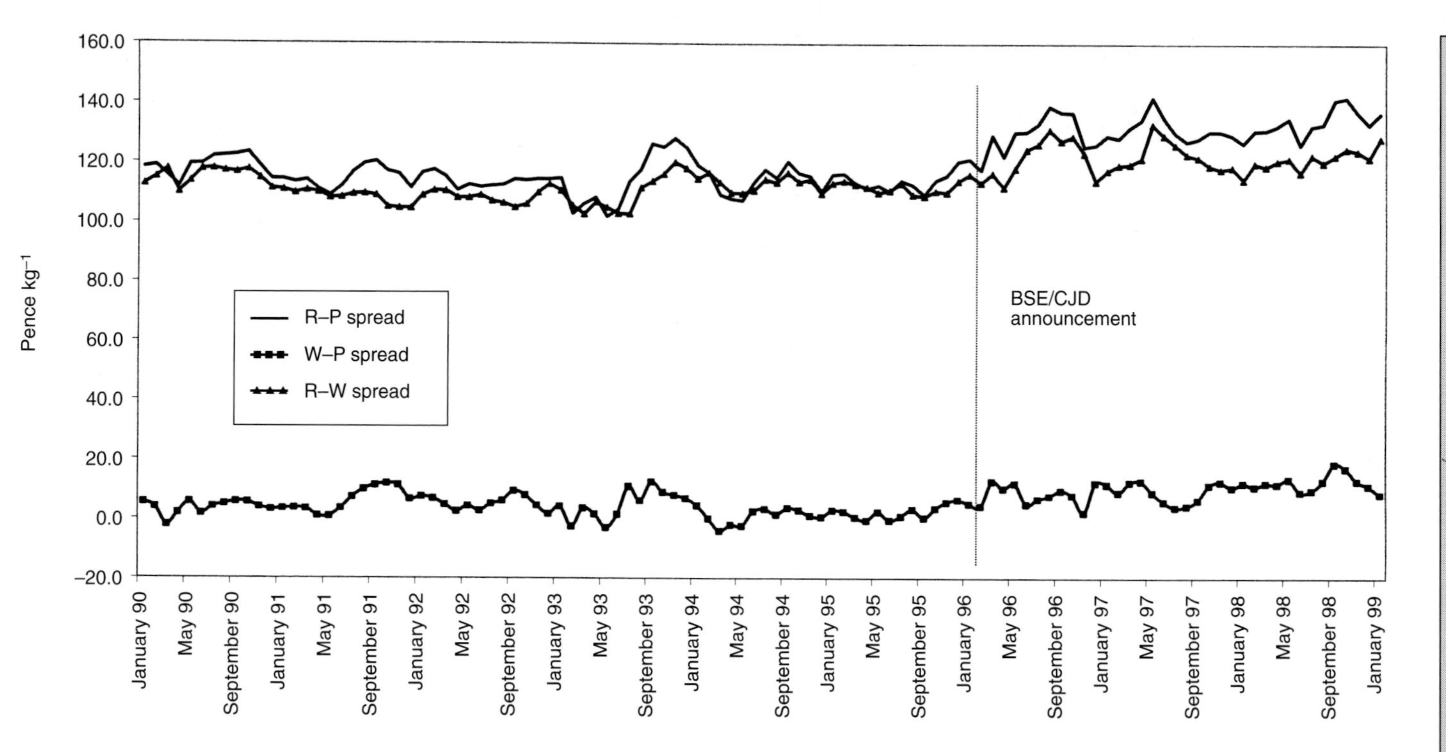

Fig. 14.2. UK beef price spreads. R, retailer; P, producer; W, wholesaler.

now notorious and disturbing televised image of an infected cow helplessly stumbling in the later stages of the disease shocked the viewing public instantly. Repeated broadcasts fixed the image in the national psyche and became a potent emblem of the frightening potential of BSE for human health.

By any measure, the publicity was staggering. Prior to the ministerial announcement on 20 March 1996 linking BSE in cattle to its human variant, variant CJD, there were around 14 articles per month relating to food safety in the national broadsheet press. In the same month, the European beef-export ban took effect (European Commission Decision 96/293) and the count peaked at 333, averaging 93 per month for the remainder of the 1990s. Other key points in the chronology of food safety in Britain are identifiable in the series plotted (in logs) in Fig. 14.3, such as: the peak in the summer of 1990 reflecting advice relating to beef-on-the-bone products for export; the announcement relating to the breeding from the offspring of BSE cases; and the 'Specified Bovine Offal (Amendment) Order 1995 (SI 1995/3246)' prohibiting the use (and export) of certain offals destined for human consumption.

As a preliminary to the statistical analysis, all the data series are tested for non-stationarity, using standard Dickey–Fuller procedures. Model selection criteria are presented in Table 14.2, with augmented Dickey-Fuller (ADF) test statistics under the null of a unit root. The results clearly demonstrate that the three prices are non-stationary in levels and stationary in first differences, although inference regarding the order of integration of the food-publicity index (i_t) is less clear-cut. Since, both the Akaike information criterion (AIC) and the Hannan–Quinn criterion (HQC) tests select a model with two lags, in which the null of a unit cannot be rejected at the 5% level, we conclude that the index is also I(1), as visual inspection of Fig. 14.3 suggests.[5]

Methodology

The formal analysis of price transmission is conducted in a vector autoregressive (VAR) framework to exploit the properties of integration and co-integration that appear to exist in the data. VAR methods offer a tractable framework for the investigation of dynamic relations, particularly when the variables are co-integrated. If co-integration is ignored, estimation and inference are, at best, impaired and, at worst, invalidated (Harris, 1995). VAR models also readily facilitate investigation of the dynamic response path of variables to exogenous shocks, using what is called impulse response analysis.

Given the familiarity of these methods, we merely offer a sketch of

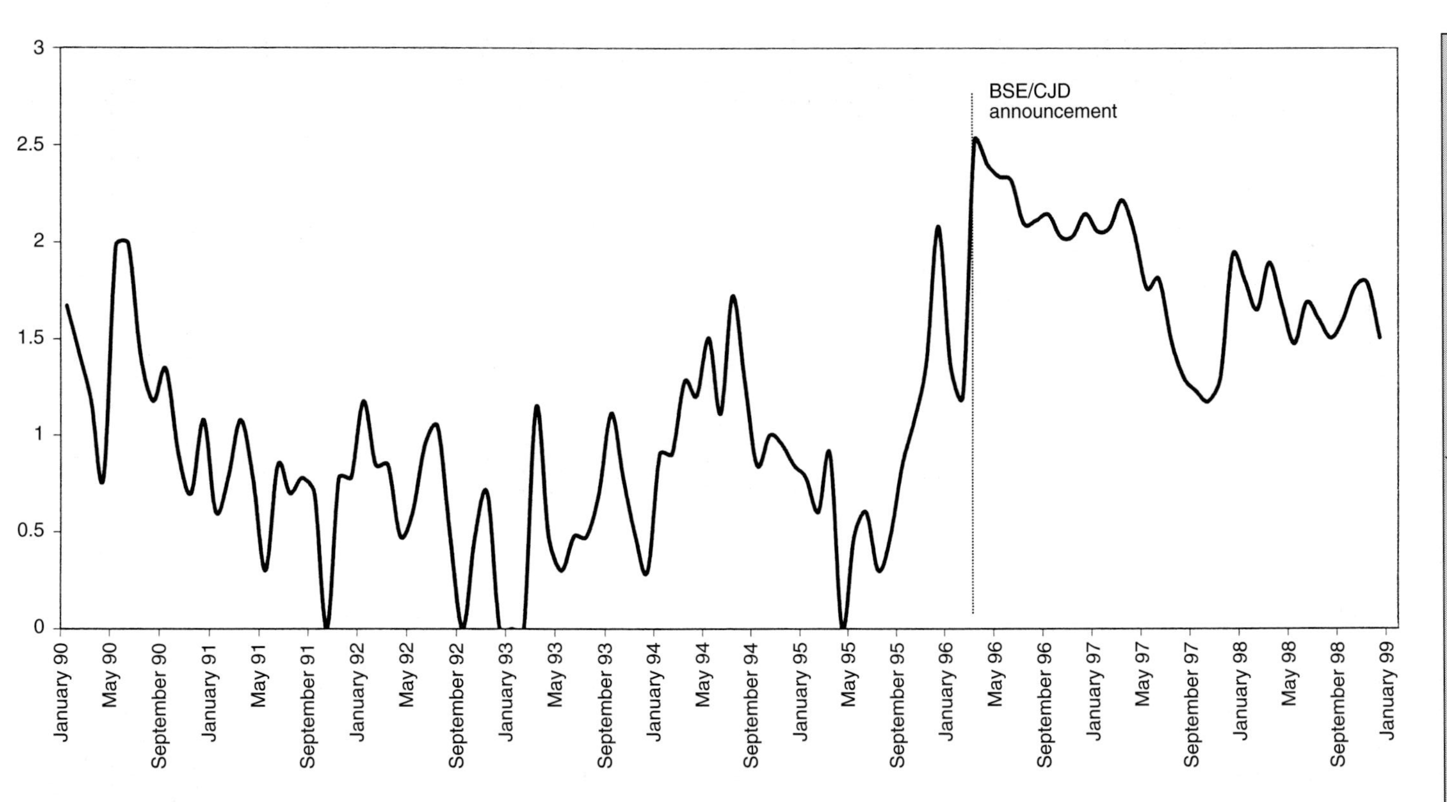

Fig. 14.3. The food-publicity index (logs).

Table 14.2. ADF tests for the order of integration.

	Lag length	ADF test statistic	AIC	SBC	HQC
R_t	0	−1.54	−281.57	−285.60	−283.21
	1	−1.53	−282.57	−287.93	−284.74
	2	−1.78	−281.90	−288.61	−284.62
	3	−1.89	−282.55	−290.59	−285.81
ΔR_t	0	−10.40	−281.78	−284.46	−282.86
	1	−6.28	−281.53	−285.56	−283.17
	2	−5.07	−282.40	−287.76	−284.57
	3	−4.22	−283.14	−289.85	−285.86
W_t	0	−1.31	−309.45	−313.47	−311.08
	1	−2.31	−297.16	−302.52	−299.33
	2	−2.06	−297.92	−304.63	−300.64
	3	−2.07	−298.86	−306.91	−302.13
ΔW_t	0	−6.36	−297.86	−300.54	−298.95
	1	−6.01	−298.11	−302.13	−299.74
	2	−5.20	−299.10	−304.47	−301.28
	3	−4.99	−299.78	−306.48	−302.50
P_t	0	−1.36	−293.87	−297.90	−295.51
	1	−2.13	−285.63	−291.00	−287.81
	2	−2.14	−286.57	−293.27	−289.29
	3	−2.22	−287.37	−295.41	−290.63
ΔP_t	0	−6.96	−285.94	−288.62	−287.03
	1	−5.79	−286.93	−290.96	−288.56
	2	−4.95	−287.92	−293.29	−290.10
	3	−4.48	−288.92	−295.62	−291.64
i_t	0	−3.87	−144.08	−146.76	−145.16
	1	−3.28	−144.16	−148.18	−145.79
	2	−2.66	−142.39	−147.75	−144.56
	3	−2.44	−142.80	−149.50	−145.52
Δi_t	0	−12.90	−148.43	−151.12	−149.52
	1	−10.70	−144.95	−148.97	−146.58
	2	−8.51	−144.82	−150.19	−147.00
	3	−5.97	−145.43	−152.14	−148.15

Tests are evaluated from lag 12, although only lag 3 and below are reported. The ADF regression of the levels (e.g. R_t) includes a constant, trend and seasonals, with the 95% critical value of the ADF statistic given as −3.45. The ADF regression of the differences (e.g. ΔR_t) includes a constant and seasonals (but no trend) with a 95% critical value of the ADF statistic of −2.88. The highest numerical value of the model-selection criteria (AIC (Akaike Information criterion), SBC (Schwarz Bayesian criterion) and HQC (Hannan–Quinn criterion)) indicates the preferred model.

the relevant aspects here.[6] Consider the general polynomial distributed lag, or VAR(k), model:

$$\mathbf{x}_t = \mathbf{\Pi}_1 \mathbf{x}_{t-1} + \ldots + \mathbf{\Pi}_k \mathbf{x}_{t-k} + \boldsymbol{\mu} + \boldsymbol{\varepsilon}_t \quad (14.1)$$

where $t = 1, \ldots, T$; $\mathbf{x}_t$ is an $(n \times 1)$ vector variables, $\boldsymbol{\mu}$ an $(n \times 1)$ vector of deterministic components (such as constant, seasonal factors and dummy variables) and $\boldsymbol{\varepsilon}_t$ an $(n \times 1)$ vector of normally distributed disturbances of zero mean and non-diagonal variance–covariance matrix $\mathbf{\Omega}$, i.e. $\boldsymbol{\varepsilon}_t \sim$ n.i.d.$(0, \mathbf{\Omega})$. The order (k) of the VAR is also determined by the data, and here we adopt standard model-selection criteria (AIC, Schwarz Bayesian criterion (SBC) and HQC) for this purpose. In cases where the variables co-integrate, Johansen (1988) shows that Equation 14.1 has an equilibrium-correction representation given by:

$$\Delta \mathbf{x}_t = \boldsymbol{\alpha\beta}' \mathbf{x}_{t-k} + \sum_{i=1}^{k-1} \Gamma_i \, \Delta \mathbf{x}_{t-i} + \boldsymbol{\mu} + \boldsymbol{\varepsilon}_t \quad (14.2)$$

where $\boldsymbol{\beta}'$ is the $(r \times n)$ matrix of co-integrating vectors, $\boldsymbol{\alpha}$ is an $(n \times r)$ matrix of loading (or error correction) coefficients and r denotes the number of linearly independent combinations of $\mathbf{x}_t$ that are integrated of order zero {I(0)}. These linear combinations represent the long-run relationships between the variables in $\mathbf{x}_t$. Empirically, the rank of $\mathbf{\Pi}$ is determined using the trace and maximal eigenvalue statistics, which are compared with critical values derived by Osterwald-Lenum (1992).

Following Stock and Watson (1988), with n variables, there can be at most $n - 1$ linearly independent co-integrating relations and $n - r$ common trends. Therefore, with a triplet of prices, there can be at most two such co-integrating combinations, since, if any two pairs of prices co-move (co-integrate), then so must the third. In this case, the prices share a single common trend and may be expected to co-move over time. The presence of a single co-integrating relation among the triplet of prices implies two separate common trends among the price triplet. As a result, price pairs do not co-integrate, since any two prices will possess different trends. In this case, price co-movement between any pair will be much less apparent when 'eyeballing' the series together, but the presence of co-integration implies that the long-run behaviour of any two variables is reflected in the third. Where there are no co-integrating combinations among the triplet, each price has its own (possibly similar, but nevertheless) distinct trend. If the price triplet is to co-integrate, then at least one additional variable is required. Owing to the parallels between common trends, co-integration and co-movement, formal tests for co-integration offer insights into the economic relationships at hand, signalling not only the extent of economic integration in vertical markets but also the factors necessary to induce such market integration.

Should there be more than one co-integrating vector, as is likely to be the case here, it is often difficult to interpret the co-integrating vectors directly (Lütkepohl and Reimers, 1992). Impulse response analysis provides a tractable means of evaluating the time path of variables in $\mathbf{x}_t$ to exogenous shocks. As such, it offers insights into the dynamic behaviour of variables in the system by combining the short- and long-run behaviour embodied in Equation 14.2. Calculation of the impulse response functions for a system of co-integrated variables follows much the same approach as that used in standard (stationary) VARs (Lütkepohl and Reimers, 1992). However, the fact that co-integration implies an error-correction mechanism has implications for calculation and interpretation of impulse responses, not least since shocks to a co-integrated system do not die out, but persist in the long run, albeit via a time path that leads to a new equilibrium.

The time paths of variables to shocks, or impulse response functions, are found by imposing a recursive structure on the moving-average representation of the VAR. Lütkepohl and Reimers (1992) show that the impulse response function of Equation 14.1 is given by:

$$\mathbf{\Phi}_s = \left(\boldsymbol{\varphi}_{ij,s}\right) = \sum_{l=1}^{s} \mathbf{\Phi}_{s-l}\, \mathbf{A}_l \quad \text{for } s = 1, 2, \ldots \tag{14.3}$$

where $\mathbf{\Phi}_0 = \mathbf{I}_n$, $\mathbf{A}_l = 0$ for $l > k$. A plot of $\boldsymbol{\phi}_{ij,s}$ is the impulse response function of variable i with respect to a unit shock to variable j, s periods ago, all other variables at the time of the shock (and earlier) being held constant. While this is a useful measure of the dynamics, it ignores the contemporaneous correlation that may exist between the variables (expressed in the off-diagonal elements of $\boldsymbol{\varepsilon}_t$) and the fact that, in general, the impulse response functions are not unique. Here, we use the generalized impulse response function developed by Pesaran and Shin (1998), which takes account of these problems, although even here a caveat is warranted since, as with any dynamic simulation, the issue of exogeneity is relevant. In the current application, given that changes in the food-publicity index are likely to drive price changes and not vice versa, the index is treated as exogenous to prices in the impulse response analysis.

Co-integration

As an initial step, Equation 14.1 is estimated for the price triplet (R_t, W_t and P_t). An unrestricted VAR(12) model, augmented by seasonals and four impulse dummy variables[7] gives a good approximation, such that residuals conform to the stated assumptions for $\boldsymbol{\varepsilon}_t$ (see Table 14.3). Given that prices are I(1), the model is examined for the presence of co-integration implied by the co-movement of prices apparent in

Table 14.3. Co-integration test results.

	H_0	Maximal eigenvalue	95% CV	Trace	95% CV
(a)	R_t, W_t and P_t				
	$r = 0$	20.4	21.0	30.7*	29.7
	$r \leq 1$	10.0	14.1	10.3	15.4
	$r \leq 2$	0.3	3.8	0.3	3.8
	Diagnostic tests Vector AR 1–2 F(18,136) = 1.4219 [0.1307] Vector normality $\chi^2(6)$ = 7.1989 [0.3028]				
(b)	R_t, W_t, P_t and i_t				
	$r = 0$	44.3**	27.1	105.0**	47.2
	$r \leq 1$	37.0**	21.0	60.8**	29.7
	$r \leq 2$	22.3**	14.1	23.7**	15.4
	$r \leq 3$	1.4	3.8	1.4	3.8
	Diagnostic tests Vector AR 1–2 F(32,123) = 1.1678 [0.2696] Vector normality $\chi^2(8)$ = 11.008 [0.2012]				

* Denotes rejection of H_0 at the 5% significance level; ** denotes rejection of H_0 at the 1% significance level. Critical values are asymptotic and do not take account of dummy variables. Re-estimation of the models without these variables introduces residual non-normality but does not affect the inferences regarding co-integration that are reported in the text.

Fig. 14.1. Panel (a) of Table 14.3 reports the co-integration test statistics for this model.

While there appears to be an indication of co-integration among the triplet, the formal evidence is at best weak. Specifically, the trace-test statistic rejects the null hypothesis of no co-integration at the 5% significance level, but the maximal eigenvalue test does not. In addition, the test statistics do not provide any substantive evidence of the multiple co-integrating relationships suggested by the pair-wise co-movement of the data. One literal, albeit bald, interpretation of this result is that beef markets are poorly integrated. Alternatively, the explanation might lie in the role of omitted variables, in particular, given the preceding discussion, the impact of BSE and related health concerns on price movements.

Augmenting the price-transmission model with (the natural logarithm of) the food publicity index i_t has a marked effect on inference. As the results in panel (b) of Table 14.3 show, evidence in favour of co-integration is now much stronger: the trace and maximal eigenvalue test statistics now reject the null of no co-integration at the 1% significance level. Moreover, both tests suggest the presence of three co-integrating vectors, a result consistent with pair-wise co-movement in combination with the food-scares variable. The clear conclusion of the co-integration analysis is that the food-publicity index plays a key role in the long-run evolution of UK beef prices and that, once the effect of the index is taken into account, prices co-move in a manner consistent with market integration.

Impulse Response Analysis

The above result begs a number of questions, not least those relating to the precise role that the food publicity index plays in price formation. To investigate this issue empirically, consider Fig. 14.4, which shows the generalized impulse response functions of the three beef prices to a one-standard-error shock in the equation describing the food publicity index. They indicate that heightened publicity regarding food safety leads to an instantaneous fall of 0.39 p kg^{-1} (pence per kg), 0.66 p kg^{-1} and 1.34 p kg^{-1} in retail, wholesale and producer prices, respectively. Allowing the shock to work through the system, the impacts grow such that the long-run impact is three to four times greater. The point estimates suggest that, in the long run, prices at the three stages fall by 1.80 p kg^{-1}, 2.80 p kg^{-1} and 3.11 p kg^{-1}, respectively.

These results indicate that, first, UK beef prices were responsive and negatively related to the public's awareness of food safety issues (principally BSE) in the 1990s, but that, secondly, the impact was not common across stages in the marketing chain. This second point suggests that price spreads also move systematically in response to publicity about the safety of food. Shocks to the food publicity index cause the wholesale–producer price spread to expand more than the retail–wholesale price spread. Moreover, the difference between retail and producer prices, the measure that receives most attention in the public debate on this issue, rises by an even larger amount in response to BSE publicity. Specifically, the same one-standard-error shock in the log of the publicity index induces a 1 p kg^{-1} increase in the retail–wholesale spread, a 0.3 p kg^{-1} increase in the wholesale–producer spread and thus a 1.3 p kg^{-1} increase in the retail–producer spread. Given that media interest has generally risen over the sample period, it is not surprising that price spreads have risen over time.

The observation that the food scares index should lead to a decline in prices at each marketing stage is, to a large degree, expected and consistent with the dominance of the demand relative to supply sources (e.g. due to increased regulation) of exogenous shocks in the beef sector over the 1990s. That this price decline should vary between stages is a little more surprising, particularly so, given the nature of the data, which, being consistent with the fixed-proportions technology, might

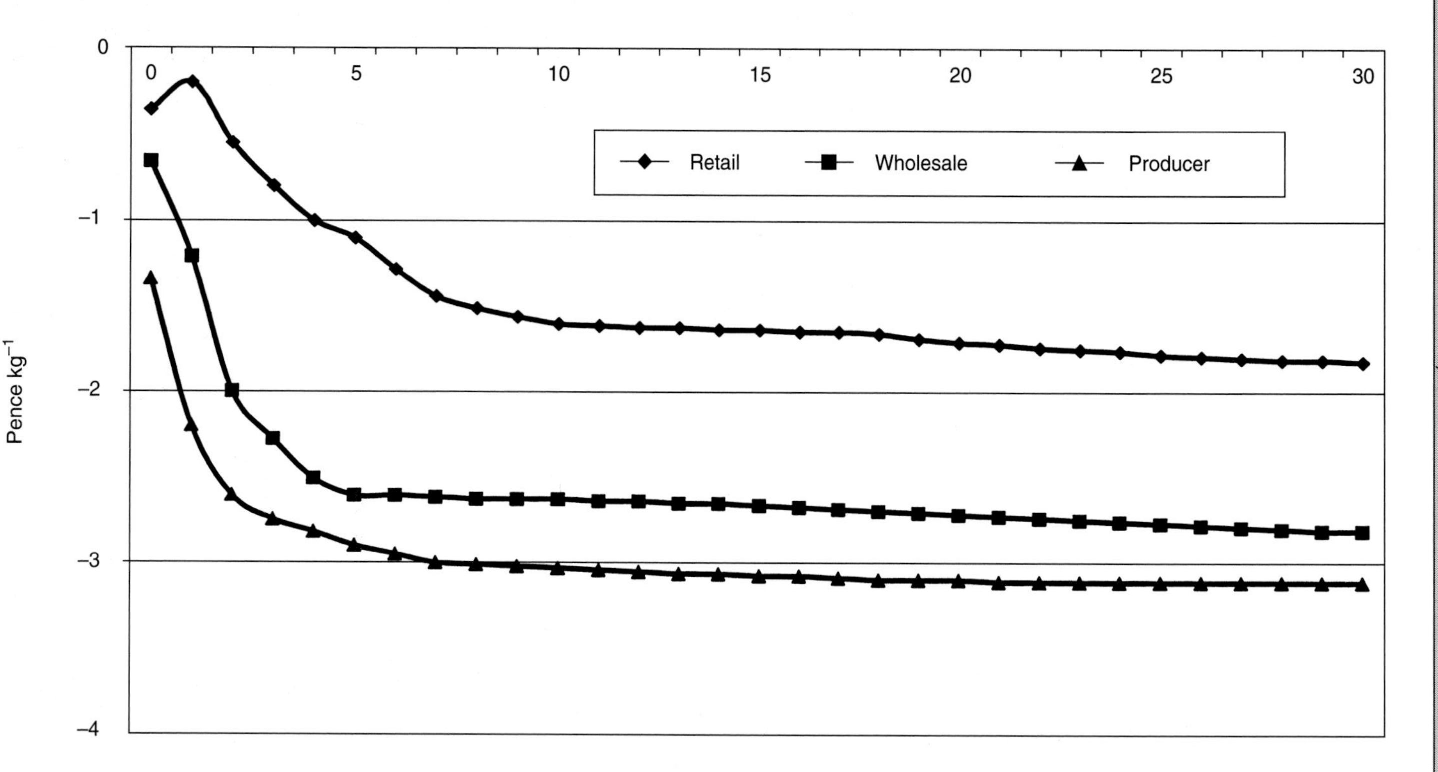

Fig. 14.4. Generalized impulse response functions of beef prices to a one-standard-error shock in the food publicity index.

otherwise suggest that the price declines would be equal. The fact that these price declines vary between stages, implying a widening of the price spreads, is possibly (although not necessarily) indicative of a food chain characterized by some degree of market power at each vertical stage.

Conclusions

BSE and its link to variant CJD in humans has been the most significant food scare to have affected the UK and Europe more generally in recent years. This chapter has focused on the impact of publicity, predominantly concerns relating to the emerging BSE crisis, on the development of prices in the UK beef sector during the 1990s. Acknowledging the co-movement that exists between prices in the meat marketing chain, we use a co-integrating framework, the results of which show the importance of information as embodied in the food publicity index in price transmission. Prices at all levels have tended to decrease during the 1990s, a result that is consistent with inward shifts in the demand function. Perhaps more interesting is the fact that the extent of price adjustment varies between marketing stages. In particular, price changes at the retail (wholesale) sector decline but less so than price changes at the wholesale (farm) sector. While supply-side shocks may also have been relevant to the development of prices in the UK beef sector over the 1990s due to increased regulation, the fact that widening margins are observed suggests the dominance of demand-side shifts in a vertically related market, which does not correspond to the assumptions of a perfectly competitive model. Although the results presented in this chapter do not constitute a formal test of market power, they lend support to the UK's Competition Commission investigation into the degree of market power in the food sector.

In sum, the focus of this work has been on the impact of food scares on the UK beef market, highlighting specifically the fact that the incidence of the BSE crisis has not fallen equally on the various sectors in the beef marketing chain. While prices at each stage have fallen, marketing margins between stages have increased, suggesting that the incidence of the BSE crisis has fallen primarily on farmers, less so on the processing sector and least of all on retailing. Given the nature of the technology that links these sectors, the relative changes in the marketing margins are consistent with some degree of market power at each stage. While the approach adopted here uses a reduced form representation to examine the impact of BSE on price relationships, future work aims to develop a more structural framework for assessing the impact of food scares that explicitly allows for varying degrees of competition at each stage of the marketing chain. In addition, since the BSE crisis would have affected the consumption not only of beef but of meat products more generally, the BSE crisis would have had an impact on a wider range of interrelated markets than were identified in this chapter. Whether consumers substituted beef for other products beyond the structural changes that were already occurring or decreased consumption of meat products more generally and the impact these effects had on prices and price spreads between vertically related markets are the subject of our ongoing research.

Acknowledgements

The authors are grateful to Tony Rayner and the editors of this volume for comments on an earlier draft. This research was undertaken with finance provided by MAFF.

Notes

[1] For example, compensation was received by abattoirs for the slaughter of cattle aged 30 months and over, which was known as the 'Over 30 Month Scheme' (DTZ Pieda Consulting, 1998).
[2] For details on data construction, see MAFF (1999).
[3] The newspapers are *The Times* and the *Sunday Times*, *Telegraph*, *Independent* and *Guardian*.
[4] Although unknown, the probability of contracting CJD from BSE-infected meat was considered to be minuscule, despite the frenzied media interest in the acknowledgement of a possible link between the two. Some media reports at the time went so far as to claim that one was more likely to find a mislaid winning lottery ticket in the street on 2 successive weeks than to die from eating beef (reported on BBC Radio 4, 17 May 1996).
[5] This discrepancy is relatively common, since Schwarz Bayesian criterion penalizes overparameterization more severely than the other model-selection criteria.
[6] Those seeking an introduction are referred to Enders (1995) or Johnston and DiNardo (1997).
[6] Dummies are for 1993(3), 1995(9), 1996(3) and 1996(4). While they have a negligible impact on parameter estimates, they are included to satisfy the normality assumption. Dummies correspond to the timing of key announcements in the BSE/variant CJD (vCJD) crisis, namely the Chief Medical Officer's statement that beef was safe, the high occurrence of vCJD in farmers and the ministerial announcement of the link between BSE and vCJD, respectively.

References

Bhuyan, S. and Lopez, R.A. (1997) Oligopoly power in the food and tobacco industries. *American Journal of Agricultural Economics* 79, 1035–1043.

Brown, D. and Schrader, L. (1990) Cholesterol information and shell egg consumption. *American Journal of Agricultural Economics* 72, 548–555.

Burton, M. and Young, T. (1992) The structure of changing tastes for meat and fish in Great Britain. *European Review of Agricultural Economics* 19, 165–180.

Burton, M. and Young, T. (1996) The impact of BSE on the demand for beef and other meats in Great Britain. *Applied Economics* 28, 687–693.

Burton, M., Tomlinson, R. and Young, T. (1995) Consumers' decisions whether or not to eat meat: a double hurdle model of single adult households. *Journal of Agricultural Economics* 45, 202–212.

Chang, H. and Griffiths, G. (1998) Examining long-run relationships between Australian beef prices. *Australian Journal of Agricultural Economics* 42, 369–387.

Competition Commission (2000) *Supermarkets. A report on the Supply of Groceries from Multiple Stores in the United Kingdom*, 3 vols. Stationery Office, London.

DTZ Pieda Consulting (1998) *Economic Impact of BSE on the UK Economy.* A Report Commissioned by the UK Agricultural Departments and HM Treasury, HMSO, London, 59 pp.

Enders, W. (1995) *Applied Econometric Time Series.* Wiley Series in Probability and Mathematical Statistics, Wiley, New York, 433 pp.

Gardner, B. (1975) The farm–retail spread in a competitive industry. *American Journal of Agricultural Economics* 57, 399–409.

Goodwin, B. and Holt, M. (1999) Price transmission and asymmetric adjustment in the US beef sector. *American Journal of Agricultural Economics* 81, 630–637.

Harris, R. (1995) *Using Co-integration Analysis in Econometric Modelling*. Prentice Hall, London, 176 pp.

Johansen, S. (1988) Statistical analysis of co-integration vectors. *Journal of Economic Dynamics and Control* 12, 231–254.

Johnston, J. and DiNardo, J. (1997) *Econometric Methods*, 4th edn. McGraw-Hill, New York, 531 pp.

Kinnucan, H.W., Xiao, H., Hsia, C.-J. and Jackson, J.D. (1997) Effects of health information and generic advertising on US meat demand. *American Journal of Agricultural Economics* 79, 13–23.

Lund, P. (1998) Eating out: statistics and society – presidential address. *Journal of Agricultural Economics* 49, 279–293.

Lund, P.J. and Derry, B.J. (1985) Household food consumption: the influence of household characteristics. *Journal of Agricultural Economics* 36, 41–58.

Lütkepohl, H. and Reimers, H.E. (1992) Impulse response analysis of co-integrated systems. *Journal of Economic Dynamics and Control* 16, 53–78.

MacDonald, S. and Roberts, D. (1998) The economy wide effects of the BSE crisis: a CGE analysis. *Journal of Agricultural Economics* 49, 458–471.

McCorriston, S., Morgan, C.W. and Rayner, A.J. (1998) Processing technology, market power and price transmission. *Journal of Agricultural Economics* 49, 185–201.

McCorriston, S., Morgan, C.W. and Rayner, A.J. (1999) Market structure, demand shifts and pass-back elasticities. Mimeo, University of Exeter.

Meat and Livestock Commission (1999) Retail price spreads and indicative retail price margins for beef, pork and lamb. MLC mimeo, Milton Keynes.

Ministry of Agriculture, Fisheries and Food (MAFF) (1999) *Report of an Investigation into the Relationship between Producer, Wholesale and Retail Prices of Beef, Lamb and Pork*. MAFF, London, 29 pp.
Ministry of Agriculture, Fisheries and Food (MAFF) (various years) *National Food Survey. Annual Report on Food Expenditure, Consumption and Nutrient Intakes*. Formerly, *Household Food Consumption and Expenditure: Annual Report of the National Food Survey Committee*. Stationery Office, London.
Murcott, A. (1998) *The Nation's Diet: the Social Science of Food Choice*. Longman, London, 384 pp.
Muth, R. (1964) The derived demand curve for a productive factor and the industry supply curve. *Oxford Economic Papers* 16, 221–234.
New Scientist (4 November 2000) The human tragedy may be just beginning. *New Scientist* 148 (2263), 9.
Osterwald-Lenum, M. (1992) A note with quantiles of the asymptotic distribution of the maxiumum likelihood co-integration rank test statistics. *Oxford Bulletin of Economics and Statistics* 54, 461–472.
Pesaran, H. and Shin, Y. (1998) Generalised impulse response analysis in linear multivariate models. *Economics Letters* 58, 17–29.
Phillips, N. (2000) *The BSE Inquiry: Report, Evidence and Supporting Papers of the Inquiry into the Emergence and Identification of Bovine Spongiform Encephalopahy (BSE) and variant Creutzfeldt–Jakob disease (vCJD) and the Action Taken in Response to it up to 20 March 1996*, 16 vol. Stationery Office, London.
Stock, J.M. and Watson, M.W. (1988) Testing for common trends. *Journal of the American Statistical Association* 83, 1097–1107.
Thomas, W.J. (1972) *The Demand for Food: an Exercise in Household Budget Analysis*. Manchester University Press, Manchester, 136 pp.
von Cramon-Taubadel, S. (1998) Estimating asymmetric price transmission with the error-correcting representation: an application to the German pork market. *European Review of Agricultural Economics* 25, 1–18.
Warde, A. and Martens, L. (1998) *Eating Out*. Cambridge University Press, Cambridge, 189 pp.

Assessment of Findings and Future Research 15

WEN S. CHERN[1] AND
KYRRE RICKERTSEN[2]

[1]*The Ohio State University, Columbus, Ohio, USA;*
[2]*Agricultural University of Norway, Ås, Norway*

The impact of health information on food demand is difficult to assess, as demonstrated in several chapters, and it is not always easy to draw clear-cut conclusions and make policy recommendations. Nevertheless, this book has made several contributions, which are summarized and discussed below.

First, information on the historical trends of food consumption and summaries of food-demand studies covering many countries have been collected. Such summaries are not easily available and will make the book a useful reference. Secondly, the book shows how different measures of health information can be constructed and discusses some of the problems associated with constructing these measures. Thirdly, applications of various methods for demand estimation incorporating health-information variables have been demonstrated. Fourthly, even though the book shows the difficulties with regard to drawing general conclusions about the effects of health information, it has provided evidence on the impacts of health information on food demand in the USA and, to some extent, in Europe. Fifthly, the government can influence the diet through various policy instruments, such as changing the relative prices of various foods, income support for various types of households, provision of information concerning the diet and health, education and support for research on health and nutrition. Some of these policy measures have been discussed in this book, providing input for further discussion. This book marks only another milestone, not the end of the journey, in this field of enquiry, and suggestions for future research will be given.

Evidence from Historical Trends

Chapter 2 and the European chapters include data on the per capita food consumption during the last 20 years of the 20th century. The data show that there are similarities as well as differences in food-consumption patterns between the USA and the selected countries in Europe. There are also many notable differences among countries in Europe. Even though Americans tend to eat more meat than their European counterparts, the per capita meat consumption in Germany has actually been higher than that in the USA. In 1999, Americans consumed 91.5 kg per person, compared with Germany's 93.3 kg in 1998 or France's 88 kg in 1998. Other Europeans consumed less meat and the per capita figures were 72 kg in 2000 in the UK, 65.3 kg in 1998 in Spain, 60.3 kg in 1998 in Norway and 47.3 kg in 1998 in Scotland. However, it should be noted that the Scottish data are household survey data, which are not directly comparable with the data for the other countries, as discussed in Chapter 10. Americans also eat notably more beef than Europeans. However, overall, the German diet looks more like the American diet in terms of the quantities of major food groups consumed, except fish, than the diets observed in other European countries. The US per capita fish consumption of 6.9 kg in 1999 is the lowest among the countries covered in this book. Norway had the highest per capita fish consumption of 25.2 kg in 1998. This is undoubtedly one reason why Rickertsen and von Cramon-Taubadel (Chapter 3) could not find a European pattern in the impact of health information on food demand. Indeed, there are many European patterns of food consumption.

Nevertheless, there are also some notable trends. During the last 20 years, Americans have been eating more meat, fish, cereals, fruits, vegetables, and fats and oils, but less milk, eggs and sugar. In the European Union (EU), the changes have also been notable, but not uniform. While the per capita meat consumption steadily increased in the UK and Norway, it remained relatively constant in France and Spain, but declined somewhat in Scotland. The data from Germany show a decreasing trend. The declining trend of fluid-milk consumption seems relatively universal in the USA and the EU. However, the increasing trend of fruit and vegetable consumption is much more profound in the USA than in Europe. In fact, the per capita fruit and vegetable consumption declined during the 1985–1998 period in Spain. On the other hand, the per capita consumption of fats and oils steadily increased in the USA and Germany, but declined in Spain, Norway and Scotland, and remained relatively constant in France. The changing patterns of food consumption observed in the EU and the USA are probably caused by many factors, such as income, relative prices, demographic changes and health-risk concerns. Therefore, it is not possible to pinpoint the precise causes for these changes.

Measures of Health Information

The Medline database was used to construct several nutrition–health-information indices used in studies in the USA (Chapters 2 and 11) and Europe (Chapters 3, 7, 8, 10 and 13). The indices are based on the number of medical-journal articles related to specific diet–health relationships, such as those related to fat and cholesterol, using specific keywords in the database search. Similar indices based on the number of published articles in newspapers and other printed media were also developed and used in Chapters 2, 12 and 14. Undoubtedly, authors in this book have made significant progress in using the Medline database in constructing the health-information indices over time. The advantages of this approach are: (i) this index could be constructed over a long time period; (ii) the observations can be constructed on a monthly basis or aggregated to a quarterly or annual basis; and (iii) we can deal with well-specified aspects of the diet–health relationships. For example, we can construct the indices related to fat and cholesterol used in several chapters in this book. We can also construct indices related to 'good' nutrients, such as calcium and fibre or other health issues related to obesity. While the applications of the health-information indices have generated substantial evidence regarding the impacts of health-risk concerns about fat and cholesterol on food demand in general and specific foods in particular, the econometric results of these indices are far from perfect. Of course, there are many reasons why these indices may not perform satisfactorily in some models in some countries. Furthermore, some methodological issues remain unsolved.

First, as discussed in Chapter 2, there are two general approaches in using the number of published medical-journal articles. One way is to include the raw data series of the number of medical articles in a demand system and let the demand system determine the lag structure of the impacts of health information. The other is to assume the lag structure of the carry-over effects and the duration of the effects of an article published and to use this assumed lag structure to construct the health information prior to its inclusion in a demand system. Either of these approaches has its pros and cons. The former approach may appear to be less restrictive in letting the demand-system estimation simultaneously determine the demand parameters, as well as the lag structure of the health information. However, the variation on the structure of the carry-over and duration of a published article among different foods may raise questions about the ability of a demand system to distinguish the dynamics of food-consumption behaviour and the diffusion of health information from any published article.

Secondly, the choice of keywords for any data search is difficult and, to some extent, arbitrary. The number of articles identified by a search in Medline varies substantially depending on the choice of keywords, and apparently minor changes may change the resulting health-information index in a major way, as discussed in Chapter 3.

Thirdly, the construction of any index over time is complicated by the fact that the discussion of possible linkages is changing over time. For example, when it comes to dietary fat, the focus of medical research has gradually changed from dietary fat and cholesterol to 'good' and 'bad' cholesterol and, further, to beneficial omega-3 fatty acids found in fish oils and the *trans* fats found in certain vegetable oils. Given the changing focus of the health information, the effects on demand for specific food items may not be stable over time, as demonstrated for eggs in Chapter 12.

Other measures of health information were also created and used by various authors in the book. In the cross-sectional context, health information may be directly measured in a household survey by asking the survey respondents questions reflecting knowledge on nutrition and diet–health relationships. Answers to these questions may be used as a measure of consumer health information and knowledge, as in Chapter 6. Health information may be approximated by the level of education, as in Chapter 4, and by food-label use, as in Chapter 5. The latter two approximations are relatively *ad hoc*. Even though higher education may be directly related to more effective use of information, education itself may capture other factors affecting consumption behaviour. Similarly, the correlation between health information and label use remains, at best, an approximation.

Due to the lack of direct measures of health information and diverse aspects of diet–health relationships, this book shows that it is necessary to create proxies for measuring changing patterns of health information over time and across individuals. The use of these proxies has produced some successful econometric results in quantifying the impacts of health-risk concerns on the demand for various foods in Europe and the USA. More validation is needed on how best to use the indices based on the number of published medical-journal articles and on how to improve the household survey to collect more direct measures of health information and knowledge from survey respondents.

Methods of Demand Estimation Incorporating Health-information Variables

Most authors in this book employed a complete demand-system approach to estimate the impact of health information on food demand. There are, of course, many choices of demand systems. The most used model in this book is the linear approximate almost ideal demand system (LA/AIDS). In all cases, the problems associated with the Stone index in the original LA/AIDS have been taken care of by using alternative price indices. For the most part, the health-information variables were incorporated by demographic translation, along with other demographic variables. The use of the LA/AIDS rather than the original non-linear almost ideal demand system (AIDS) may be necessary in cases where the model includes many food items and there are many demographic variables. Several authors in the book used other demand systems, such as the quadratic AIDS (QUAIDS) and the generalized addilog demand system (GADS). The models were estimated by either time-series or cross-sectional data. When time-series data are used, one needs to examine the autocorrelation structure. As noted in Chapter 10, these econometric procedures were not uniformly applied, possibly contributing to some of the mixed results obtained in Chapter 3. As usual, we are unable to reconcile all inconsistencies on econometric ground or as a true reflection of consumer behaviour. Therefore, the econometric results generated from these demand systems need to be further validated in the future. Also, other methods of incorporating health-information and demographic variables, such as demographic scaling and the Gorman method, need to be explored.

Other single-equation or simultaneous-equations approaches were also employed in various chapters in the book. Quantile regression (Chapter 4) and the ordered polychotomous response model (Chapter 6) are seldom used in food-demand analysis. Quantile regression can be effectively used to estimate the structural relationship for each segment of the distribution, such as fat intake, enabling researchers to target the analysis to a certain high-risk population group. This is particularly pertinent for nutrition analysis. Thus these are good additions to the literature. Two US studies (Chapters 5 and 6) extended a single-equation approach to a simultaneous-equations system in order to treat the health information/knowledge as endogenous variables. While the treatment of health information as an endogenous variable may appear to be intuitively plausible, the underlying microeconomic foundation has yet to be developed. We note that many demand analysts use Becker's household-production theory or Lancaster's characteristics model to incorporate the health-risk concerns or health information/ knowledge into consumer-choice behaviour. However, in these models, health information or knowledge is assumed as given and these microeconomic models will determine the optimal choices of goods given the information or knowledge. It is not clear how to specify an economic model in which the consumer will simultaneously determine both the choice of attributes (in Lancaster's model) or commodities (in Becker's model) and the amount of health information or nutrition knowledge.

It is important to point out that the authors in this book all used structural econometric modelling as their methodology. There are other marketing research tools, such as using focus groups based on psychological profiling, which can be used to uncover the consumer perception of foods with different health and nutrition attributes. Other methods, such as stated-choice methods and experimental economics, can also be employed to estimate the consumer's willingness to pay for certain desirable health and nutrition attributes of foods.

Major Findings

Despite the imperfection in measuring health information, there is overwhelming evidence of the impacts of health-risk concerns on food demand in the USA and the EU, as shown by the studies covered in this book. Overall, the evidence is stronger in the USA than in Europe. The US studies show that concerns about the diet–health relationship have induced people to eat healthier diets, with less meat (beef, pork and poultry), eggs, and fats and oils, but more fruits, vegetables and dairy products. It is somewhat surprising that the health concerns caused a reduction in the consumption of poultry but an increase in the consumption of dairy products. Other studies have shown that, over the years, poultry consumption has increased at the expense of 'red meats' due to health concerns. The net impacts on dairy products can be relatively uncertain because the consumer preference for individual dairy products varies when it comes to dealing with health-risk concerns. This is consistent with the historical trends of increasing consumption of ice cream, cheese and low-fat milk, but decreasing consumption of whole milk. The US studies also show that consumers have responded strongly to negative information, such as the risks related to fat and cholesterol, in the consumption of meat and eggs. In fact, there is robust evidence of the effectiveness of negative health–diet information relative to commercial advertising programmes, as discussed in Chapter 11. Furthermore, nutrition labelling has enhanced the quality of American diets as measured by the healthy eating index, as discussed in Chapter 5.

The European studies provide mixed evidence on the impacts of health-risk concerns related to fat and cholesterol on food demand. As the same fat- and cholesterol-information index was applied to more or less the same model structure in the five European countries, the estimated impacts of health information varied from country to country, as discussed in Chapter 3. Therefore, there exists no unified 'European pattern' regarding the estimated impacts of health-risk concerns. However, the country studies in France (Chapter 8), Germany (Chapter 7), Norway (Chapter 10) and Spain (Chapter 9) show stronger effects of concerns over health and nutrition, using varying sources of data. We also note that not all the estimated impacts with respect to individual food items are plausible. For example, it is shown in France that more health information would decrease the consumption of beef, pork, other meats, eggs and butter, but rather surprisingly increase the consumption of oils and dairy products. These findings are very similar to those found in the US studies. In Germany, the impacts of health information are stronger in West Germany than in East Germany, which is plausible. In West Germany, the results show, however, that an increase in health information would induce increases in the consumption of fish and fish products, decreases in the consumption of pork, but have no impacts on the consumption of beef and other food products. In Norway, the estimated impacts of health information are surprisingly negative for fish, but significantly positive for poultry, fats and oils, and other foods. In the UK, the information about the 'mad-cow' scare is shown to have significantly negative impacts on the prices of beef.

Policy Implications

As described in Chapter 5, four of the top ten causes of death in the USA – heart disease, cancer, stroke and diabetes – are associated with poor diets. The economic costs of diet-related illnesses are also huge. In Chapter 4, the total costs related to medical treatment, premature deaths and lost productivity are reported to be $71 billion per year (in 1995 dollars) in the USA, and an even higher estimate is reported in Chapter 5.

Coronary disease is also the most common cause of deaths in Europe. More than 40% of all deaths in the Nordic countries were caused by cardiovascular diseases in 1996. Even in France, which together with Japan has the lowest occurrence of heart diseases, they are the primary cause of mortality, accounting for 32% of the deaths (Chapter 8).

Fat and cholesterol are an important factor in causing heart as well as other diseases, and it is hardly surprising that the intake of fat and cholesterol was discussed in all the chapters (except for Chapter 14). Governments across the world are also implementing various policies aimed at reducing the intakes of fat and cholesterol.

It is rather uncontroversial that the public should finance medical research providing the basis for the health information that is discussed in this book. It is also clear from the findings obtained in this book that continuing efforts to conduct medical research related to diet and health and to produce medical-journal publications and thus more health information are needed in the future. Health information based on published medical articles cannot be held constant without new research and discoveries. The government's role in providing information, such as the *Dietary Guidelines for Americans* and the *Food Guide Pyramid*, should also be enhanced, and the studies in this book, as well as many other studies, show that various types of nutritional information affect the diet in a healthy direction.

An increased government role through mandatory nutrition-labelling regulations in the USA was more controversial. The Nutrition Labeling and Education Act that was passed in 1990 requires mandatory nutrition labelling for almost all packaged foods. The labelling is standardized, for example by using standardized serving sizes, to make the information easily available for consumers, and there are strict regulations concerning health claims. Even though mandatory labelling schemes induce costs, the expected positive effects of label use (also found in Chapter 5) give support to the introduction of such standardized and 'easy to understand' labelling schemes.

Information is not always successful in changing the diet, as discussed in Chapter 13, and a more radical approach, with taxation of 'unhealthy' and subsidies for 'healthy' foods, is possible. However, we see several problems with such an approach. First, as discussed above, what is 'healthy' cannot be established once and for all by medical research and the evidence is changing over time, as for eggs, discussed in Chapter 12. Secondly, taxation and subsidy schemes are typically not very flexible and it may be difficult to lower taxes or remove subsidies when new knowledge is established. Thirdly, even if products such as chicken, fish and vegetables are relatively healthy, it does not mean that processed products based on these food items are healthy. Potato chips and deep-fried chicken are hardly healthy products. Given these problems, we are reluctant to propose taxation as a strategy for improving diets.

Suggestions for Future Research

More work is needed to construct measures of health information. The existing indices constructed from the Medline database need further validation and they will also need periodic updating. Since the Medline database is being continuously expanded, the updating of these indices has to be carefully done. One cannot simply conduct a similar keyword search for additional months/years. In fact, one needs to cover the time periods for the existing indices, as well as seeing whether or not the existing indices can be replicated with the new Medline database. The period covered by the existing indices may also need updating to ensure consistent coverage to include all expanded journals in the current database.

Methods of collecting health information or nutrition knowledge from household survey also need improvement. At the very least, the information on negative and positive attributes of nutrient contents in food needs to be distinguished. Other methods using focus groups, contingent valuation and experimental economics can be used to enhance the measurement of health information.

All the models were estimated by using either time-series or cross-sectional data. Panel data are more informative and have more variability in health information than cross-sectional data and less collinearity among the variables than time-series data. Extensions of the analysis to the use of panel data would therefore allow us to study the effects of health information on specific households over time and to study the dynamics of change at the household level.

The functional form of the complete demand system remains an unresolved issue. There are other candidate models, such as the Rotterdam or translog models, which were not used by authors in this book. Since the Rotterdam model has been widely used in the analysis of the impacts of advertising, its extension to analysing the impacts of health information should be pursued, with the use of time-series data. Related to time-series modelling, a more careful specification of the autocorrelation structure and co-integration analysis is needed by future researchers in the field.

Extension of demand analysis to incorporate the supply side in estimating the impacts of health information on both the consumer and the producer represents an important area of future research. Another example of the impact of health concerns on the food economy is given by the study of the 'mad-cow' scare in the UK. From the policy perspective, it is important to know the ultimate impacts of the consumer's changing health concerns on the prices of various foods. Linking the demand analysis to the supply side and the market will provide useful insights into the potential structure changes in our food production and marketing systems in the future.

Index

Figures in **bold** indicate major references. Figures in *italic* refer to figures and tables.